Special Publication No. 77

Industrial Applications of Surfactants II

The Proceedings of a Symposium organised by the North West Region of the Industrial Division of the Royal Society of Chemistry

University of Salford, 19th—20th April 1989

Edited by
D. R. Karsa
Harcros Chemicals (UK) Limited, Manchester

ROYAL
SOCIETY OF
CHEMISTRY

British Library Cataloguing in Publication Data

Industrial applications of surfactants II.
1. Surface — active agents
I. Karsa, D. R. (David Robert), *1943-* II. Royal Society of Chemistry, *Industrial Division, N. W. Region* III. Series
668'.1 ISBN 0-85186-617-4

Published by the Royal Society of Chemistry,
Thomas Graham House, Cambridge CB4 4WF

Printed in Great Britain by
Whitstable Litho Printers Ltd., Whitstable, Kent

Introduction

The market for surfactants in Western Europe, USA and Japan is 6,000K tonnes, of which approximately 50% are used in areas outside the household detergent and personal care areas; in the so-called 'industrial' applications.

The continued search for higher value added, more performance-related applications and raw material pressures in the commodity area of the surfactant market have resulted in a continued growth of interest in the properties and applications of speciality surfactants; areas where products are not sold for their chemical structure against a specification but where products are formulated and sold for the effect they produce in a particular process. Many of these applications fall within the industrial sector.

The proceedings of this conference are complementary to the first industrial conference 'The Industrial Applications of Surfactants' held at the University of Salford, Greater Manchester in April 1986. The following papers describe a wide range of chemical species not featured in the first conference. These include acetylenic glycols and derivatives, alkanolamides, phosphate esters, amine oxides, sulphosuccinates, ether carboxylates, naphthalene derivatives and others. Further selected application areas, including agrochemical formulations, the construction industry, pulp and paper production, mineral flotation and the uses of cationic and ampholytic biocides, are described in depth.

Surfactant technology continues to be an area for innovation and topics including biosurfactants, the formation and application of 'ultra-thin films', polymeric surfactants and the application of cationics as phase transfer catalysts are examples which are featured in this monograph.

It is hoped that the reader will find these topics equally stimulating and the monograph a useful addition to the knowledge of surfactant properties and applications.

D R KARSA

Gratitude is expressed to the co-sponsors of this conference,

Akzo Chemicals BV

Harcros Chemicals UK Limited

ICI Chemicals & Polymers Limited

Shell Chemicals UK Limited

once more for their generous support which helped ensure the success of this meeting.

Contents

Innovative Approaches to Surfactant Development

David R. Karsa
HARCROS CHEMICAL GROUP, SPECIALITY CHEMICALS DIVISION, MANCHESTER M30 0BH, UK

The diverse applications of surface active agents across the industrial spectrum and continuing changes in performance, environmental and legislative requirements ensures an on-going need for product innovation. Although sometimes described as a 'mature' technology, there are many avenues still unexplored. Specific end-uses still await the commercial availability of surfactants of a particular structure and performance and there are many classes of surfactants, often synthesised many years ago, which remain relatively unknown and unexploited.

In this paper, it is hoped to illustrate that performance demands in the industrial sector still offer considerable scope for product innovation.

In Western Europe, Japan and the United States, at least 45% of surfactants are utilised in what may be broadly classified as "industrial applications", representing some 2,600K tonnes of products.[1] This figure has been quoted as high as 58% and to some extent depends where products such as industrial and institutional cleaning products are classified. The household detergent and personal care market represents the largest single usage of surfactants and product development and acceptance in this sector has significantly influenced the major raw materials available for industrial applications.

Hence the larger volume products available are those based on raw materials from petrochemical and oleochemical majors, such as fatty alcohol and alkylphenol ethoxylates and alkanolamides in the nonionic area and alkylbenzene sulphonates, fatty alcohol and fatty alcohol ether sulphates in the anionic area. The lower volume, more specialised products are often derivatives of these base materials, for example phosphate esters, sulphosuccinates and propoxylates derived from fatty alcohol or alkylphenol ethoxylates.

Raw Materials

In the industrial sector, the criteria which govern product selection are often quite different to those which dictate choice in the personal care and household detergent areas. For example, there are many applications where the surfactant's contact with the environment is minimal and biodegradability is not a pre-requisite.

Both success and failure of a product in the detergents sector can preclude exploitation in non-detergent areas. Major raw material producers are primarily geared to producing a limited range of high volume material to supply to a limited number of detergent manufacturers and toiletry companies, with sometimes little interest in other uses. Examples would be the oleochemical-derived alpha-sulphonated fatty acids and esters and paraffin (alkane) sulphonates which have been almost entirely targeted at the domestic area, but which, for example, can be utilised as highly effective emulsifiers in the emulsion polymer process. Likewise, the failure of alpha-olefin sulphonates to gain wide acceptance in the toiletry, liquid 'soap' and detergent areas in Western Europe, in part due to concern over sultone formation on sulphation, has precluded their availability for other areas where their performance has been established. Such uses include agrochemical formulations and, again, in emulsion polymer processes. Hence existing raw materials are capable of further exploitation, subject to their cost-performance advantages in specific areas.

Looking to the future, a number of readily available raw materials offer interesting prospects. Those derived from re-newable resources such as natural oils and carbohydrates afford the basis of new surfactant molecules. In the case of natural oils, the high molecular weight, highly hydrophilic oil can be reacted with hydrophilic moieties under controlled conditions to build-in sufficient hydrophilicity to acquire the desired surface active properties. Such hydrophobic species can be envisaged for use in defoamer formulations and low foam surfactant systems.

Carbohydrates on the other hand afford part of a hydrophilic moiety which by alkoxylation and subsequent alkylation afford surface active species. Starch, sucrose, glucose, dextrins etc. all offer the potential to give readily biodegradable surfactants, potentially of a low order of toxicity.

Well established materials include products such as phosphated maize starch, which is used, for example, as a dispersant for inorganic pigments, and alkyl glycosides which exhibit excellent solubility in highly built liquid cleaner formulations.

The polyhydroxyl-functionality of the carbohydrates may be used to advantage to build multiple poly(ethylene oxide) or poly(ethylene oxide)/poly(propylene oxide) side chains followed by further reaction to increase hydrophobicity. It is an area of synthesis where product degradation and prevention of the formation of highly coloured species can sometimes be difficult to control. Remarkably it has been found that water soluble carbohydrates may be alkoxylated in aqueous solution with very little polyglycol formation when reacted at 55°C (or below)[2]. The process yields almost colourless products. (See Table 1 below).

TABLE 1

Ethoxylation of Carbohydrates in Aqueous Solution

	Charges (g)		
Sucrose	1,197	513	-
Polysorbitol (*66% syrup/4,900 MW)	-	-	699*
Water	616	264	*
32% Caustic Soda	45	32	21
Ethylene oxide	308	2,640	2,418
HCl (conc).	38	26	18.6
Moles EO/sucrose unit	1.9	40	20
Reaction temperature	35	50	50
% MEG (by g.l.c.)	0.4	0.2	0.2
% DEG (by g.l.c.)	0.5	0.5	0.2
% TEG (by g.l.c.)	-	-	1.3
% higher glycols (by g.l.c.)	-	-	4.2
% EO unreacted	4.9	-	0.6
% EO charge as glycol	0.9	0.7	6.5
% EO as ethoxylated derivative	90.2	99.3	92.9

Alkoxylated derivatives can be further reacted with propylene oxide or with esters or epoxides to increase hydrophobility. (Figure 1).

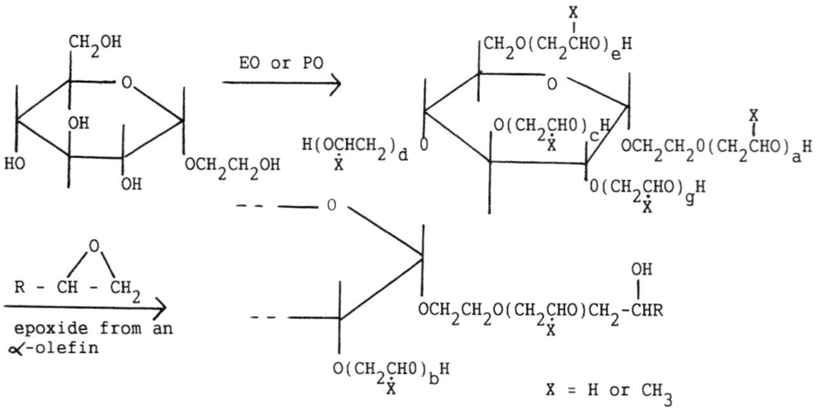

Figure 1

As the search for alternative biodegradable, mild, low order toxicity surfactants grows, carbohydrate chemistry must offer a range of potentially low cost raw materials capable of many conversion routes to effective surfactants.

Sometimes it appears in surfactant production, unit processes are limited to the various techniques of sulphation/sulphonation, alkoylation plus amidation, phosphorylation, esterification etc. Polymerisation is one unit process which is totally under exploited. 'Polymeric', or more accurately 'Oligomeric' surfactants result from the marriage of polymer and surfactant chemistry. Many surfactant producers have no involvement with polymer chemistry and hence are not exposed to the possibilities which may be available. Oligomeric surfactants are effective emulsifiers and dispersants which generally depend on steric stabilisation for their effectiveness. In general, they exhibit low foam and relatively poor wetting properties. Basic types include:

1 "Comb" surfactants, for example pendant polyalkylene glycol chains attached to an acrylic or methacrylic backbone.
2 Random polymeric surfactants, for example aromatic or aliphatic polycarboxylic acids or anhydrides reacted with polyols, alkoxylated polyamines etc.

Chapter 4 of this mongraph describes some of the newer species which have been commercialised in this area. Applications range from oil spill dispersion, metal working, hydraulic fluids, alkyl resin emulsification through to paper deinking and agrochemical dispersions. Species such as alkoxylated alkyl phenol-formaldehyde condensates, commonly used as emulsifiers and de-mulsifiers in the oil industry and naphthalene sulphonic acid-formaldehyde condensates, widely used as dispersions and concrete super plasticisers, are representative of oligomeric species which are already well established in particular niche markets, i.e. Figure 2

Figure 2 Examples of well known oligomeric surfactants

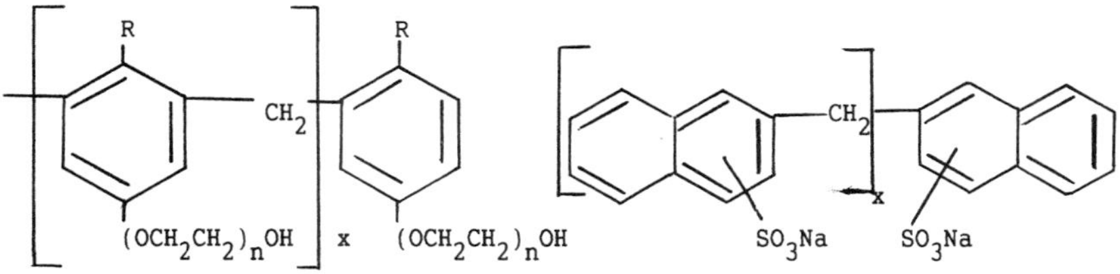

Alkylphenolethoxylate-formaldehyde condensates.

Sodium salt of naphthalene sulphonic acid formaldehyde condensate.

Biosurfactants

Longer term, newer technologies may provide sources of raw materials. Biotechnology and biosurfactants have featured widely in the literature of the last decade. These tend to be high molecular weight materials e.g. α-dodecyl-β-hydroxy arachidic acid (Figure 3).

Figure 3

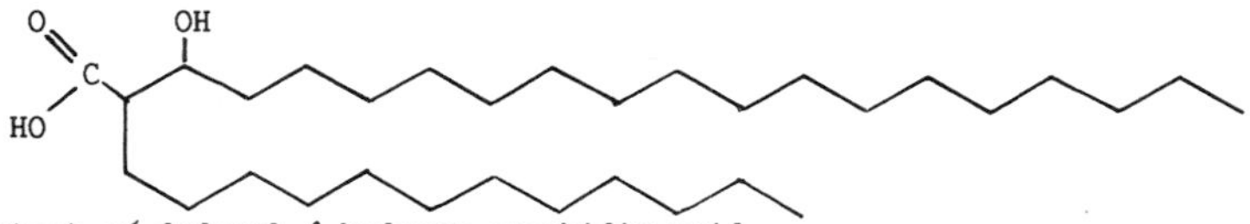

Biosurfactant α-dodecyl-β-hydroxy arachidic acid.

The main practical use described in the literature appears to be in oil field chemicals, particularly in the oil field recovery area but the question remains whether cost-effective materials can ever be produced in bulk from biotechnology processes. An up-date on the current status of this class of surfactant appears in Chapter 2.

On the petrochemical side, it is more difficult to envisage the longer term movements in raw materials. On the anionic front, fatty alcohols and alkylate will continue to be major feedstock for the foreseeable future, particularly with recent positive reports on alkylates and the ultimate fate of alkylbenzene sulphonates in the environment.[3] No doubt paraffin sulphonates may see some growth although they can only be produced on high capital, dedicated plants. Alpha-olefins seem to have gained little acceptance in Europe, even after the increased availability of alpha-olefins over the past decade.

On the nonionic front wider use of fatty alcohol derivatives and continued use of alkylphenol ethoxylates in industrial applications appears to be the medium term picture.

Stereochemistry:

The orientation of a surfactant at air/liquid, liquid/liquid or liquid/solid interfaces is governed by both the nature of the hydrophobe and the number and position of the hydrophilic group(s) in simple terms, the effectiveness of the surfactant in surface coverage and packaging. More effective surface coverage can be achieved in two ways; by an oligomeric or polyfunctional surfactant or by a surfactant with a sterically large hydrophobe (Figure 4).

Conventional Surfactant

Oligomeric Surfactant

Surfactant with a 'bulky' hydrophobe

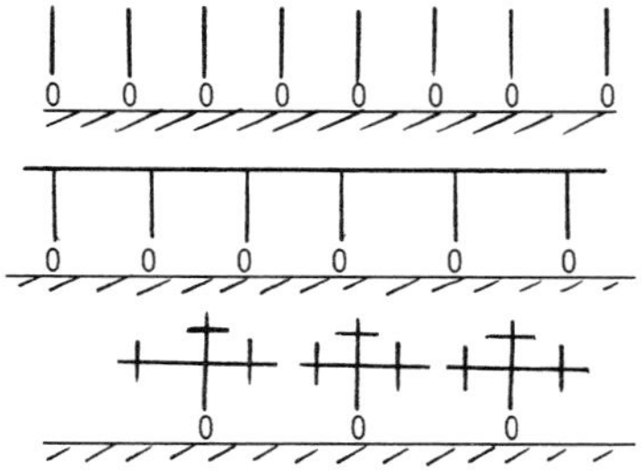

Figure 4

0 = hydrophilic group

This simplistic picture is borne out in practice, and although oligomeric surfactants or surfactants based on very large hydrophobes will always be much more expensive on a weight/weight basis, they can be far more cost effective than conventional surfactants in more demanding applications.

For example 26 weight percent of nonylphenol ethoxylate is required to produce a maximum 35% solids, low viscosity phthalocyanine blue pigment paste. A higher solids content pigment dispersion cannot be achieved with a nonylphenol ethoxylate. However an addition of only 15% by weight of a polysubstituted phenol ethoxylate ('bulky' hydrophobe) will achieve a solids content of 50% for the same pigment.

Microemulsion work, particularly in agrochemical formulations has also illustrated the benefit of both anionic and nonionic species based on a sterically large hydrophobe, e.g.

Synthetic pyrethroid	10.0% w/v
Ammonium salt of sulphated distyryl phenol ethoxylate (5 EO)	7.0%
Tristyryl phenol ethoxylate (7 EO)	5.0%
70% alcoholic solution of calcium dodecyl benzene sulphonate	3.0%
Antifreeze (glycol)	5.0%
Antifoam	0.7%

Synthetic Pyrethroid Microemulsion

Reactivity and Stability

Both surfactants with a reactive group and those which will chemically or thermally degrade can offer performance advantages. In both cases, the surface active properties of the surfactant are essential for the functioning of a particular process but at a later stage of processing, or under 'in use' conditions, those residual surface active properties are unwanted or even detrimental. One example would be the residual emulsifier in an emulsion polymer leading to foam problems during paint formulation, paint can filling and application, plus poor scrubability of the dried paint film. In principle, the following general concept can be applied to advantage

Surfactant —— Reactivity or Degradation ——→ Total loss of surface active properties or Different, residual surface active properties.

The main application for "reactive surfactants" is as primary emulsifiers in emulsion polymer processes. Such 'polymerisable' surfactants must have both a reactive (usually vinyl- or allyl-) end-group with a hydrophilic moiety at the other extremity of the the molecule. The chemical structure between these two functional groups must be of a sufficient molecular weight and hydrophobicity that the molecule is surface active and capable of affecting an emulsion polymerisation reaction without additional external emulsifier. Such products are not readily available in Europe and should not be confused with readily available 'hydrophilic monomers', such as sodium vinyl sulphonate, which are not true surfactants.

Typical examples of reactive surfactants primarily targeted at emulsion polymer production are given below:

(i)

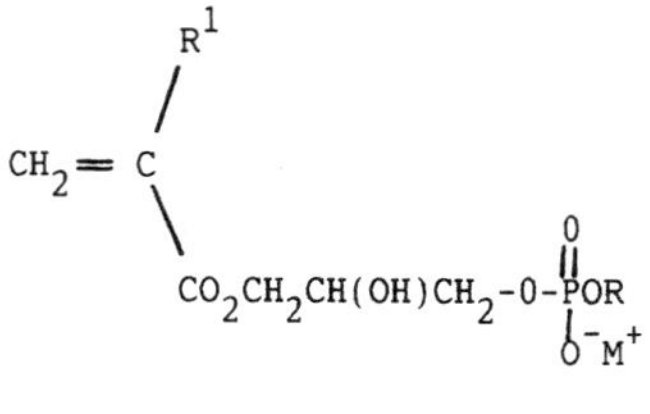

R = Alkyl or Fluroalkyl
R^1 = H, CH_3
M = H, Alkali metal, NH_4, Alkylamine or Alkanolamine Salt.

(ii)

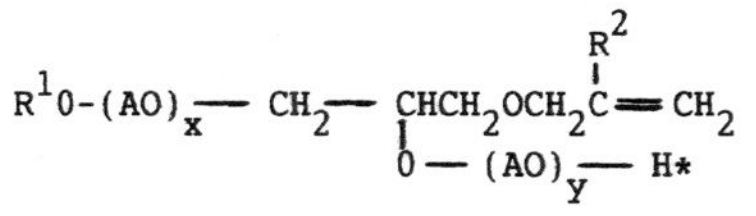

R^1 = Alkyl, Alkenyl, Alkylaryl (C_8-C_{30})

A = Alkylene of 2-4 C atoms

x = 0 or 1-100

y = 1-200

*Nonionic or sulphate, phosphate, sulphosuccinate derivatives.

and

(iii)

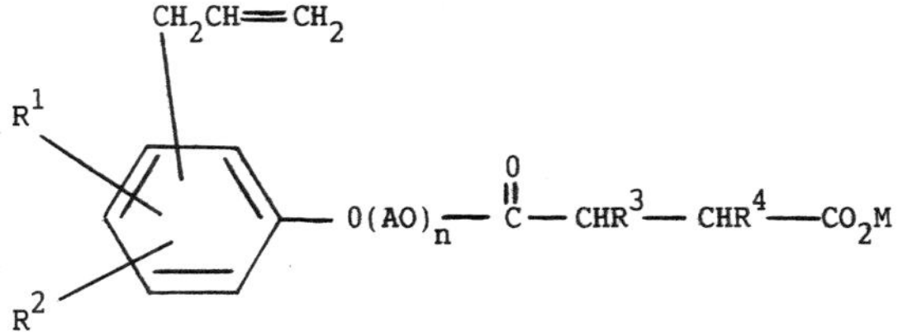

Where R^1 = C_{4-18} Alkyl, Alkenyl <u>etc</u>.

R^2 = H or R^1

R^3 and R^4 = SO_3M and the other equals H

A = C_{2-4} Alkylene

M = Alkali metal, NH_4 <u>etc.</u>

For example, product (ii) finds its main application as the sole emulsifier in acrylate-based and styrene-butadiene co-polymer emulsion production.

Here the emulsifier reacts into the polymer as it is formed affording a "self-stabilising" polymer emulsion. This is of particular interest in the area of paints and surface coatings as the resultant dried films exhibit excellent scrub resistance and superior water resistance to many conventional water-based systems.

Degradable surfactants

Thermal and/or chemical degradation can also be employed to advantage to eliminate or alter surface active properties during processing. Ammonium salts are a simple, early example of this concept.

Ammonium salts of soaps or alkyl sulphates can, under thermal conditions lose ammonia and minimise the residual surface activity. For example mechanically foamed acrylic or styrene-butadiene-based co-polymer emulsions can be used to produce foamed polymer coatings for carpet and upholstery backing, shoe interlayers and for the production of crushed foam backings for curtain and drapery materials. Once the foam is cured, the surfactant is no longer required and can for example, in the case of a wet foam-backed floor covering, foam excessively and even cause delamination of the foam from the substrate. Use of an ammonium salt as a foam stabiliser can help overcome these problems after the curing stage.

Fatty alcohol and alkylphenol ether sulphates will degrade in highly acidic conditions, the degradation being accelerated at elevated temperatures and becoming almost autocatalytic, the liberated ether sulphonic acid further assisting the degradation. With longer chain length ethoxylates, the degradation product will still be a water-soluble surface active nonionic, most likely with good wetting and emulsifying properties.

Thermal degradation of Sodium Salt of Nonylphenol ether (4 EO)sulphate at 90°C and pH 10 and 4

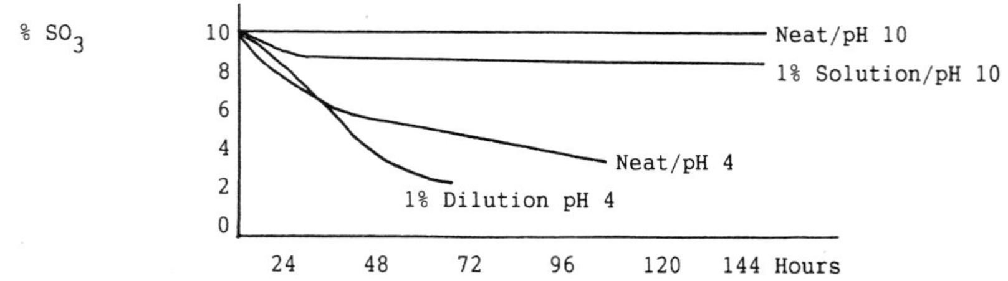

Sodium and ammonium salts of alkylphenol ether (4 to 30 EO) sulphates are used to produce acrylic and styrene-acrylic emulsion polymers for paints and surface coatings. It could be envisaged that where the desired pH/temperature conditions could be applied, a coating with poor water sensitivity and good scrub resistance could be produced.

In a similar fashion, ester hydrolysis can be employed to yield different surface active properties at the further stage of an industrial process. This is best illustrated by considering a di-ester sulphosuccinate surfactant which possesses excellent wetting properties. On hydrolysis in acid conditions the ester groups hydrolyse and generate free fatty alcohol (or alkoxylates) some of which by careful selection can have effective defoaming properties. In the production of phosphoric acid from phosphate rock, using concentrated sulphuric acid, a sodium di-octyl sulphosuccinate provides the required wetting of the sulphuric acid onto the phosphate rock and also rapidly liberates free octanol which then acts as an essential defoamer in the system. Addition of octanol alone to the system is totally ineffective as combined wetting and then defoaming is required.

In the textile area, such combinations of wetter/dye leveller and the need to control process foam could be approached without the need for expensive defoamer addition. Likewise this approach may be applicable to automatic clothes washing powders and liquids where "controlled foam" properties are needed.

Bünte salts, organo-thiosulphates combine the properties of reactivity and degradation (both thermal and chemical). Although first developed over a hundred years ago they have only begun to be exploited over the past decade.

The organo-thiosulphate consists of an $-S-SO_3Na$ group which exhibits an instability which has been exploited in recent years, for example, in the shrink-proofing of wool to produce machine-washable garments.

Bünte salts are readily prepared from compounds possessing hydroxy-groups by the two-step process described below:

$$R\text{-}OH \xrightarrow[\text{(\textit{para}-toluene sulphonic acid catalyst)}]{ClCH_2CO_2H} R\text{-}O\text{-}\overset{O}{\overset{\|}{C}}\text{-}CH_2Cl$$

$$R\text{-}O\text{-}\overset{O}{\overset{\|}{C}}\text{-}CH_2Cl \xrightarrow[\text{pH 5 to 6}]{Na_2S_2O_3} R\text{-}O\text{-}\overset{O}{\overset{\|}{C}}\text{-}CH_2S_2O_3^{-}\ Na^{+}$$

Polyfunctional Bünte salts are used as shrink-proofing agents for wool. Hence, treated garments become machine washable. Shrinkproofing is achieved by breaking the disulphide linkages residing on the surface of the wool and reforming them through this shrinkproofing agent. The subsequent structural changes in the outer fibres increase the stability of the disulphide groups during washing. In the pad-batch process <u>meta</u>-bisulphite ions are used to cleave the disulphide linkages in the wool in a reversible process, and the Bünte salts react faster with the sulphide end-groups than do the thiosulphate-groups formed concurrently <u>e.g.</u>

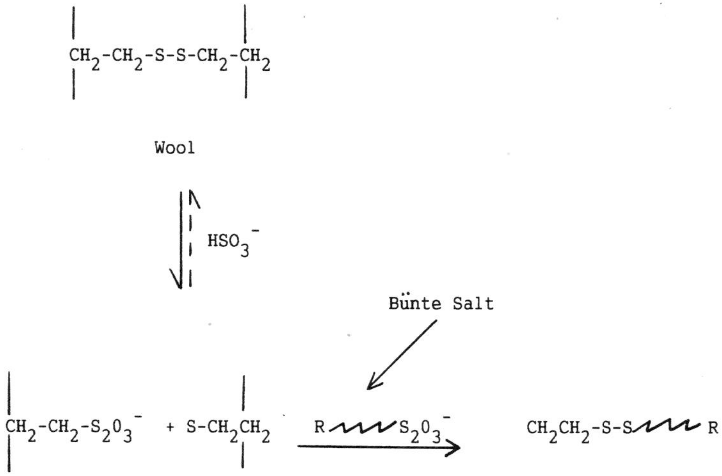

Electromicroscope studies of wool fibres confirm the shrinkproofing process is a combination of both chemical reaction and physical bonding of the fibres.

Very recently, this process has been applied to the treatment of human hair in the field of silicone surfactants.

Dimethicone thiosulphates (Polysiloxane polyorgano thiosulphates) of the general types:

$$CH_3-\underset{\underset{CH_3}{|}}{\overset{\overset{CH_3}{|}}{Si}}-O{\left[\underset{\underset{CH_3}{|}}{\overset{\overset{CH_3}{|}}{Si}}-O\right]}_n {\left[\underset{\underset{\underset{\underset{\underset{CH_2S_2O_3Na}{|}}{\underset{CH-OH}{|}}}{\underset{CH_2}{|}}}{\underset{O}{|}}}{\overset{\overset{CH_3}{|}}{\underset{(CH_2)_3}{|}\,Si}}-O\right]}_m \underset{\underset{CH_3}{|}}{\overset{\overset{CH_3}{|}}{Si}}-CH_3$$

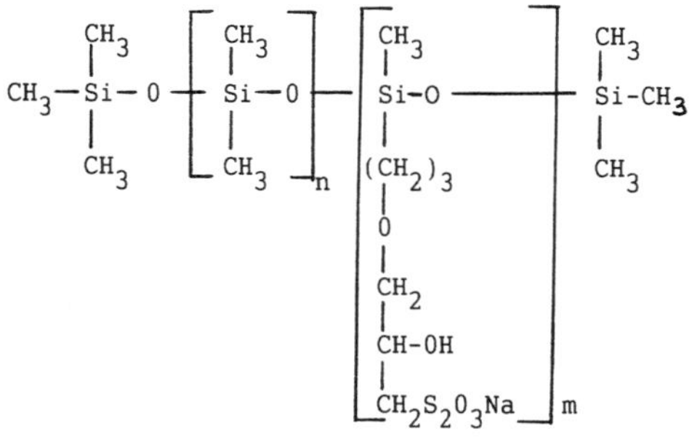

and

$$NaO_3S_2CH_2-\overset{\overset{OH}{|}}{CH}-CH_2O(CH_2)_3{\left[\underset{\underset{CH_3}{|}}{\overset{\overset{CH_3}{|}}{Si}}-O\right]}_n \underset{\underset{CH_3}{|}}{\overset{\overset{CH_3}{|}}{Si}}-(CH_2)_3-O-CH_2-\overset{\overset{OH}{|}}{CH}-CH_2S_2O_3Na$$

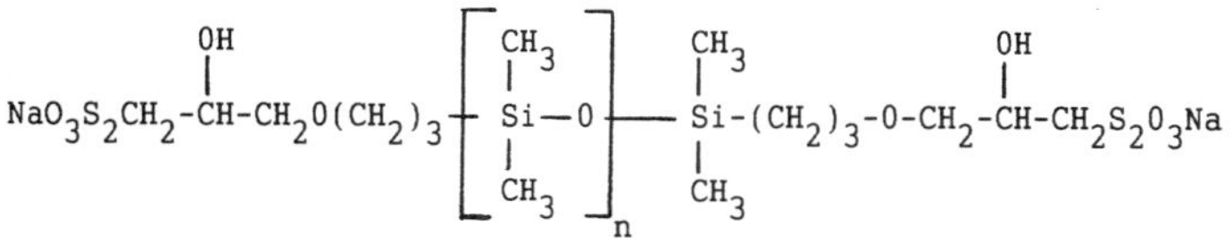

have potential application as film forming components in cold-wave hair preparations. As can be seen above, the first example is also typical of the 'comb' surfactant polymeric structure previously described.

It should also be noted that mono-functional Bünte salts based on fatty alcohol ethoxylates afford a thiosulphate analogue of conventional fatty alcohol ether sulphates and exhibit comparable foaming, wetting and detergency.

<u>NONIONICS</u> - Requirements for primary and ultimate biodegradation.

In the European Community, the use of biodegradable nonionics in cleaning products has been hastened by EC legislation and the proposed ending of a number of derogations permitting the use of non-biodegradable species such as ethylene oxide-propylene oxide block co-polymers. However, in areas such as industrial rinse aids and certain high pressure cleaning processes, the biodegradable replacement products usually offer inferior wetting and defoaming characteristics. In products such as fatty alcohol EO/PO co-polymers, the number of moles of propylene oxide is generally a compromise to maintain biodegradability at the expense of the product's defoaming capability. The formulation and blending in of other components can improve performance but rarely reaches the standard achieved by the block co-polymers. In fact, there appears to be no biodegradable replacements for particularly high molecular weight ethylene oxide-propylene oxide co-polymers based on primary amines, where protein soil defoaming in high agitation conditions coupled with excellent wetting properties are a pre-requisite. Hence the area of low foam biodegradable nonionics affords an area where innovation is necessary to meet performance criteria. This may possibly be achieved by more subtle formulations of biodegradable products with other minor components, by alternative 'end-blocking' of nonionics with hydrophobic groups or considering more innovative products, such as "bridged nonionics" (one or more ethoxylates bridged via their poly(ethylene oxide) chains).

When ultimate biodegradation is considered, then alkylphenol ethoxylates are being examined more closely. Although alkylphenol ethoxylates biodegrade somewhat more slowly than fatty alcohol ethoxylates of a similar HLB or ethylene oxide content, nevertheless they pass the EC test for primary biodegradation. Concern has been expressed regarding the fish toxicity of degradation products such as the nonylphenol + 1 EO, + 2 EO <u>etc</u>. although recent work has illustrated that these components are less toxic than previously thought.[4] Having said this, there is considerable pressure and action, particularly in continental Europe, to replace alkylphenol ethoxylates in industrial cleaning applications. This is not proving to be a simple substitution process with the aromaticity of the alkylphenol ethoxylates and their narrower poly(ethylene oxide) distribution distinguishing them from the fatty alcohol ethoxylates with a choice of alkyl chain lengths which influence performance and their broader spread of poly(ethylene oxide), with high free alcohol levels in the lower EO content products.

More subtle blending of fatty alcohol ethoxylates of varying chain lengths and poly(ethylene oxide) contents and other components is necessary in some cases to approach the performance of alkylphenol ethoxylates, which still remain the most cost-effective materials in many cleaning formulations. It will be interesting to see whether narrow distribution fatty alcohol ethoxylates commercialised in the United States particularly for use in liquid laundry detergents offer any performance advantages in any industrial formulations.[5,6]

Cationics and amphoterics

Greater awareness of the unique properties of amphoteric and cationic surfactants has been matched by growing usage in particular market sectors. The biocidal activity of cationics and ampholytes will be discussed in a later chapter, but it is clear that, for example, particular ampholytes are gaining wide acceptance for use in food processing plants as the bacteriacides of choice. Again the potential for novel species, such as polymeric betaines, have yet to be evaluated in this important area.

The growing use of cationics as phase transfer catalysts will be discussed in a future chapter of this monograph. It is clear that low molecular weight cationics and their ability to carry counter ions across a phase boundary have still to be explored in many industrial processes.

The elementary 'surfactant rule' which states that cationic and anionic surfactants are incompatible must be modified when applied to anionic/ cationic pairs where one species is of a relatively low molecular weight. In such cases, these soluble complexes offer formulation and property advantages which could be further developed and exploited. Such 'soluble complexes' are also formed if either or both species have high poly(ethylene oxide) contents.

Under-exploited surfactant species

The literature is full of surfactants which have been developed on a laboratory scale and usually, for understandable reasons, never been commercialised. This has sometimes been due to lack of required raw materials in bulk, to more cost-effective species being available and sometimes due to the inventor not being clear where unusual surface active properties could be usefully applied. The Bünte salts described earlier are a classic example of where the latter comment applies.

The following products are just a few examples where the author has direct experience, but only represent a small cross-section of species which should be re-appraised in the light of changing circumstances.

(a) <u>Sodium lauryl sulphoacetate[7,8] and analogs</u>

This class of product can be prepared by producing the chloracetyl ester of a fatty alcohol, followed by reaction with sodium bisulphite, <u>i.e.</u>

$$\underset{(R=C_{12,14}\ \text{alkyl})}{R\text{-}OH} + ClCH_2\text{-}CO_2H \longrightarrow R\text{-}O\text{-}\overset{\overset{O}{\parallel}}{C}\text{-}CH_2Cl + H_2O$$

$$\downarrow (Na_2SO_3)$$

$$R\text{-}O\text{-}\overset{\overset{O}{\parallel}}{C}\text{-}CH_2SO_3Na + NaCl$$

or alternatively, by esterification of the alcohol directly with sulpho acetic acid,

$$ROH + HOOC\text{-}CH_2\text{-}SO_3^-Na^+ \longrightarrow R\text{-}O\text{-}\overset{\overset{O}{\parallel}}{C}\text{-}CH_2SO_3^-Na^+ + H_2O$$

In practice, these processes are more complex with the use of co-solvents and an ester-washing stage being necessary in some processes.

Similar materials can be produced by reaction of a fatty acid chloride with sodium isethionate, <u>i.e.</u>

$$R.COCl + HOCH_2CH_2SO_3Na \longrightarrow R\text{-}\overset{\overset{O}{\parallel}}{C}\text{-}OCH_2CH_2SO_3Na + HCl$$

Sodium alkyl β-sulphopropioniate

and there are many other references in the literature to similar classes of sulpho-esters.

Sodium lauryl sulphoacetate itself is an extremely low toxicity, low irritancy surfactant which exhibits excellent foaming, wetting, emulsification and detersive properties in both hard and soft water.

Aqueous solutions exhibit interesting rheological properties - a 10% aqueous solution is a clear gel at ambient temperature. When produced in commercial quantities in the past, this material has been sold as a free-flowing powder which is relatively non-hygroscopic. The material also has a number of FDA approvals.

The 'mildness' of this lends it be considered for toiletry and cosmetic applications, including bubble baths, shampoo, dentifrices, creams and lotions and in speciality 'soaps'.

(b) Sulpho N-alkyl propionamides (sulphated Ritter acrylamides)[9,10]

Reaction of an alpha-olefin with acrylonitrile under highly acidic conditions produces an N-alkylacrylamide:

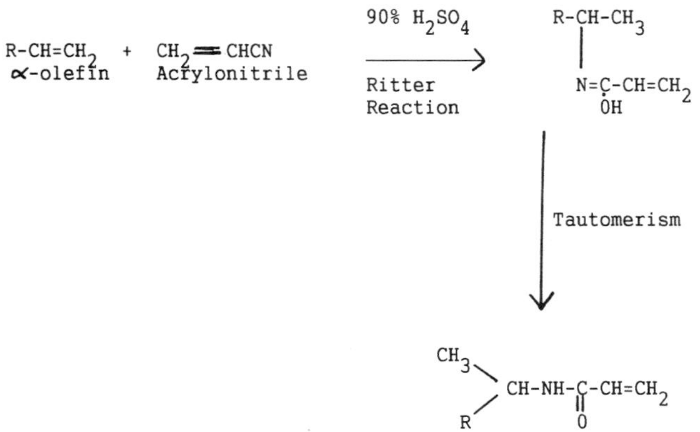

N-alkylacrylamide
(Ritter acrylamide)

The N-2alkylacrylamide may then be reacted with sodium bisulphite under a variety of conditions, which avoid polymer formation, to give the corresponding sulphonate in high yield.

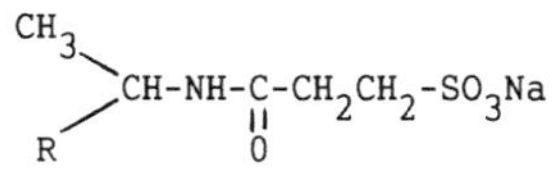

Alpha-olefin purity is critical to the final performance of the sulphonate and the carbon chain length of R is critical with respect to surface active properties. In the region of R = C_8-C_{10} alkyl, very low foaming properties and poor/moderate wetting properties are observed. At higher chain length, R = C_{12} to C_{14}, high foaming in both hard and soft water and excellent wetting results are observed. Replacement of sodium $C_{12,14}$ alkylether (3EO) sulphate in simple detergent blends with alkylbenzene sulphonates confirms similar detergency and their effectiveness as a primary emulsifier in emulsion polymer processes has been demonstrated for a wide range of co-polymers.

(c) Mannich base derivatives

The reaction between an alkylphenol, formaldehyde and diethanolamine to yield a Mannich base is well known

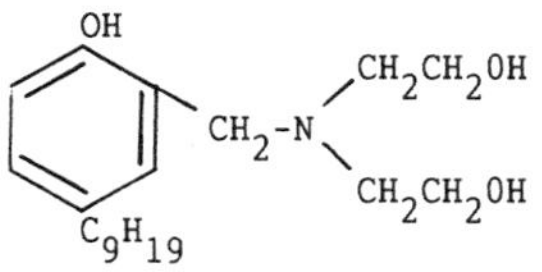

or with excess diethanolamine

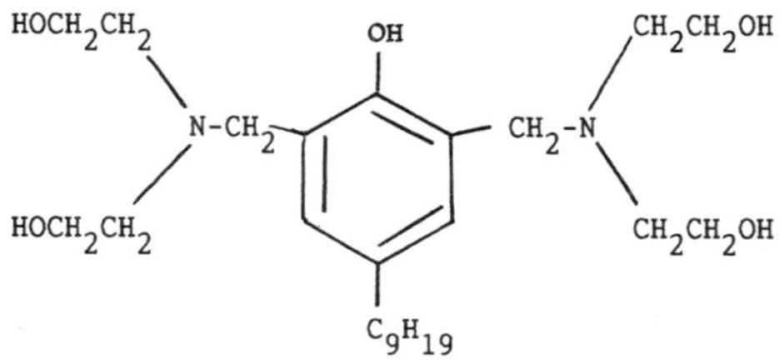

Propoxylated derivatives find application in the formulation and production of polyurethane systems, whereas further ethoxylation with optional sulphation or quaternisation affords ranges of surface active agents. Quaternised Mannich bases have been used in the production of fluff pulp as a de-bonding aid,[11] the extension of these and similar products [12] to uses such as softeners and antistatic agents in the textile area and as an antistat in thermoplastics are obvious extensions of their use.

<u>Conclusions</u>

The major outlets for surfactants, the detergent and personal care areas, and their close links with the major raw material suppliers understandably have great influence on the main raw materials, products and derivatives available. However, industrial applications can place quite different and often stringent demands on the surfactant required in a particular process. The evaluation in more depth of alternative raw material sources such as carboxylates or natural oils for example, will lead to cost-effective intermediates from renewable sources. Likewise many species of surfactants have never been commercialised and in changing circumstances re-appraisal of their merits may be a profitable exercise. This must be viewed in the context of the growing need for more environmentally and toxicologically acceptable materials coupled with the greater availability of certain starting materials.

Further appreciation of the importance of steric stabilisation opens up the whole field of polymeric or oligomer surfactants and mono-functional species with large 'bulky' hydrophobes. Reactive and degradable surfactants afford an under-exploited concept of destroying or altering surface active properties after the surfactant has completed its prime function in a process, and new processes, such as phase transfer catalysis or ultrathin film formation, offer new and developing uses for surface active agents.

It is not possible to do justice to so large a topic in a single chapter, but it is hoped that the selection of items included in the preceding pages illustrates that, although described as a 'mature technology', the field of surface activity still affords vast areas where 'innovation' is the key word.

References

1. 'The Industrial Applications of Surfactants, Chapter 1., Proceeding of Royal Society of Chemistry Conference; University of Salford, April 1986. Edited by D R Karsa, Special Publication No 59 (1987), ISBN 0 85186 636 0.

2. Unpublished work, A.T. Pugh et al., Lankro Chemicals Limited.

3. Proceeding of the International Status Seminar "Alkylbenzene Sulphonate (LAS)" held in Aachen/FRG, November 9th-10th, 1988; Tenside, Surfactants Detergents 26 (1989) 2, 85-179.

4. Biodegradation of Nonylphenol ethoxylate (APEO), A, Neufahrt, K Hofmann and G Täuber (Hoechst AG), Communicaciones presentadas a las XVIII Journadas del Comite Espanol de la Detergencia. Barcelona, March 25th-27th, 1987, pp 183-198.

5. New Formulating Potential: Control of the Ethoxylate Distribution of Nonionic Surfactants, K W Pillan, G C Johnson and P A Siracusa (Union Carbide Corp), Soap/Cosmetics/Chemical Specialites, March 1986, pp 34-40, 70-77.

6. Polyoxyethylene Oligomer Distribution of Nonionic Surfactants, T. Sato, Y. Saito and I Anazawa (Nihon University, Japan). J.A.O.C.S., 65(6), 1988, pp 996-999.

7. Preparation and Properties of Sodium Alkyl Sulphoacetates, T.Hikota and K Meguro (University of Tokyo), J.A.O.C.S., 46(2), 1963, pp 579-582.

8. 'Surface Activity' (2nd Edition), edited by Moilliet, Collie and Black (pub. Spon), p26 ('Sulphoacyl compounds').

9. US Patent 3,170,951 (American Cyanamid Co), 1965.

10. US Patent 3,396,153 (American Cyanamid Co), 1968

11. Canadian Patent 1042466 (Berol Kemi AB), 1978.

12. UK Patent 2,070 040 (Berol Kemi AB), 1980.

Biosurfactants — an Overview

I. D. Robb
UNILEVER RESEARCH LABORATORY, PORT SUNLIGHT, WIRRAL L63 3JW, UK

INTRODUCTION

Biosurfactant is a term that normally refers to surface active molecules that are produced by a variety of micro-organisms such as yeasts, fungi or bacteria. As with most surface active materials, they have a hydrophilic and a hydrophobic portion, both of which can be synthesised from a variety of simple feedstocks, including hydrocarbons, sugars or oils.

The role of these biosurfactants in the functioning of the microorganisms is not clear. The biosurfactants occur in two principal places - either (i) extracellular in which case they may be used to aid emulsification of their hydrocarbon feedstock or (ii) they may reside within the cell wall, in which case they possibly assist in transport of hydrocarbon across the cell wall. The extracellular surfactants may play a part in cell recognition, particularly when the specific and delicate nature of the interactions between polysaccharides and proteins are considered.

Whilst biosurfactants are no doubt well suited to perform their function in the microorganism, they offer few specific performance advantages over synthetic surfactants when used in many industrial processes or products. In addition they tend to be more expensive than the most common synthetic surfactants, and much wider use would be made of bio-surfactants if their cost could be reduced.

Types of Surfactant

The hydrophobic part of biosurfactants are normally linear, saturated hydrocarbons, though unsaturated chains and hydroxy-substituted chains are known, as well as the terpenes, which include sterols and cholesterol derivatives. The hydrophilic portion is commonly composed of a sugar or amino acid, the many variations in the spatial arrangements of the hydrophilic groups on this portion of the molecules making an important contribution to various properties displayed by the surfactants. Classification of the surfactants can be made according to the microorganism from which it is obtained.

Bacterial Surfactants

a) Glycolipids. Typical glycolipds are produced by Pseudomonas bacteria, e.g. Pseudomonas aeruginosa No.141 grown on a peptone/glycerol substrate produces(1) a surfactant comprising of 2 rhamnose units joined by an α-1,2 linkage, with a hydrophobic tail of two β-hydroxydecanoic acids, as shown in Fig.1.

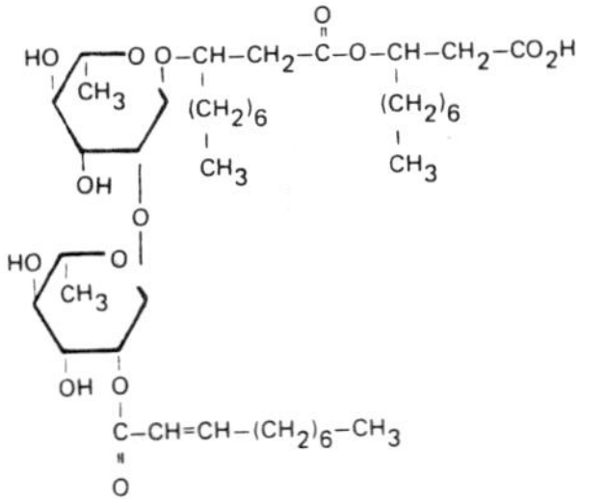

GLYCOLIPID WITH 2 RHAMNOSE UNITS

FIGURE 1

Similarly Pseudomonas Aeruginosa KY 4025, when grown[2] on 10% n-paraffin produced a surfactant with only one rhamnose unit and two β-hydroxydecanoic chains.

Similar molecules have been reported by other workers[3,4]. Rhodococcus erythropolis produced[5] on α,α-trehalose-6-coryno-mycolate (Fig.2), where the sugar and hydrocarbon are linked by an ester group.

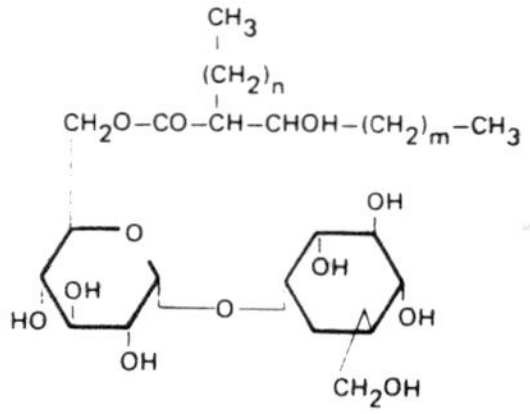

GLYCOLIPID WITH TREHALOSE UNIT
m+n=27---31

FIGURE 2

b) Amino acid lipids. One of the better known molecules is surfactin[6,7], a crystalline peptide lipid produced by Bacillus subtilis IFO 3039 and having a structure shown in Fig.3.

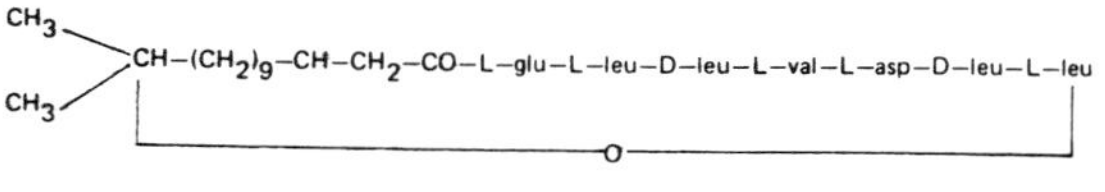

AMINO ACID CONTAINING BIOSURFACTANT, SURFACTIN.

FIGURE 3

Similar molecules have been reported elsewhere[8,9].

Yeast Surfactants

Among the surfactants produced by yeasts are those where the hydrophilic group can exist in a lactone or acidic form. Torulopsis magnoliae when fed on fatty acids and hydrocarbons(10) or glucose, yeast extract and urea(11), produces sophorolipids that can exist as a lactone or acid, as shown in Fig.4.

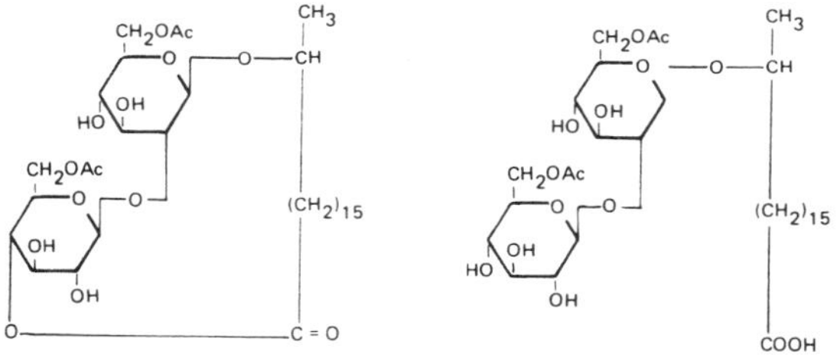

LACTONE FORM(a), ACID FORM(b)
OF SOPHOROLIPID.

FIGURE 4

The link between the hydrocarbon and sophorose unit is an ether group, giving the molecule more chemical stability than those having ester links. The lactone is susceptible to cleavage by alkali in the usual way. The lactone form of the sophorolipid has the unusual property of forming a separate phase in the growing culture. This facilitates the recovery of the surfactant, and brings the price closer to those of synthetic actives.

Fungal Surfactants

The corn smut fungas Uspilago zeae produces a bio-surfactant when grown on glucose cultures[12,13,14]. The surfactant has a di-glucose structure with β-OH alkanoic acids as the hydrophobic portion of the molecule.

Properties of Biosurfactants

The physical and chemical properties of biosurfactants are simply what would be expected from their chemical structure. Sugar groups form strong hydrogen bonds, either with water, other polar solvents or themselves (note the insolubility of cellulose in water in contrast with the high solubility of sugar in water). Thus biosurfactants tend to be soluble in water and polar solvents; in non polar hydrocarbon, solubility can be achieved through the sugar groups interacting with one another, leaving the hydrophobic tail in the solvent and thus often forming a mesophase. Surfactin dissolves more readily in alkali than pure water, mainly because of the higher charge on the molecule.

In their natural environments, the biosurfactants are naturally quite stable though under different conditions, they can rapidly decompose. In common with polysaccharides the glycosidic links are susceptible to cleavage by acids, and under alkaline or acid conditions, ester groups can be broken.

<u>Production of Biosurfactants</u>

Microorganisms can use a variety of substrates to produce either sugars or hydrocarbons. Where the starting material is glucose, glucose 6-P is produced, which in turn can be converted to other saccharides such as trehalose, sophorose or rhamnose or can be split via glyceraldehyde 3-P and pyruvate to acetyl-CoA. This in turn can add to oxaloacetate to produce malonyl-CoA and thence to higher fatty acids. This extension of acetyl-CoA to longer fatty acids is a fundamental reaction in prokaryotes and eucaryotes. At each stage in the cycle an acetyl group is added to the growing chain to produce a slightly longer fatty acid. From this mechanism can be understood why naturally occurring fatty chains have an even number of carbon atoms.

Alternatively hydrocarbon substrates can be utilised to produce fatty acids or saccharides. Alkanes can be oxidised via the alcohol and aldehyde to the fatty acid. Oxidation at the β position produces acetyl-CoA, from which various fatty acids can be synthesised. Thus sugars, alkanes or mixtures of both can be utilised to synthesise biosurfactants, either from the feedstock directly or via intermediate compounds.

A major factor contributing to the cost of biosurfactants is their recovery from solution. Methods of recovery based on solvent extraction or precipitation are commonly used, though the final cost of the material is strongly correlated with its concentration in solution. This can be seen in Fig.5 where the cost of biomaterials in $US is related to their solution concentration.

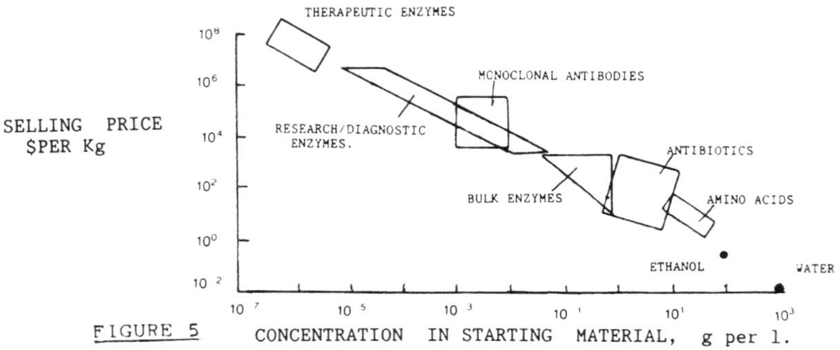

FIGURE 5

Adsorption and Aggregation Properties of Alkyl Polyglycosides

The adsorption and aggregation behaviour of bio-surfactants, especially alkyl carbohydrates can be understood from a study of the closely related alkyl polyglycosides (APG). APG's are produced synthetically but have similar structures to some biosurfactants and because of their lower cost, their application and solution behaviour have been more widely studied.

APGs can be made by reacting short chain alcohols with glucose in the presence of a catalyst, then exchanging the short alcohol for a longer, desired one, a process called trans-glucosidation. This gives surfactants with hydrophilic groups from 1-5 glucose units in length. These molecules are nonionic and can be used in products involving detergency, oil solubilisation, emulsification etc. Nonionic surfactants are already widely used in industry, and are often of the type containing an alkyl ethoxylate.

These and APGs have different properties as a result of their different hydrophilic groups. The ethoxylate group's hydration is both temperature and electrolyte sensitive, whereas the glucose group's hydration tends to be insensitive to both. With increasing temperature, the ethoxylate group dehydrates, thus becoming more insoluble and giving rise to phase separation (cloud point) of ethoxylated nonionic surfactants. In the presence of certain anions, e.g. $SO_4^=$, F^-, OH^- or $CO_3^=$, ethoxylate groups become less soluble due to the negative adsorption of these anions from the ether group. Addition of these anions does not lead to dehydration of the ether groups, though the net effect is the same. The glucose group, by contrast, does not dehydrate with increasing temperature nor are anions negatively adsorbed from sugars. This may be a result of the hydration being to a proton rather than the oxygen of an ether group. Thus the sugar based biosurfactants and APGs are normally more soluble than the ethoxylated nonionics, do not form a separate phase with increasing temperature, are milder to the skin than many nonionic surfactants and are rapidly biodegraded.

In dilute solutions, biosurfactants adsorb at the air/water interface and reduce surface tension in the same way as synthetic surfactants. A comparison of data from biosurfactants and synthetic ones is given in Table 1.

TABLE 1

Surfactant	Critical micelle conc.mol.ℓ$^{-1}$	γ-surface tension mNm^{-1}	Area/ molecule Å^2	Ref.
Octyl glycol ether	$4.9x10^{-3}$	26	32	15
Octyl glycerol ether	$5.8x10^{-3}$	25	32	15
Octyl β-D-glucoside	$2.5x10^{-2}$	31	41	15
C_{12} β-D-maltoside	$1.5x10^{-4}$	36	50	16
$C_{12}EO_8$	$8x10^{-5}$	36	64	16
Rhamnolipid		29[a]		17
Trehalose (tetraester)		26[a]		18

[a]Data obtained in electrolyte solution.

The data show that the bulky glucose head group retards the packing of the molecules at interfaces compared to the single ether groups. Increasing the hydrocarbon chain length allows micelles to form more easily, though the larger maltose head group shown in Fig.6 retards packing at the A/W interface compared to the smaller glucose head group. By manipulating the nature of both the hydrophilic and hydrophobic group, various properties of solubilisation and surface tension lowering can be obtained.

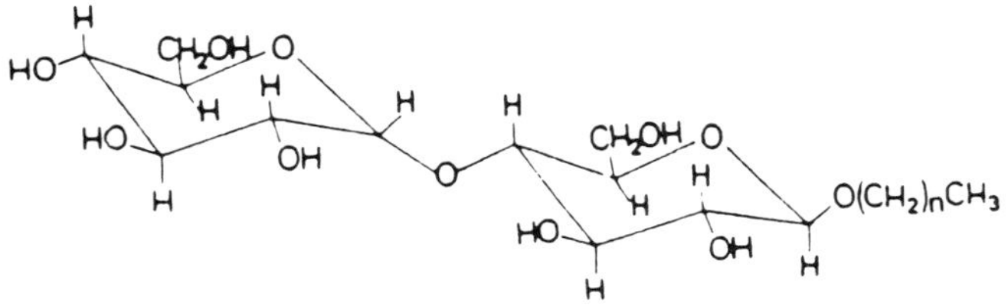

ALKYL MALTOSIDE

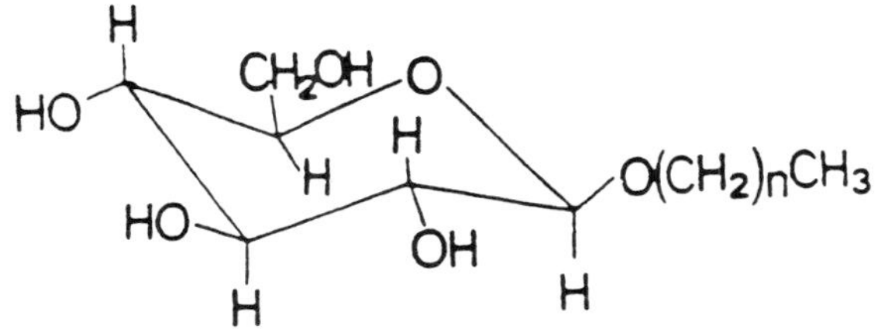

ALKYL GLUCOSIDE

FIGURE 6

In more concentrated systems, biosurfactants can form liquid crystal structures in a similar way to synthetic surfactants through the phase diagrams tend to be more simple than for equivalent ethoxylated nonionic surfactants. This is shown in data of Warr et al.[19] in Fig.7, where n-dodecyl β-D-maltoside shows the simple transitions of isotropic[4] --> lamellar (L_1) --> solid (S). More recently, Marcus et al.[21] have demonstrated a H_2 phase between L_1 and L_α. In contrast[20], the phase diagram for $C_{12}EO_8$ is also shown in Fig.7 and shows in addition to the isotropic, hexagonal and solid, other phases such as hexagonal (H_1), liquid surfactant (L_2) and bicontinuous cubic phase.

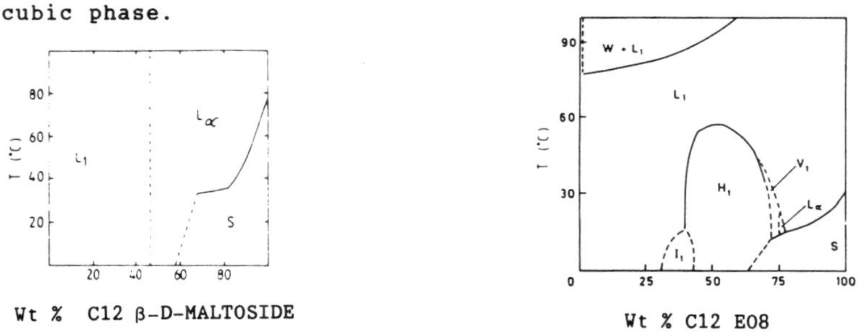

FIGURE 7

This difference is mainly due to the decrease in solubility of the ethylene oxide group in water with increasing temperature, allowing the repulsion between head groups within the micelle to decrease with increasing temperature. The delicate balance of forces, attractive between the hydrophobic group, repulsive between the headgroups and usually repulsive between micelle surfaces largely determines the phase diagram.

In other systems(22) n-alkyl β-D-glucopyranosides have been shown to exhibit smetic A phases, with some form of bilayer structuring. One possible structure is shown in Fig.8.

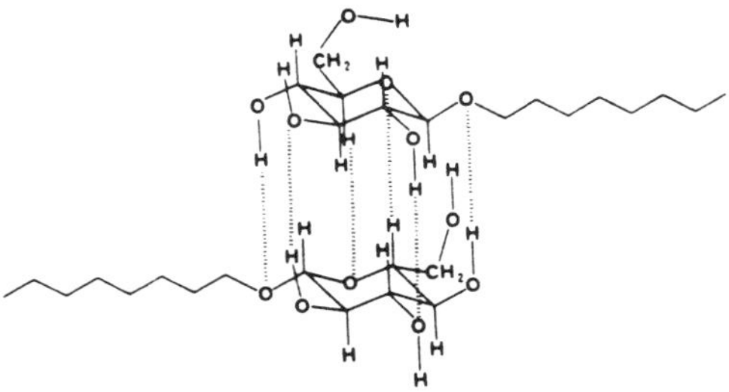

POSSIBLE HEAD GROUP INTERACTIONS.

FIGURE 8

A review of the liquid crystal phases of alkyl carbohydrates has been made by Jeffrey(23). Most of the thermotropic liquid crystals have been found with mono-alkyl surfactants, and they show two "melting points" - one where the alkyl chains melt and another at higher temperatures where the sugar groups dissociate. The X-ray studies(24) of liquid crystals of heptyl α-D-mannopyranoside and octyl 1-S-β-D xylopyranoside show only one powder diffraction spectrum at 22Å and 30Å respectively. These periodicities are between 1 and 2 times the respective chain lengths, indicating tilting of the bi-layers. Although liquid crystals have mostly been found with monalkyl carbohydrates van Doren et al.[25] have recently found dialkyl carbohydrates that form liquid crystal phases.

The aggregation between the hydrophilic groups can lead to extended structures in solution. Long 'whisker-like' structures were found[26] with n-octylaldonamides in solution; the length; diameter ratios ($\sim10^4$) depended on the solvent used.

Industrial Applications of Biosurfactants

The main reported use of biosurfactants is in the recovery of oil. The recovery of oil from wells in the second stage can require assistance either from water pressure or surfactants to lower oil/water interfacial tensions. Water or viscous polymer solutions are injected on one side of the oil field and used to push oil out of the rock. Lowering interfacial tensions allows oil droplets to be deformed more easily thus facilitating their movement through the porous rock. However the cost of using synthetic surfactant to achieve this increased oil recovery is only justified if the oil price is high (~$30/b). Microorganisms injected down oil wells can produce both gas and surfactants, both of which can assist the recovery of the oil. Significant problems exist in keeping the microorganisms alive and producing surfactants under the high temperatures and salinities sometimes encountered in wells. Nevertheless, this application obviously does not require the expensive step of separating the biosurfactant from the medium in which it was produced. A detailed review of the application of bio-surfactants to oil recovery has been written by Brown et al.[27].

REFERENCES

1. J R Edwards and J A Hayashi, Arch.Biochem.Biophys., 1965, 111, 415.
2. S Itoh, H Honda, F Tomito and T Suzuki, J.Antibiotics, 1971, 24, 855.
3. M Yamaguchi, A Sato and A Yukuyama, Chem.Ind., 1976, 4, 741.
4. C Syldatk, S Lang, F Wagner, V Wray and L Witle, Zeit. Naturforschung, 1985, 40C, 51.
5. A Kretschmer, H Bock and F Wagner, Appl.Environ.Microbiol., 1982, 44, 864.
6. A Kakinuma, M Hori, H Sugino, I Yoshida, M Isono, G Tamura and K Arima, Agr.Biol.Chem., 1969, 33, 1523.
7. K Arima, A Kakinuma and G Tamura, Biochem. Biophys.Res. Commun., 1968, 31, 488.
8. F Peypoux, G Michel and L Delcambe, Eur.J.Biochem., 1976, 63, 391.
9. F Besson, F Peypoux, G Michel and L Delcambe, Eur.J. Biochem., 1977, 77, 61.
10. A P Tulock, A Hill and J F T Spencer, Can.J.Chem., 1968, 46, 3337.
11. C Asselineau and J Asselineau, Prog.Chem.Fats, Other Lipids, 1978, 16, 59.
12. R U Lemieux, Can.J.Chem., 1951, 29, 415.
13. R U Lemieux and R Charanduk, Can.J.Chem., 1951, 29, 759.
14. S S Bhattacharjee, R H Haskins and P A J Gorin, Carbohyd. Res., 1970, 13, 235.

15. K Shinoda, T Yamanaka and K Kinoshita, J.Phys.Chem., 1959, 63, 648.

16. C J Drummond, G G Warr, F Grieser, B W Ninham and D Fennell Evans, J.Phys.Chem., 1985, 89, 2103.

17. L H Guerra-Santos, O Kaeppeli and D Fiechter, Appl.Environ. Microbiol., 1984, 48, 301.

18. F Wagner, J-S Kim, S Lang, Z-Y Li, G Marwede, U Matulovic, E Ristau and C Syldatk, Proc III Eur.Congr.Biotechnol. Verlag Chem., 1984, I_{1-3}.

19. G G Warr, C J Drummond, F Geiser, B W Ninham and D F Evans, J.Phys.Chem., 1986, 90 4581.

20. D J Mitchell, G J T Tiddy, L Waring, T Bostock and M McDonald, JCS Faraday Trans I, 1983, 79, 975.

21. M A Marcus and P L Finn, Liquid Crystals, 1988, 3, 381.

22. J W Goodby, Mol.Cryst.Liq.Cryst., 1984, 110, 205.

23. G A Jeffrey, Acc.Chem.Res., 1986, 19, 168.

24. D C Carter, J R Rubble and G A Jeffrey, Carbohydrate Research, 1982, 102, 59.

25. H van Doren, T J Buma, R M Kellogg and H Wynberg, JCS Chem. Commun., 1988, 460.

26. J H Fuhrhop, P Schneider, E Boekema and W Helfrich, J.Amer.Chem Soc., 1988, 110, 286.

27. M J Brown, V Moses, J P Robinson and D G Springham, CRC Crit. Rev. Biotechnol., 1986, 3 159.

Formation and Commercial Applications of Organic Ultrathin Films

J. H. Harwell and E. A. O'Rear
INSTITUTE OF APPLIED SURFACTANT SCIENCE AND SCHOOL OF CHEMICAL ENGINEERING AND MATERIALS SCIENCE, THE UNIVERSITY OF OKLAHOMA, NORMAN, OKLAHOMA 73019, USA

Introduction

This paper describes a new technique, using surfactants, which may be used to engineer the properties of a surface or interface through the formation of an ultrathin organic film. The technique is quite versatile, is applicable to a variety of surfaces, and can be used to form films of widely varying characteristics; this versatility makes the concept quite exciting. On the other hand, this technique is very new, the first published reports having appeared only in 1987[1,2]. Though the paper will describe the technique in some detail, it should be realized that only a few systems have been investigated to date, and that the first successful commercial applications are still in the future.

Background

The new surface modification technique described here is based on an extension of two of the most well know aspects of surfactant behavior to a solid/solution interface. These two aspects of surfactant behavior are micelle formation and solubilization; we refer to their interfacial analogues as admicelle formation and adsolubilization. Just as micelle formation and solubilization form the basis for emulsion polymerization, their interfacial analogues form the basis for this new surface modification technique.

The term "admicelle" is intended to convey the idea of an aggregate of adsorbed surfactant molecules with properties very much like those of a micelle[3]. One of these properties is the ability of the surfactant aggregate to incorporate non-surfactant molecules into the aggregate; in this case, the non-surfactant molecules will be monomers of the polymer with which the surface is to be coated. Because the surfactant aggregate is at an interface, we refer to the incorporation of the monomers into the aggregate as adsolubilization.

The formation of the organic thin film can be thought of as occurring in three steps[1]. In step 1 the surface to be coated with the film is exposed to a surfactant solution under conditions where the chosen surfactant will adsorb on the chosen surface to an extent sufficient to form a bilayer of surfactant on the surface. In step 2 the bilayer is exposed to monomers, which must have sufficiently low solubility in the supernatant solution so that they preferentially partition into the adsorbed surfactant layer. The organic interior of the bilayer acts like a two-dimensional solvent for the monomers. In step 3 polymerization is initiated; if the components of the system have been chosen properly, there will be insufficient monomer outside of the bilayer to sustain a polymerization reaction in the supernatant; thus, the reaction will occur in the bilayer, resulting in formation of a thin film of polymer on the surface.

As of this writing we have published work with only one system: sodium dodecyl sulfate (SDS) as the surfactant, alpha-alumina as the surface, and styrene as the monomer. Most of our work has been done using porous alumina powder as the substrate, simply because it is easy to work with. In an aqueous slurry, alumina exhibits an electrophoretic mobility of zero at pH 9.5; at lower pH values the surface is positively charged and SDS readily forms admicelles on the surface, at surfactant concentrations well below its critical micelle concentration (CMC). At pH 4, SDS forms a complete bilayer on the alumina at a concentration just below the CMC. If the polymerization reaction is to be confined to the bilayer, it is important that there be no micelles in which the monomer can be solubilized; solution conditions, surface, and surfactant must be matched so that the bilayer is formed at a concentration below the CMC.

Because styrene has a very low solubility in water, it partitions readily into the surfactant bilayer; what is difficult is getting sufficient styrene to the bilayer to sustain the polymerization reaction. In our initial attempts to form a thin film by our technique, we found that a saturated solution of styrene would be stripped of styrene by exposure to the bilayer, so that the equilibrium styrene concentration in the supernatant was almost undetectable. While this was no problem in and of itself, even large quantities of styrene-saturated solution gave us very low loadings of the bilayer. We tried putting excess styrene in contact with the supernatant, but this had two drawbacks. Because the amount of styrene taken up by the bilayer was so small, we had no effective way of determining the equilibrium amount of styrene in the bilayer, since the excess styrene kept the chemical potential of the styrene constant in the whole system. Even more important, when we tried to run the polymerization reaction in the presence of the excess styrene, the supernatant solution became cloudy, which lead us to believe that we were probably making polystyrene by emulsion polymerization; we were not restricting the polymerization reaction to the bilayer. By adding ethanol to the supernatant[2], however, in concentrations up to 0.5 M, we were able to make enough styrene available to the bilayer to saturate it with styrene, while not leaving enough styrene in the supernatant to sustain polymerization outside of the bilayer. Fortunately, the presence of the ethanol did not disturb the surfactant bilayer. While we did not make an extensive study of the behavior of the styrene/SDS/alumina/ethanol system, it is interesting that, for most of the systems we measured, the ratio of surfactant to styrene in the bilayer was 2:1, suggesting a structure something like a styrene sandwich, with the SDS molecules serving as the slices of bread. In no system that we measured was the proportion of styrene in the bilayer ever found to be be greater than this.

After equilibration of the alumina surface with the solution of SDS, styrene, and ethanol, polymerization was initiated by adding sodium persulfate and heating the solution above 50°C. Persulfate is a water soluble initiator; in emulsion polymerization systems it is generally believed that styrene radicals originating in the supernatant must partition into the surfactant aggregate, where there is a sufficient density of styrene monomers to sustain polymerization. We have been able to model the production of polystyrene in our system by making this assumption[1].

We tracked the course of polymerization in the bilayer by two separate methods[1,2]. In the first method, we ran the reaction in closed vials for fixed periods of time, quenched the reaction in ice water, then separated the alumina from the supernatant by filtration, using ice water to rinse the powder. The amount of polystyrene formed was determined by extracting the polystyrene from the surface of the alumina with tetrahydrofuran (THF), then determining the amount of polystyrene in the THF using UV spectroscopy. We found that, as we had expected, the polymerization kinetics were identical to those found in emulsion polymerization of styrene, corresponding to the case were mass transfer of monomer from the supernatant to the surfactant aggregate was the rate controlling step. While the amount of styrene in the bilayer was at equilibrium before polymerization began, additional styrene began to partition into the bilayer after the polymerization reaction had been initiated. This probably has at least two causes, one being the high solubility of styrene in polystyrene, the other being that as the polymerization reaction proceeded, styrene was consumed in parts of the bilayer, making room for additional styrene to be adsorbed from the supernatant.

The other method we used for tracking the film-forming process involved ellipsometry[4]. A layer of aluminum was vacuum deposited onto a glass slide; an oxide layer was formed on the aluminum by aging the aluminum layer in an oven at 85°C. The slide, with the layers of aluminum and aluminum oxide, was then placed in contact with the surfactant/ethanol/styrene solution at concentrations and at a pH at which the system had equilibrated when in contact with the alumina powder. After a suitable period for equilibration, the polymerization reaction was initiated as before, allowed to run for a preselected period of time, then quenched. After quenching, the slide was removed from the vial in which the reaction was run, rinsed in distilled, deionized water, and allowed to dry. We then used ellipsometry to look for evidence of formation of a thin film on top of the oxide layer at each of the different reaction times. The results confirmed that a film was forming and that its dimensions were consistent with those we expected from the mechanism we postulated: For very short reaction times--on the order of a few minutes--no film was detected on top of the oxide layer. We interpreted this to indicate that polymerization had not been allowed to continue in these cases for sufficient time to produce a film with enough integrity to adhere to the oxide layer in the presence of the distilled water. For intermediate reaction times we detected films of less than 1 nm thickness; this is less than the thickness of a monolayer of SDS; we interpreted this to indicate that patches of the surface were now covered with a film and that the ellipsometer was giving us an average value for the surface. Then, for reaction times greater than 20 minutes, we found films of 3.2 nm to 3.4 nm thickness; this corresponds to the thickness of an SDS bilayer with the SDS tails fully extended and not interpenetrating. This was precisely the results we were hoping to obtain. At very much longer reaction times we began to find films of increasing thickness, some approaching 20 nm. We believe that films of this thickness are consistent with the transfer of additional styrene from the supernatant to the surface which we observed occurs as the polymerization reaction proceeds.

Applications

We stumbled onto one of the possible uses of the process very early in our feasibility studies. A natural question to ask about the films concerns how well they adhere to the oxide layer. To test this we decided to pack a liquid chromatography column with the filmed alumina, then pass distilled, deionized water through the column for several times the void volume in the column, then extract the polystyrene with THF to see how much remained. When we tried to make an aqueous slurry for pumping into the column, we were surprised to find that the alumina now floated on the water; indeed, if some of the alumina was forced down below the surface of the water with a spatula, it immediately popped back to the surface, remaining completely dry. We hypothesized that the upper layer of the bilayer had been leached out of the film by the washing process to which we subjected the alumina after the polymerization reaction had been quenched. After washing the alumina was dried in an oven at 60°C to drive off any unreacted styrene. With the polystyrene film now applied to the surface, the surface had become hydrophobic, and the pores of the alumina would no longer take up water; the air which was now trapped in the pores of the alumina made it buoyant.

To confirm this hypothesis we performed and analyzed a number of nitrogen sorption studies on the bare and on the filmed alumina samples[5]. If the film had been applied uniformly to the surface of the alumina, including the internal pore surface, we expected that the surface area determined from the nitrogen sorption would be only slightly reduced and the pore size distribution would remain qualitatively the same, only shifted to narrower pores. If the buoyancy was produced by particles of polymer blocking the pores, we expected a possibly sharp reduction in surface area and a dramatic change in the pore size distribution. The results of our study suggested that the polymer film was uniformly distributed over the internal surface of the alumina. Application of the film reduced the surface area from 96 m^2/g to 79 m^2/g. More importantly, the pore size distribution curves were shifted to smaller diameter pores, but the distribution curves maintained the same shape.

The film coated alumina is only hydrophobic if it has been washed between the polymerization and drying steps. When the washing step is skipped, the alumina stays hydrophilic. Additionally, if surfactant is added to the water on which the hydrophobic alumina is floating, the alumina eventually wets, the pores fill with water, and the powder particles sink. All of this points to the surfactant in the second layer of the bilayer being reversibly adsorbed on the polymer layer. It should be possible to replace the second layer with any surfactant of interest, in order to give the surface whatever functional groups or mixture of hydrophobic and hydrophilic character desirable; all the while the substrate will have all the mechanical and chemical properties of the inorganic material on which the film was formed.

We have used the filmed alumina as an ion exchange material[6]. After washing off the SDS layer, we flushed a column packed with the alumina with a solution of sodium tetradecylsulfate (TDS); TDS has a Krafft point near 50°C, so it was applied to the film surface above this temperature. When used as an ion exchange material below 50°C, the TDS did not leach off of the surface. Sodium ions associated with the TDS groups were readily exchanged for calcium ions when a solution of calcium chloride was run through the column. Though the sample we prepared had a low ion exchange capacity, it was on an inorganic support.

We also used the filmed alumina as an adsorbent for low molecular weight inorganic compounds[6]. When an aqueous solution of tert-butyl-phenol was passed through a column packed with the filmed alumina, it had a capacity for the phenol which was comparable to that of activated alumina. While these are "low-tech" uses of the material created by forming the film on the alumina, the more important lesson to be learned from this is the ease with which an inorganic powder can be tailored so that it has surface properties to accomplish whatever purpose is intended. By varying the amount of adsorption in step 1 of the film forming process the ratio of hydrophobic patches to hydrophilic patches on the surface can also be varied.

Another potential area of application for films formed by this process is in the manufacturing of electronic devices. Various electronic devices assist or entertain us daily at our homes, offices, factories and even our cars. Over the span of a single generation, these gadgets have become interwoven with modern economies and cultures. Essential to this proliferation is the development of microelectronic fabrication techniques and particularly, microlithography. Microlithography is the method by which circuit patterns are miniaturized and reproduced on silicon wafers.

Driven largely by the demand for more powerful and faster computers, researchers in microlithography endeavor to produce higher densities of component circuit element. Tandem research efforts are underway in various radiation sources(e.g. deep UV, electron beam, and X-ray) and in new resist materials for higher resolution and improved contrast and sensitivity properties. Advances have been so rapid that computers become obsolescent long before their physical demise.

Organic thin films are utilized in microelectronics fabrication as a radiation sensitive layer spread over wafers of silicon and its oxide (Figure 1). Patterned exposure, such as with a mask, changes the chemical nature of radiated regions of the organic resin so that, during a subsequent developing step, exposed regions are selectively removed or retained. If the exposed section remains after development, then the material is said to be a negative resist; if the unexposed region is retained however, then the resin is called a positive resist[7]. Polystyrene is an example of a negative resist. It forms crosslinks on exposure to UV or e-beam radiation with the higher molecular weight polymer being less soluble in developing solvent. Polymethylmethacrylate, on the other hand, undergoes bond scissions and functions as a

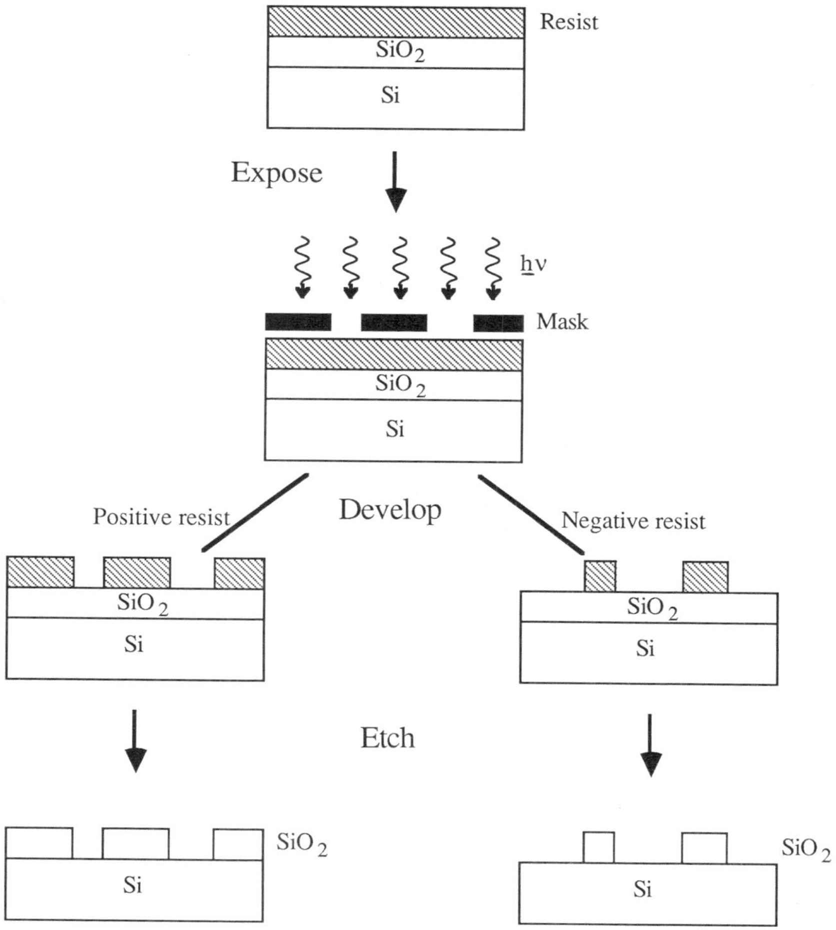

Figure 1. How a resist is used in microelectronic fabrication.

positive resist. More common positive resists, such as the Novolac resins, contain solution inhibitors that rearrange photochemically through the Wolff reaction to a water-soluble carboxylic acid form and thereby permit dissolution of the polymer matrix[8].

Among other factors, certain physical phenomena contribute to loss of resolution and place practical limits on circuit element densities. A key parameter in helping to control these processes so that results can approach theoretical limits is film thickness of the organic resin. Film thickness affects different exposure methods through different mechanisms. Where light is the exposing radiation, for example, diffraction occurs at the boundaries between the transparent and opaque regions of the mask. The thicker the photoresist layer, the greater the light beam will become diffuse or spread out at the resin-silicon dioxide interface. In the case of exposure by e-beam, interactions of the electrons with the atoms in the resist layer cause lateral scattering. For the ideal case of a monoenergetic beam, Fermi derived the following probability relationship for forward beam scatter at depth z into the material and radial distance r from the incident beam axis[9]:

$$H(r,z) = 3\lambda \exp(-3\lambda r^2/4z^3)/(4\pi z^3)$$

in which λ is the transport mean free path in nanometers and is given by

$$\lambda = 5.12 \times 10^{-4} E^2 A/(\rho z^2 \ln(0.725 E^{1/2} Z^{1/2}))$$

where A and Z are the mass number and atomic number of the target,

ρ the density, and E is the electron beam energy. By setting r=0, we see that the probability that an electron will not be deflected from the beam path strongly decreases with penetration depth. In short, the cumulative effect with increasing thickness of the resist is greater dispersion at the substrate-resist boundary.

Conventional films are applied in an empirical process called spin coating. Polymer in an organic solvent is applied to a silicon wafer that is then rotated at a prescribed angular velocity and acceleration under vacuum. It is generally thought that this method is limited to preparing films of 0.5-1.0 microns due to the presence of pinhole defects in the thinnest films[10], though some recent results bring this limit into question[11].

Certain frontier investigations in microlithography have explored the use of Langmuir-Blodgett(LB) deposited films as resist materials. These studies on LB films might be broadly categorized in two groups, either deposition of small amphiphilic molecules or of macromolecules. In the LB method, a monolayer of amphiphilic material is first formed at a liquid-gas interface in a special trough. Subsequently, the material is deposited as a monolayer onto the substrate by sweeping across the surface with an inert barrier as the vertically oriented substrate is slowly removed from or lowered into the solution. For multilayer structures, dipping is repeated sequentially[12-14].

Various single component, deposited Langmuir-Blodgett films have been demonstrated as high resolution e-beam lithographic resists. For instance, Barraud and coworkers in France[15-17] polymerized 30 monolayers(90 nm) of α-tricosenoic acid on an aluminum substrate in a scanning electron microscope at 5 kV accelerating voltage. Very high resolutions of 60 nm, sensitivities up to 0.5

μC/cm2 and also contrasts of 0.7 to 2.4 were obtained. Lando's group in the United States pre-stabilized 69 nm thick multilayers of ω-octadecylacrylic acid with a small dose of UV radiation prior to e-beam exposure[10]. Any subsequent electron irradiation produced a negative resist while high doses completely polymerized the film. In this case, the film could be utilized as a positive resist material. Still other researchers have shown vinyl esters[18-20], long chain fumaryl monoesters[21] and acrylics[22-25] among other molecules[26,27] to be polymerizable either as monolayers or deposited Langmuir-Blodgett films.

The contributions of LB technology have been significant, yet LB layers suffer from some serious limitations[28]. Among these are film stability, requirements for a vibration-free environment, limited production rate, and difficulties of scale-up. Although the formation of LB films has a number of serious obstacles to routing implementation, the technique can help to overcome pinhole defects, address inherent limitations of resolution in conventional resists due to polymer size, and can form thinner films to reduce the effects of diffraction of electron scattering.

The most notable alternatives to the LB technique have been those categorized as self-assembling films that utilize surface active molecules. Though not as old as the LB approach nor as well-studied, self-assembling approaches may help to address the limitations of LB deposition. Several adsorption systems have been used to form organic monolayers and could serve as the basis for polymerized films. For instance, aliphatic carboxylic acids that are known to self-assemble on the oxide layers over aluminum and silver[29-31] are similar to polymerizable LB films. As another possible stepping stone, Nuzzo and Allara have reported the interesting discovery that organic sulfur compounds spontaneously form monolayers on gold surfaces[32-34]. To a large degree, the above systems are tied philosophically to early successful attempts to form polymer films at the air-liquid interface. That is, the underlying approach is to incorporate polymerizable functional groups into surface active molecular species. The new approach described in this paper centers on the use of adsorbed surfactant aggregates not (necessarily) as a reactive species, but as an interfacial polymerization solvent for separate monomeric species. Given the highly technical research competitiveness of the electronics industry and its expected market growth, the importance of surfactants through the LB approach or alternative methods will increase into the next century.

We have made some preliminary attempts to modify the system we used to place films on alumina for making films on silicon oxide. The most obvious change that needs to be made is the replacement of the SDS by a cationic surfactant. We have been working with cetylpyridinium chloride (CPC). Though there is substantial adsorption of the CPC on the silicon oxide, we have not yet confirmed that we actually have a bilayer; the silicon oxide is considerably more difficult to manipulate through varying the pH of the solution than is the alumina.

Conclusions

These are not the only potential applications of the thin films we have described in this paper. The great similarity between the surfactant bilayer and lipid bilayers such as make up the membranes of living cells suggests that these films may find application as a means of immobilizing hydrophobic enzymes or synzymes. Thin films have potentially important applications in other areas, such as corrosion inhibition, solid lubricants, optical coatings, surface-modified electrodes, and membranes for chemical or biochemical separations. They may also be important in developing techniques for engineering the interface between fiber and polymer in fiber reinforced polymer composites. In certain aspects of ceramic processing and the processing of green bodies for microelectronic applications films of the type described here may offer a unique opportunity to design the interface of the particles being processed so that it has precisely the properties desired. There is no question that many of these areas will be of ever increasing economic importance in the future and that development of new generations of thin films will play a critical role in this; what remains to be seen is whether or not thin films formed by polymerization in adsorbed surfactant layers will best satisfy the demands of these emerging technologies.

References

1. J. Wu, J.H. Harwell and E.A. O'Rear, J. Phys. Chem. 1987, 91, 623.
2. J. Wu, J.H. Harwell and E.A. O'Rear, Langmuir 1987, 3, 531.
3. J.H. Harwell, J.C. Hoskins, R.S. Schechter and W.H. Wade, 1985, 1, 251.
4. J. Wu, J.H. Harwell and E.A. O'Rear, Colloids and Surfaces 1987, 26, 155.
5. J. Wu, J.H. Harwell and E.A. O'Rear, AICHE J. 1988, 34, 1511.
6. J. Wu, PhD Dissertation, The University of Oklahoma, 1987.
7. C. Jaeger, "Introduction to Microelectronics Fabrication," Addison-Wesley, Reading, Massachusetts(U.S.A.), 1988, Vol V, Chapter 2, p. 13.
8. C.G. Willson, Organic Resist Materials-Theory and Chemistry, in "Introduction to Microlithography," American Chemical Society, Washington, D.C., 1983, ACS Symp Ser 219, p. 87.
9. L.F. Thompson and M.J. Bowden, The Lithographic Process: The Physics, in "Introduction to Microlithography," American Chemical Society, Washington, D.C., 1983, ACS Symp Ser 219, p. 15.
10. G. Fariss, J. Lando and S. Rickert, Thin Solid Films, 1983, 99, 305.
11. S.W.J. Kuan, C.C. Fu, R.F.W. Pease and C.W. Frank, Proc. SPIE(in press).
12. V.K. Agarwal, Physics Today, 1988, 41,40.
13. A.W. Adamson, "Physical Chemistry of Surfaces," 4th ed., Wiley-Interscience, New York, 1982, Chapter IV.
14. M. Breton, J Macromol Sci-Rev Macromol Chem, 1981, C21, 61.

15. A. Barraud, C. Rosilio and A. Ruaudel-Teixier, J Coll Interface Sci, 1977, 62, 509.
16. A. Barraud, C. Rosilio and A. Ruaudel-Teixier, Thin Solid Films, 1980, 68, 91.
17. A. Barraud, C. Rosilio and A. Ruaudel-Teixier, Thin Solid Films, 1980, 68, 99.
18. A. Cemel, T. Fort and J.B. Lando, J Polym Sci Pt A-1, 1972, 10, 2061.
19. S.A. Letts, T. Fort and J. B. Lando, J Coll Interface Sci, 1976, 56, 64.
20. V. Enkelmann and J.B. Lando, J Polym Sci, 1977, 15, 1843.
21. D. Naegele, J.B. Lando and H. Ringsdorf, Macromolecules, 1977, 10, 1339.
22. N. Beredjick and W.J. Burlant, J Polym Sci Pt A-1, 1970, 8, 2807.
23. V.R. Ackerman, O. Inacker and H. Ringsdorf, Kolloid Z. u. z. Polymere, 1971, 249, 1118.
24. A. Banerjie and J.B. Lando, Thin Solid Films, 1980, 68, 67.
25. A. Dubault, C. Casagrande and M. Veyssie, J Phys Chem, 1975, 79, 2254.
26. K.C. O'Brien, C.E. Rogers and J.B. Lando, Thin Solid Films, 1983, 102, 131.
27. S.L. Regen, P. Kirszensztejn and A. Singh, Macromolecules, 1983, 16, 335.
28. L. Netzer, R. Iscovici and J. Sagiv, Thin Solid Films, 1983, 99, 235.
29. D.L. Allara and R.G. Nuzzo, Langmuir, 1985, 1, 45.
30. D.L. Allara and R.G. Nuzzo, Langmuir, 1985, 1, 52.
31. N.E. Schlotter, M.D. Porter, T.B. Bright and D.L. Allara, Chem Phys Lett, 1986, 132, 93.
32. R.G. Nuzzo and D.L. Allara, J Am Chem Soc, 1983, 105, 4481.
33. M.D. Porter, T.B. Bright, D.L. Allara and C.E.D. Chidsey, J Am Chem Soc, 1987, 109, 3559.
34. E.B. Troughton, C.D. Bain, G.M. Whitesides, R.G. Nuzzo, D.L. Allara and M.D. Porter, Langmuir, 1988, 4, 365.

Polymeric Surfactants — Properties and Applications

G. Bognolo
ICI SPECIALTY CHEMICALS, EVERSLAAN 45, 3078, EVERBERG, BELGIUM

INTRODUCTION

The last few years have witnessed a continuous growth in the development of surface active agent technology especially for the industrial (i.e. the non-detergent) sector resulting from :

- the increasing availability on an industrial scale of a broader and more sophisticated range of products. In particular molecules that only a few years ago were at the stage of quasi-curiosities of unclear industrial interest, are now becoming an integral part of major manufacturing processes and finished goods.

- the improved knowledge of the application aspects and in particular the structure/performance relationship.

The stimulus for this growth originates from two facts :

- practically any product that we use in our daily life requires the use of surface active agents in its manufacturing history.

- the production technologies become more and more sophisticated and demand higher performance and specificity of effects from the surfactants.

To the surfactants producers and technologists it should be particularly rewarding to realise that the industry is moving away from the classical, oversimplified equation :

Surfactant = soap

There is indeed an increasing appreciation of a more complete and useful concept that a surfactant is a structural and functional entity, often of considerable complexity that is an integral and essential part of a production process or finished product formulation and whose specific effects (resulting form interfacial phenomena) can be optimised by a scientific approach to the molecule design and to the application technology.

This conceptual evolution from the surfactants users and the need for extending and enhancing the range of effects provided has enabled the introduction and commercial exploitation of products designed according to structural and application principles different from the classical surfactants based on the conventional sulfonation, sulfation, esterification, ethoxylation or quaternisation chemistry.

These molecules are surface active in the broadest meaning of the word i.e. they are amphipathics in nature and are active at the interface of immiscible phases. However, the effects provided are not necessarily related to parameters like surface or interfacial tension or critical micelle concentration, but result rather from an extensive interaction between the moieties in the amphipathic molecule and the immiscible phases.

The evidence from basic and applied research supported by results from industrial applications, confirm that this interaction can be best exploited when the amphipathic molecule is of polymeric nature, because of :

- steric stabilisation effects against flocculation and coalescence
- higher solvation/adsorption energy for molecule positioned at the interface
- higher solubility of the polymer amphipathic moieties in "difficult" media like higher molecular weight paraffinic oils or concentrated solutions of inorganic salts.
- better stability of the amphipathic molecule at the interface.

Certain conventional surfactants for example the higher molecular weight ethoxylated derivatives or the ethylene oxide/propylene oxide block copolymers have already a polymeric structure, that is often adequate for binding with the water in aqueous systems. However their interaction with oils or other hydrophobic phases is limited by the chain length of the lipophilic moieties, 18 carbon atoms being the maximum available and economically acceptable for the majority of applications.

It is a key feature of polymeric surfactants that their structure provides a much better interaction with these hydrophobic media, which allows unique effects to be obtained in both aqueous and non-aqueous systems of the emulsion or dispersion type.

It is some of these structural aspects and their use in industrial application that I would like to discuss in some details.

POLYMERIC SURFACTANTS STRUCTURES

There are 2 types of basic structures that we have found especially functional :

A) The "random structure" type of polymers, prepared by reacting aliphatic carboxylic acids, aliphatic and/or aromatic polycarboxylic acids or anhydrides, polyalkylene glycols (usually polyethylene glycol) and polyols.

The resulting products are statistical mixtures, with a wide molecular weight distribution, that can be described in first approximation as containing loops of hydrophilic moieties (polyoxyethylene) bonded through ester linkage to the lipophilic moieties in a random, tridimensional network.

By changing the proportion and type of the reaction components it is possible to produce a broad range of surfactants with different characteristics, with respect to water or hydrocarbon solubility, cloud point (for the water soluble polymers), emulsion stabilisation at high temperature, ionic strength and shear dispersion properties in aqueous and non-aqueous media.

The key features of the random type of polymeric surfactants is that they exhibit both emulsification and dispersion properties, and that they impart good resistance to emulsion coalescence even at low interfacial coverage and at high shear.

These properties have been used in many different industrial applications, ranging from cleaning formulations to dispersions of high molecular weight water-soluble polymers in hydrocarbon media, to paper de-inking in the flotation process. Hypermer A-394, A-409 and A-109 are the surfactants most commonly used. Two applications that deserve special mention are :

a) the emulsification/dispersion of crude oil residues, as for example in tank cleaning or oil spill dispersant formulations.

b) emulsion polymerisation. The high interfacial activity of the random polymer allows superior emulsion stability to be achieved during and after polymerisation whilst significantly improving the water resistance characteristics of the latex.

One interesting variation in the design of random polymeric molecules involves the use of high molecular weight alkenyl succinic anhydrides as the anhydride component and as total or partial replacement of the aliphatic carboxylic acids.

The compounds produced are very soluble in aliphatic hydrocarbons even at comparatively high polyethylene glycol content and the chain length of the alkenyl moiety provides high stabilisation of the dispersed phase in both aqueous and non-aqueous systems.

It is a peculiar feature of these surfactants that the same molecule can often be used to produce emulsions of the oil-in-water and water-in-oil type, that are stable in demanding conditions of temperature and ionic strength of the aqueous phase.

Similarly to the compounds described before, the higher molecular weight species tend to provide better stabilisation for the emulsions but require more energy input in the emulsion preparation. This can be overcome by the use of synergistic blends of polymeric surfactants produced with alkenyl succinic anhydrides of different molecular weight or of polymeric and conventional surfactants.

Polymers from this class and in particular Hypermer A-60 have found numerous applications in the metal working and hydraulic fluids industry principally because of solubility in low polarity oils and the superior emulsion stabilisation properties.

B) The "ordinate structure" type of polymers resulting from the esterification of poly (12-hydroxystearic acid), (PHSA) with polyalkylene glycols to give block copolymers of the A-B-A type. The structure is conceptually similar to the classical ethylene oxide/propylene oxide condensates, however has the advantage that the differences in the polarity and solubility characteristics of the A and B segments are much more pronounced and that the hydrophobic PHSA moiety is soluble in a variety of oily phases, including aliphatic hydrocarbons. The hydrophilic moiety can be tailored by an appropriate choice of the polyalkylene glycol to achieve the desired degree of interaction with aqueous phases of different composition.

The key features of this class of polymeric surfactants, which are discussed below, can be illustrated by its most representative compounds (Hypermer B-246). These compounds are covered by the European Patent EP000424.

In the Hypermer B-246 molecule, the hydrophobe is poly (12-hydroxystearic acid) and the hydrophyle is polyethylene oxide, the molar ratio between these two moieties being 2/1. The resulting block copolymer has schematically the structure

PHSA - PEG - PHSA

Although the concept of HLB has less significance in the case of polymeric surfactants than for conventional non-ionic surfactants, for descriptive and comparative purposes Hypermer B-246 can be assigned a nominal HLB value (based on the percentage of PEG) of 5-7.

The product is insoluble in water, but soluble in esters, ether esters and ether alcohols, and in aromatic and paraffinic hydrocarbons. In concentrated solution it forms large "lamellar" liquid crystals, and many of these solutions are gels at higher concentrations.

Hypermer B-246 is a very effective, broad spectrum water-in-oil emulsifier, but can form oil-in-water emulsions with heavy paraffinic oils. It is also a dispersant for a wide range of solids, both organic and inorganic, in both water and paraffinic hydrocarbons.

Being non-ionic, Hypermer B-246 as well as other ordinate polymers in the series function by providing a steric barrier against coalescence or aggregation. When the continuous phase is oil, the PHSA chains must be solvated by the oil and extend to form the steric protective layer. The length of the extended PHSA chain has been measured with different techniques (viscometric, capacitance) and found to be dependent on the oil phase :

Decane	115 Å
Dodecane	70-90 Å
Tetradecane	70 Å

This compares with 20-22 Å for C_{18} sorbitan esters and clearly proves the degree of stabilisation achievable with this type of molecule. The stabilisation properties are further enhanced by the PEG moiety either because of the solubility in the aqueous phase, or of the affinity with the particle surface or of the rejection from the oil phase. The rejection of PEG from oil phases is particularly important with paraffinic solvents and, is best illustrated by the formation of ethylene glycol emulsions in mineral oil despite the fact that PEG is insoluble in ethylene glycol.

The most relevant industrial applications for the "ordinate" polymers commercially available are in the formulation and stabilisation of water-in-oil emulsions. Three of these deserve a particular mention, namely

a) Emulsion explosives

b) Acrylamide polymerisation through the "inverse emulsion" process

c) Water/oil/water emulsions

a) Emulsion explosives represent the most advanced development in terms of safety, ease of handling and detonation effectiveness.

They are water-in-oil emulsions where :

- 90 to 92 % by weight of the total emulsion is the oxidiser in the form of a water phase consisting of a supersaturated solution of inorganic nitrates.
- 5 - 7 % by weight is a blend of fuel (hydrocarbon of different composition depending on the explosive type) and emulsifier.
- hollow glass microspheres (sensitisers) make the balance to 100 %.

The emulsion is formulated in extreme conditions of ionic strength and dispersed phase ratio and must be stable over a wide temperature range and thermal cycles for periods of up to 24 months or more. Even the slightest instability will cause crystallisation of the supersaturated oxidiser solution and consequently loss of detonation properties.

Products based on the Hypermer B-246 technology are particularly suitable to provide the required stabilisation.

b) The inverse emulsion process enables production, at fast polymerisation rates, of high molecular weight polymers of constant quality from water soluble monomers like acrylamide, acrylic acid, dimethyl diallyl ammonium chloride, dimethyl aminoethyl methacrylate. The process consists of polymerising a water soluble monomer in the water phase of a water-in-oil emulsion stabilised by low HLB emulsifiers like sorbitan esters alone or combined with low ethoxylate derivatives.

As shown in the table below, Hypermer B-246 in combination with sorbitan monooleate can be used either to make

- low solid emulsions with reduced level of surfactant on polymer

or

- high solid emulsions with no increase in surfactant level on polymer compared to sorbitan esters in low solid formulations.

MODEL SYSTEM
Anionic polymer acrylamide/acrylic acid 70/30 at pH 5.0, oil phase low aromatic kerosene

Polymer solids (%)	Emulsifier type and level required for emulsion stability	
	Sorbitan monooleate (Span 80) alone	Hypermer B-246 Sorbitan monooleate (Span 80) 1/1 ratio
26 %	7.1 %	4.3 %
33 %	Emulsion collapses	5-4 %
40 %	Emulsion collapses	7.0 %

The stabilisation provided by the Hypermer B-246/Span 80 mix is further enhanced when the oil phase contains reduced levels of aromatic, e.g. below 0.2 %.

c) Water/oil/water emulsions are one of the most interesting technological developments in the area of colloid chemistry.

As the name already suggests, these emulsions are three phase systems, where an initial water-in-oil emulsion is further emulsified into water whilst still retaining its structure, i.e. in the final oil-in-water emulsion, the oil phase drops contain droplets of the initial water phase.

Multiple emulsions could offer a versatile, simple and cheap tool for many applications, the most immediate ones being :

- liquid surface membranes for the separation of organic and inorganic materials
- controlled release of drugs or other active principles
- control of drug overdose

Water/oil/water emulsions have been identified and are known since a long time, but their exploitation has been prevented by the impossibility of producing them on a sufficiently large scale, with an acceptable level of reproducibility and with sufficient stability for practical use.

Recent work from Dr. T. Tadros and his team, described in EP267911 has proven that the above targets can be achieved by using A-B-A block copolymers of PHSA and polyethylene oxide (of the type referred to, for example, in UK patent application 2002400) for the first water-in-oil emulsion, and of alkyl phenols propoxylated and ethoxylated or conventional ethylene oxide/propylene oxide block copolymers for the final oil-in-water emulsification.

Stable water/oil/water emulsions have been prepared by M. Seiller and co-workers (M. de Louca, C. Vaution, Y. Bensouda, A. Rabaron, M. Seiller "Les emulsions multiples", 2nd World Surfactants Congress proceedings, volume III, pp 12-33) by using "random" polymeric surfactants (Hypermer A-60) for the initial emulsion and high HLB, high molecular weight ethylene oxide/propylene oxide block copolymers (Synperonic F-127) for the final emulsion.

It is particularly remarkable that both systems exploit the steric stabilisation principle for the first and the final emulsion, and that polymeric surfactants are essential for the preparation and stabilisation of the emulsions.

CONCLUSIONS

The surfactants technology based on polymeric amphipathic molecules as described previously is making rapid progress and enables the operation of processes or the manufacture of products in conditions not attainable with conventional surface active agents.

The applications areas covered are diversified and show clearly the versatility of these products, a versatility that is further enhanced by the possibility of optimising the effects by modifying the molecular structure or by synergistic mixtures between polymers or between polymers and conventional surfactants.

Most interesting of all however is the fact that these surfactants are opening new fields for the industrial exploitation of the emulsions technology, and the initial successes with the water/oil/water systems are an exciting example of this.

We are continuing the work to optimise additional classes of polymers for both dispersions and emulsions on the evidence that polymeric surfactants have proven their technical value and the suitability to respond to existing industrial problems and on the conviction that they will prove of increasing utility as new needs will materialise.

ACKNOWLEDGEMENT

The author is indebted to Mr. A. Baker, I.C.I. Specialty Chemicals, and Dr. T. Tadros, I.C.I. Agrochemicals for their contribution and suggestions.

Ethercarboxylates for Industrial and Institutional Applications

E. Stroink
CHEMISCHE FABRIK CHEM-Y GMBH, KUPFERSTRASSE 1, D-4240 EMMERICH, WEST GERMANY

Introduction

Ethercarboxylic acids have already been the subject of numerous publications and lectures, mainly in the cosmetic field.

In many modern cosmetic formulations ethercarboxylic acids have come to be regarded as indispensable particularly because of their good dermatological properties.

But ethercarboxylic acids are also suitable for use in industrial and institutional cleaner formulations and in household cleaners.

There are several reasons why ethercarboxylic acids are increasingly being used in industry.

First, legislation on environmental safety, degradability and other aspects is becoming increasingly strict and second the requirements placed on surfactants have changed insofar as cleaning techniques and plants are becoming more and more complicated as attempts are made to shorten the cleaning time.

Furthermore, there is a general tendency in all areas to move away from powdered cleaners to liquid formulations, which means higher demands on electrolyte tolerance.

In this paper a look at the physicochemical properties of ethercarboxylic acids will be taken and the extent to which these properties are influenced by the fat chain, degree of ethoxylation, conversion grade and neutralization.

Well, what are ethercarboxylic acids ?

Ethercarboxylic acids are a group of surfactants, which are manufactured from a fatty alcohol, ethoxylated and finally carboxymethylated.

Normally the surfactant is an acid and can be neutralized using different neutralizing agents, which gives other properties, which enables many problems to be solved.

Figure 1. shows the chemical structure of an ethercarboxylic acid :

<u>figure 1:</u>

$$R - O(-CH_2CH_2O)_n - CH_2 - \overset{\overset{\displaystyle O}{\|}}{C} - OH$$

<u>e.g.</u> $R = C_4 - C_{18}$ branched, straight

<u>e.g.</u> $n = 1{,}5 - 16$

There is also a second group, the amidethercarboxylic acid which has the formula, shown in figure 2.

<u>Figure 2.</u> amidethercarboxylic acid

$$R - \overset{\overset{\displaystyle O}{\|}}{C} - \overset{\overset{\displaystyle H}{|}}{N}(-CH_2CH_2O)_n - CH_2 - \overset{\overset{\displaystyle O}{\|}}{C} - O\,Na$$

Some of the features of ethercarboxylic acids are :

- weakly dissociated acids
- non-ionic in aqueous solution
- anionic in neutralized (alkaline) solution.

These characteristics show that an ethercarboxylic acid has a double function; this surfactant group can be used in acidic as well as in neutral and alkaline cleaners.
When we consider the formula more closely, we can distinguish one molecule with three functional groups, like:

- a hydrophobic alkyl chain
- a hydrophilic polyglycol group
- a hydrophilic carboxyl group.

(see figure 3)

Figure 3. functional groups:

R	-	$(OCH_2CH_2)_n$	-	OCH_2COOH
hydrophobic	-	hydrophilic	-	hydrophilic

Hard water stability.

When we compare the formula of an ethercarboxylic acid with that of a fatty acid, we see in fact a fatty acid in which a polyglycol group is enclosed by a hydrophobic alkyl chain and a hydrophilic carboxyl group.
It is the hydrophilic polyglycol group that is responsible for the stability of the ethercarboxylic acid in different grades of water hardness and for improving the lime soap dispersing power (depending on the fat chain).
The polyglycol group can basically be regarded as an incorporated dispersing agent.
The difference between the hard water stability of an ethercarboxylic acid and a (fatty acid) soap is illustrated by table 1.

Table 1. Hard water stability according to DIN 53905 (max.score 75)

Ethercarboxylic acid	$R-O(-CH_2CH_2O)_n-CH_2-C(=O)-ONa$		
At pH 11	$R = C_{12} - C_{14}$	n = 4	------> 75 points
	$R = C_8$	N = 5	------> 75 points
	$R = C_4 - C_8$	N = 5	------> 75 points
Fatty acid soap	$R-C(=O)-ONa$		
	$R = C_{12} - C_{16}$		------> 41 points

Lime soap dispersing power.

Since fatty acid soaps form lime soaps in hard water, the lime soap dispersing power of ethercarboxylic acids was measured taking into account the different degrees of ethoxylation.
A non-ionic surfactant was also tested for comparison.
The test was done according to DIN 53903 and gives the proportion sodium oleate/dispersing agent, squeezed out as K, at which the soap is turning to turbid (See table 2).

Table 2:
lime soap dispersing power according to DIN 53903

$R - O - (EO)_2 - CH_2 - C(=O) - ONa$	-------->	K = 5
$R - O - (EO)_5 - CH_2 - C(=O) - ONa$	-------->	K = 12
$R - O - (EO)_9 - CH_2 - C(=O) - ONa$	-------->	K = 19
$R - O - (EO)_9 - H$	-------->	K = 15
$R = C_{16} - C_{18}$		

It is easy to see that with the same fat chain and an increasing degree of ethoxylation an increase in the lime soap dispersing power is achieved, whereas by contrast the non-ionic surfactant produces lower values.
Particularly those formulations which contain fatty acid soaps and which are used particularly in hard water could be improved when using ethercarboxylic acids, for example better foaming behaviour in fatty acid soaps, less deposits on the surface at the laundry.
Also for cutting and drilling oil formulations good lime soap dispersing properties of surfactants are required.

Influence of alkyl chain

What influence the carbon chain has on the ethercarboxylic acid by examining the alkaline stability, foaming behaviour and surface tension shows figure 4.

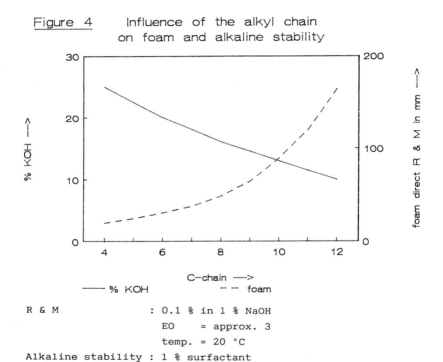

Figure 4 Influence of the alkyl chain on foam and alkaline stability

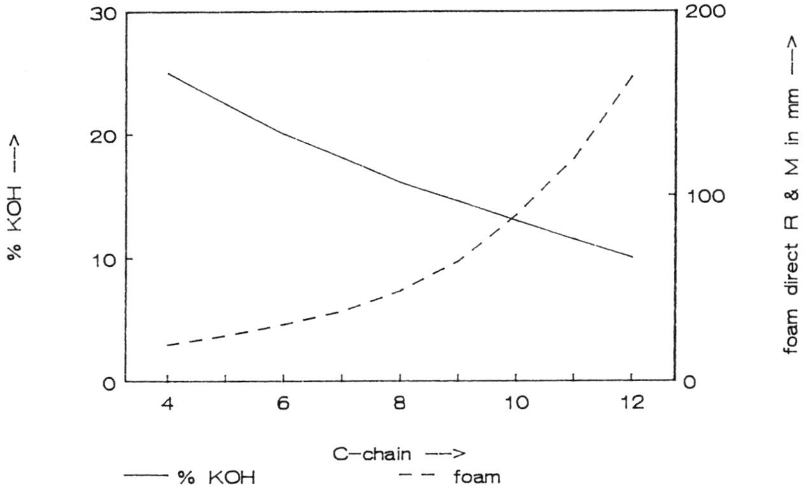

R & M : 0.1 % in 1 % NaOH
EO = approx. 3
temp. = 20 °C

Alkaline stability : 1 % surfactant

With increasing C-chain and identical degree of ethoxylation we see an increase in the foam level and a decrease in the alkaline stability and vice versa.

An ethercarboxylic acid with a C-chain of for example C_6 can therefore be described as a low-foaming surfactant which at the same time has high alkaline stability.

Another interesting characteristic of these low-foaming ethercarboxylic acids is their foaming behaviour over a wide temperature range. While many non-ionic surfactants reach there low-foaming properties at the cloud point, the ethercarboxylic acids show their low-foaming properties over a temperature range of 20 - 80 °C (See figure 5).

Figure 5 Influence of the temperature

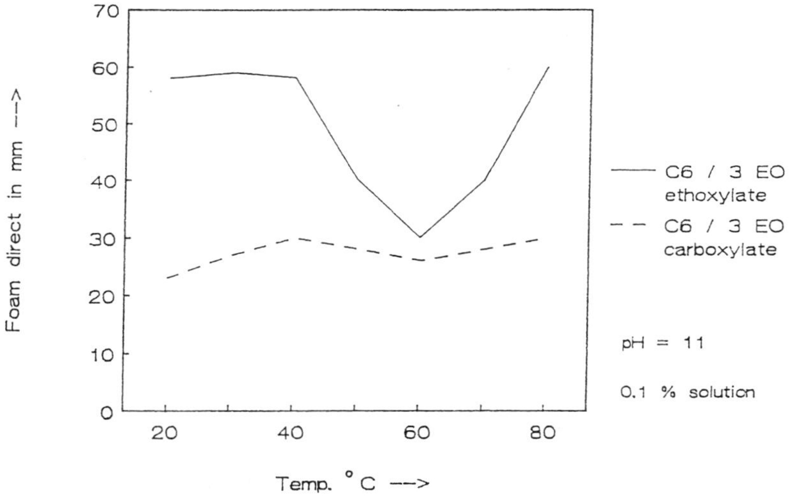

If for example a cleaning process is carried out at different temperatures, the foam must remain at the same low level at all temperatures.An ethercarboxylic acid offers favourable properties in this respect.

Surface tension.

As already mentioned, ethercarboxylic acids behave like non-ionic surfactants in aqueous solutions.

The micelles which are produced are neutral aggregates and are formed at the same concentration as with the nonionic surfactants.

In neutralized form, in other words as salt, the micelles are formed at a higher concentration which means that surface tension is generally lowest at an acidic pH as we can see on figure 6.

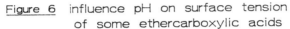

Figure 6 influence pH on surface tension of some ethercarboxylic acids

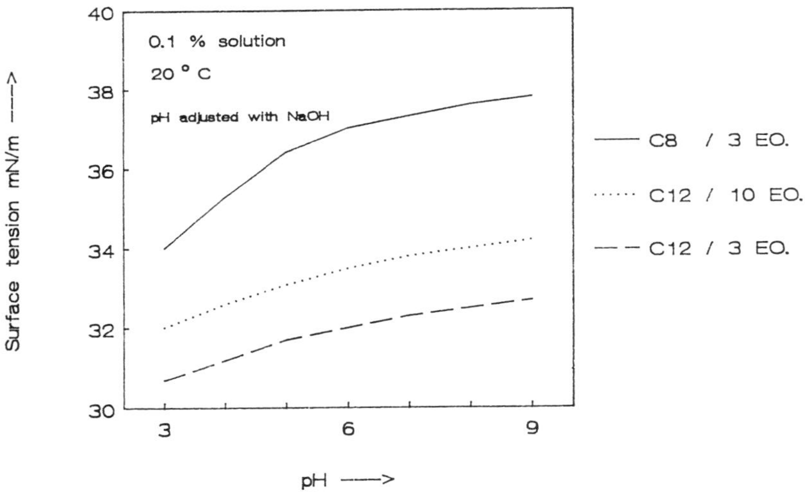

Conversion grade

An ethercarboxylic acid is manufactured from e.g. an aliphatic alcohol, ethoxylated and then carboxymethylated, but we can only speak theoretically of a 100 % ethercarboxylic acid because in practice there always will be some content of non-carboxymethylated, non-ionic material.

It is possible to vary the conversiongrade in order to obtain desired effects.

Partially neutralized ethercarboxylic acid, in fact a combination of a non-ionic and an anionic surfactant, have a particularly beneficial effect on the emulsifying power. However, the higher the conversion grade the better is the alkaline stability and oxydation stability of the ethercarboxylic acid. This is why, for example it is possible to prepare formulations based on sodium hypochlorite or hydrogen peroxide, with which chlorine and oxygen loss can be minimized.

Properties.

To obtain an impression of the physico-chemical properties of ethercarboxylic acids, following basic classification can be made :

Type I $R \geq C_{10}$

Type II $C_4 \leq R \leq C_{10}$

ad I :

Properties of ethercarboxylic acids with $R \geq C_{10}$

- low surface tension
- good cleaning action
- good emulsifying power
- hydrolysis, temperature and electrolyte stability
- acid, alkaline, and chlorine stability
- good dermatology
- hard water stability
- lime soap dispersing power.

ad II :

Properties of ethercarboxylic acids with $C_4 \leq R \leq C_{10}$

- low foaming
- highly alkaline and acid stability (up to 25 % KOH or 75 % H_3PO_4)
- electrolyte, hydrolysis, and temperature stability
- chlorine stability
- hard water stability.

So there are a large number of properties which depend on the alkyl chain (and of course of the degree of ethoxylation).
It is also possible to combine ethercarboxylic acids with different alkyl chains in order to obtain specific effects.
The combination of ethercarboxylic acids and other surfactants can create effects which cannot be obtained using a surfactant alone. Especially formulations which do not remain stable because of their high content of alkaline and/or electrolytes, the use of an ethercarboxylic acid would make such formulations stable; in such cases the ethercarboxylic acid functions as a hydrotrope.

Synergism

A combination between surfactants often produces synergism, as for example when combining an ethercarboxylic acid with a glycoside.
In certain application areas formulations are used which contain about 40 % KOH.
As is well known, glycosides have a very high alkaline stability. On the other hand, the highly foaming property of a glycoside is an effect which is not always desired.
Some ethercarboxylic acids however are very low-foaming while alkaline stability is limited to 25 % KOH.
Combinations between such low-foaming ethercarboxylic acids with a glycoside, give synergistic effects such that these ethercarboxylic acids remain stable in 40 % KOH and the highly foaming effect of the glycoside is reduced by the low-foaming ethercarboxylic acids (especially when seen on a time basis).
Table 3 illustrates the different foaming behaviour of two formulations.
The first formulation consists of 40 % KOH and 2 % glycoside; the second formulation consists of 40 % KOH and a mixture of glycoside/ethercarboxylic acid (2%). Rest is water.

Table 3:

Foam in mm according to DIN 53902.

	glycoside	mixture glycoside/ethercarboxylic acid
direct	78	70
after 1 minute	35	15
after 3 minutes	35	7

- 2 % formulation
- 0 ° GH
- 20 ° C

When formulating with the glycoside/ethercarboxylic acid mixture it is seen that hardly any foam is still present after 3 minutes although the foam level was almost identical at the start of the test.

Amidethercarboxylic acids

Although hardly interesting at all for industrial uses, amidethercarboxylic acids are interesting for institutional and household applications.

Studies have been carried out with manual dishwasher concentrates based on an amidethercarboxylic acid.

Figure 7 shows the structure of an amidethercarboxylic acid :

Figure 7

$$R - \overset{O}{\overset{\|}{C}} - \overset{H}{\overset{|}{N}} (-CH_2CH_2O)_n - CH_2 - \overset{O}{\overset{\|}{C}} - O\,Na \quad + \quad \text{glycerol derivatives}$$

R = coconutoil

n = approx. 4

In secondary alkane sulphonate/ethersuphate mixtures the influence of an amidethercarboxylic acid on foaming behaviour after addition of olive oil was investigated. Also of the same formulation the dermatological aspects was investigated. Using an amidethercarboxylic acid in a washing up liquid, the formulation becomes a stabilizing effect on the foam, especially after addition of olive oil, while the foam height in the beginning is hardly the same as without an amidethercarboxylic acid (see table 4).

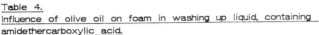

Table 4.
Influence of olive oil on foam in washing up liquid, containing amidethercarboxylic acid.

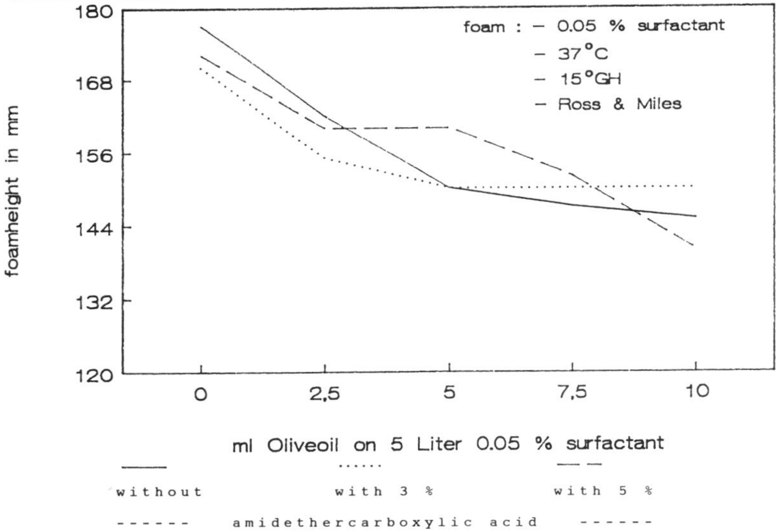

The cleaning power was not impaired by the amidethercarboxylic acid.
As regards the dermatological properties, the influence of an amidethercarboxylic acid on the washing up liquid was tested by means of a Soap Chamber Test, by an independant institute. Table 5 shows the results.

Table 5 :
Soap Chamber Test of washing up liquid.

	erythem $\bar{x}$	scaling $\bar{x}$	fissures $\bar{x}$
controls			
water	0	0	0
sodium lauryl sulphate (0.25%)	1.46	2.71	0.79
washing up liquid			
15 %	0.93	2.46	0.64
4 %	0.30	1.80	0

A zero score for fissures is obtained at 4 % concentration, while erythem shows a very low value in this hard test.

Application possibilities

Taking a look at the different applications and properties of ethercarboxylic acids both alone and in combinations with other surfactants, there is a spectrum of application areas which can be divided into three major areas :

I industrial applications
II institutional applications
III household applications

ad I : Industrial applications

application area	Properties of ethercarboxylic acid
Textile industry	- alkaline stability, low foam
High-pressure cleaning	- alkaline stability, low foam
Bottle cleaning	- low foam
Rinse aid	- low foam over a wide temperature range
Metal cleaning	- low foam
CIP cleaning	- alkaline stablity, low foam

Drilling and cutting oils	- low foam, lime soap dispersing power, non-corrosive
Electroplating	- low foam
Phosphate industry	- low foam
Conveyer belt cleaners	- low foam, lime soap dispersing power

ad II : Institutional applications

application area	Properties of ethercarboxylic acid
Deep steam carpet extraction	- low foam
Dishwashing	- low foam
Disinfectant cleaners	- chlorine stability, peroxide stability

ad III : Household applications

application area	Properties of ethercarboxylic acid
Manual dishwashing	- good dermatology, good emulsifying power
All-purpose cleaner	- good dermatology, good emulsifying power
WC cleaner(chlorine based)	- chlorine stability, viscosity increaser
Automatic dishwashing (powder)	- low foam, chlorine stability
Pre-wash	- low foam

Biodegradability and toxicity

Because of the growing environment awareness of the population and the increasingly strict legislation, the chemical industry is repeatedly finding itself in the crossfire of criticism.

This criticism is often based on ignorance of the material and people are all to quick to attach the label "toxic" or "carcinogenic" although some human habits can also be dangerous.

All ethercarboxylic acids, whether with straight or branched alkyl chain, are biodegradable in accordance with current legislation.

As regards the toxicology, it can be said that the ethercarboxylic acids are generally non-toxic that means that LD_{50} values are above 2000 mg/kg.

Trends in the Applications for Sulphosuccinate Surfactants

J. A. Milne
CYANAMID B.V., P.O. BOX 1523, 3000 BM ROTTERDAM, THE NETHERLANDS

INTRODUCTION

Sulphosuccinates, and their near relation sulphosuccinamates are comparatively unusual in that they are the only surfactants based on maleic anhydride chemistry. They are even more unusual in that some of them are synthesised from materials such as ethoxylated alcohols which are themselves surface active.

A further notable feature is that 50 years into their product life cycle, they remain a speciality; the conventional wisdom that all specialities eventually become a commodity is not observed. Cynics might say in this case that demand - used as a basis by some for defining a speciality - has not increased for whatever reason. One commonly heard response is "you can get comparable performance from other classes of surfactant at lower cost".

However, even given that this is occasionally true, it also explains why the product is a speciality. By definition, this is a surfactant type giving unique properties in a particular system which cannot be matched in all key respects by any other surfactant. Attempts to replace them with commodity surfactants invariably involve a compromise on finished product quality. Sulphosuccinates are not occupying just a few quiet backwaters of application, showing static sales development. Cyanamid as a major source of supply for these products observes very diverse outlets, uninterrupted growth for at least the last 15 years, and a performance versatility that continues to surprise us, and is still being tapped by an increasing number of chemists around the world.

This paper, in addition to providing background information on their chemistry, their market and their general features attempts to spotlight specific attributes by reference to selected applications.

CHEMISTRY

Sulphosuccinates are esters based on the reaction between maleic anhydride and an appropriate alcohol, the resultant product being then sulphonated.

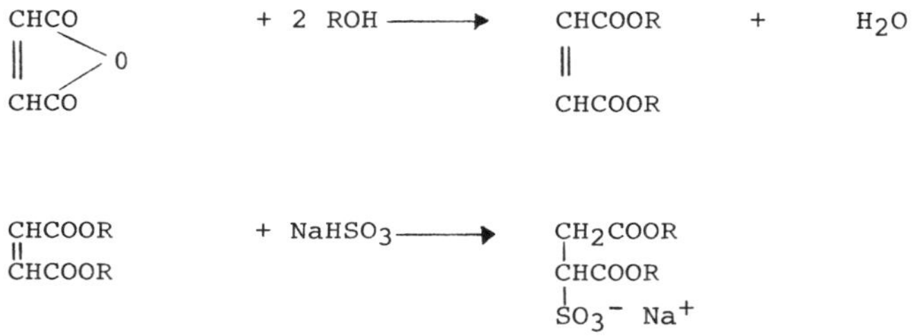

The above reaction sequence demonstrates the preparation of a diester sulphosuccinate.

This reaction is generally carried out in a stainless steel non-pressure vessel, at only mildly elevated temperatures. The sulfonation step is carried out in the aqueous phase. However, it is of great importance not to 'take short cuts' in order, for example, to raise output. Adherence to laid down procedure ensures complete conversion to a low colour (generally water white) product with very low residual electrolyte or unconverted ester content.

If the initial reaction is limited to one mole of alcohol, the esterification of only one carboxyl group takes place with the opening of the anhydride ring and a mono- or half-ester product results, which upon sulphonation gives:

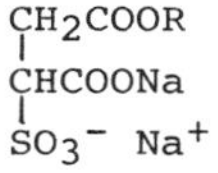

Sulphosuccinamates are produced by a similar process but incorporating generally one mole of a long chain aliphatic amine in place of the alcohol.

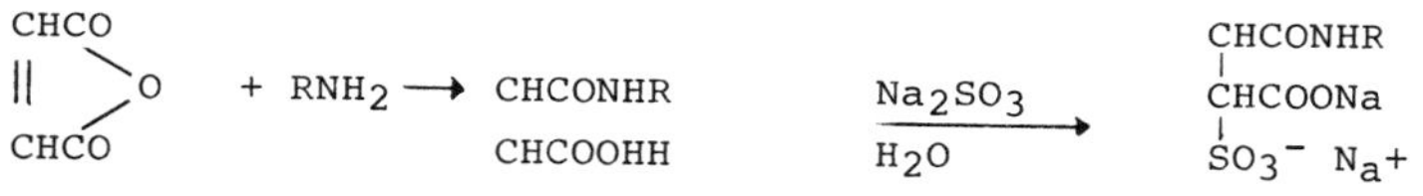

The amine can be pre-reacted with a maleate ester, in which case the final product after saponification and sulfonation has the structure:

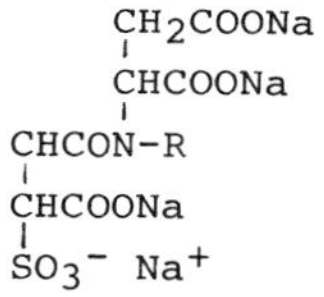

GENERAL FEATURES

Sulphosuccinates are usually supplied in liquid form. Nevertheless solid forms exist, based on for example, C_{12}, C_8, C_6, C_5 aliphatic alcohols and cyclohexanol and higher fatty acid alkylolamides. Whilst monoester types are highly soluble in water, diesters, particularly more hydrophobic types where $R > C_6$, generally require a co-solvent which is commonly alcohol. Both aliphatic and aromatic hydrocarbons and glycols are also occasionally used for C_8 diesters.

According to current EEC legislative classification, they are rated non-toxic, based on LD50 oral toxicity in rats. Ecologically speaking, their fish toxicity varies between "slightly toxic" and "relatively non-toxic". They biodegrade to varying degree from "partially" to "readily": this means between 35 and 100% by the OECD screening (28 days) test. Most commercially available grades have FDA and BGA clearance for use in compositions which come into contact with foods. It should be noted that the U.K.'s Ministry of Agriculture, Food and Fisheries recently struck off dioctyl sodium sulphosuccinate from their approved list of food emulsifiers. However, this should be reviewed again since the EPA announced in August that it was removing dioctyl sodium sulphosuccinate from List 2 (Potentially Toxic Inerts).

In general, sulphosuccinates are excellent wetters, when judged by their lowering of surface or interfacial tensions. The best is the dioctyl type which can lower the surface tension of water to 26 dynes cm -1 which is lower than most other types. Only fluorocarbon surfactants are known to be significantly better.

Sulphosuccinate emulsification performance is generally good: effective at low dosage and the more hydrophilic types are electrolyte-compatible. All have excellent (low) colour. Grades can be selected to avoid foam. For emulsion whose particle size distribution is critical, again one can select a grade to meet any requirement. This is then reproducible because consistent purity standards are met: the manufacturing process permits higher levels of purity than with, for example, products based on broad cut alcohols, or imprecise degrees of ethoxylation.

Sulphosuccinates, being esters, are not stable to extremes of pH and temperature and are prone, especially at high pH, to hydrolysis. Sulphosuccinamates, on the other hand, are not as pH sensitive.

THE MARKET (All data refers to 100% active product)

The total W. European market for surfactants excluding soap is generally held to be approaching 2 M tons per annum. Its determination with precision is impossible for the following reasons:

i) surfactant types and outlets are very diverse and no single company has activity spanning the whole range.

ii) there is inevitably a degree of double counting in that raw material for some surfactant production, for example sulphosuccinate, are themselves surface active.

iii) over the diverse range of outlets, growth rates vary enormously making quantification at any one time rather dubious.

The major consumers are household products (ca 55%) industrial and institutional cleaners 10%, personal care products 6%, and textiles 4%.

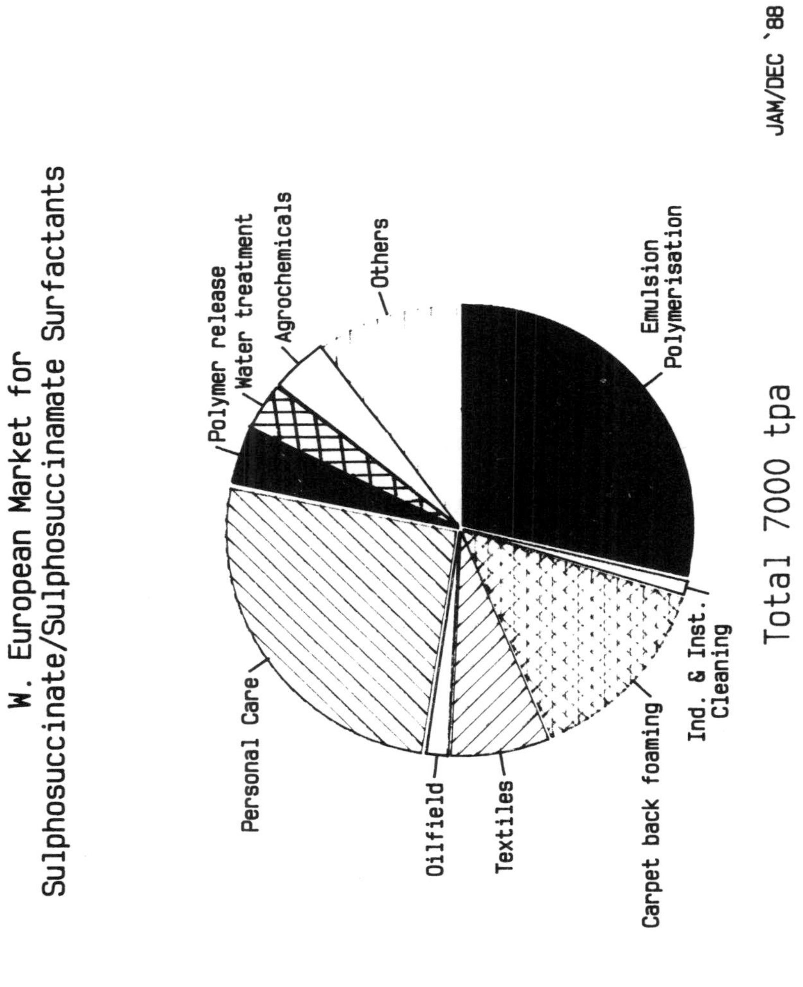
W. European Market for
Sulphosuccinate/Sulphosuccinamate Surfactants
Polymer release
Water treatment
Agrochemicals
Others
Emulsion
Polymerisation
Ind. & Inst.
Cleaning
Carpet back foaming
Textiles
Oilfield
Personal Care
Total 7000 tpa
JAM/DEC '88

However, the consumption of sulphosuccinate and sulphosuccinamate is much smaller, of the order of 7000 tpa dry weight basis. About half of that goes to emulsion polymerisation and personal care products. The major outlets are illustrated in Figure I.

Consequently, it can be seen that market penetration of these types of surfactants vary from negligible e.g. in household applications, through 0.75% in textiles, 1.5% in personal care to 6% in E.P.

To establish growth rates is even more difficult. For example, the early 80's saw consumption in oilfield surfactants grow strongly - in excess of 20% p.a. for a couple of years, whilst the textile industry was in recession and surfactant demand actually shrank. Even within an industry such as emulsion polymerisation, current growth rate of so-called pressure emulsions, ethylene/vinyl acetate copolymers, is estimated at 14% p.a. while output of acrylics, owing to acrylic acid shortage is going down. Surfactant demand for both is directly linked. Superimposing on this the fact that there is a strong trend towards lower addition rates of speciality surfactants, means surfactant consumption in E.P. is growing at 1-1,5% p.a. as a whole, yet sulphosuccinates perhaps as high as 6-7% p.a.

Microemulsions are increasingly popular in agrochemical formulating and these demand high levels of surfactant addition up to 25%. Accordingly, we see demand overall growing at 5-6% currently, but sulphosuccinates, not providing the basis for formulating microemulsions, probably lagging at only 3% p.a.

We estimate the overall growth rate between 1985 and 1990 for sulphosuccinates to be just in excess of 6% p.a. on a dry wt. basis.

Now I would like to discuss applications in more detail.

OIL SPILLAGE TREATMENT

There is a continuous need for dispersants which assist clean-up operations for oil spills at sea, in coastal water in harbour and on beaches. Of course, use of a dispersant is not the only technique available : booms to physically collect and allow pumping of surface oil are often the preferred option in harbour. Recent announcements from the chemical press cite chemicals which convert the oil to a solid film which can then be rolled in and subsequently incinerated. However, particularly where the oil slick can be sprayed from the air, a formulated dispersant is the preferred option.

The performance criteria for the active ingredient are largely met by sodium dioctylsulphosuccinate (DOSS):

1. high surface activity in terms of lowering the oil/water interfacial tension.
2. non toxicity to sea creatures
3. absence of low flashpoint solvent in the formulation
4. no low cloud point, or freezing point in the formulation
5. long shelf life (preferably min. 5 years)
6. viscosity > 50 mpas at $0^{o}C$.
7. (for type II dispersants) must be diluable with water.

Crude oil spilled at sea forms an emulsion - 'chocolate mousse' - under the wave action of the sea and is stabilised by asphaltenes and/or petrolenes in the crude. To break this sea water-in-oil emulsion, surfactant should be added at or below its CMC, otherwise emulsification not demulsification might result. In practice, 1.3 x CMC has been found an optimal addition rate for DOSS because of interfacial adsorption of the surfactant. DOSS is more effective than most other surfactants in displacing the asphaltene stabilising layer.

Various formulations are commercially viable. As a guide, 10-25% DOSS (active basis) together with 20-25% ethoxylated natural oil ,e.g., fish oil or fatty acid derivative. The latter is instrumental in breaking the emulsion once the DOSS has wet the crude oil surface, but it is acknowledged DOSS is the critical ingredient.

OIL SPILLAGE TREATMENT

MECHANISM OF DISPERSION

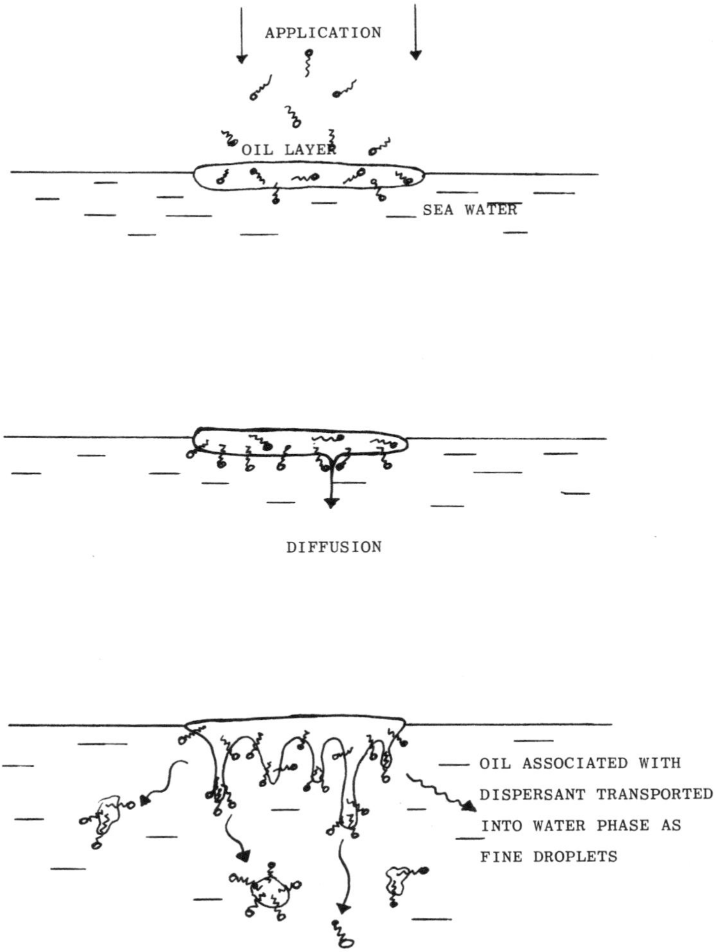

DEWATERING OF OIL

Oil recovered from the well is invariably mixed with water. It can vary from 95%+ down to about 25% content in the extractant. Over much of this range it exists as, firstly, a water in oil emulsion then when its content is below around 50%, an oil in water emulsion.

Demulsifiers used at the well-head, provide the first means of separation ensuring that 'wet' oil (upto 1% water) is piped or shipped to the refinery, and the water separated is sufficiently clean to be passed back to the land or sea.

Once in the refinery, the first need is to desalt the oil, i.e., add water to dissolve salts, caustic to precipitate Mg^{++} and Ca^{++} ions, then again demulsify the resultant emulsion in oil.

Sulphosuccinates are a major surfactant type for desalting/demulsifying. As in the case of oil slick breaking, its power to disrupt the effect of asphaltenes is of major importance. As both Na^+ and Cl^- are anathema in the refinery equipment - the cause of pitting and corrosion - sulphosuccinate in the form of ammonium salts rather than the more common sodium form, are generally preferred. It is also believed that Na^+ can cause reduced activity of hydrocracking catalyst.

Excess surfactant, carried through with the desalted oil to a heat exchanger (prior to cracking) can benefit the foam demulsification step - reduce fouling by asphaltenes, minerals and FeS.

Sulphosuccinates cannot be used to demulsify the wash water by product - this is generally accomplished by polyamine or block copolymer diamine.

Whilst many surfactants can match DOSS demulsification performance per se, the latter does have additional properties which favour its use:

1) it is generally an excellent wetter for suspended solids and often for their subsequent dispersion.

2) it is believed to contribute to corrosion protection (by a mechanism of excellent wetting - displacing air bubbles at the surface - and strong adsorption on solid metallic surfaces).

A Refinery Operation : Desalting

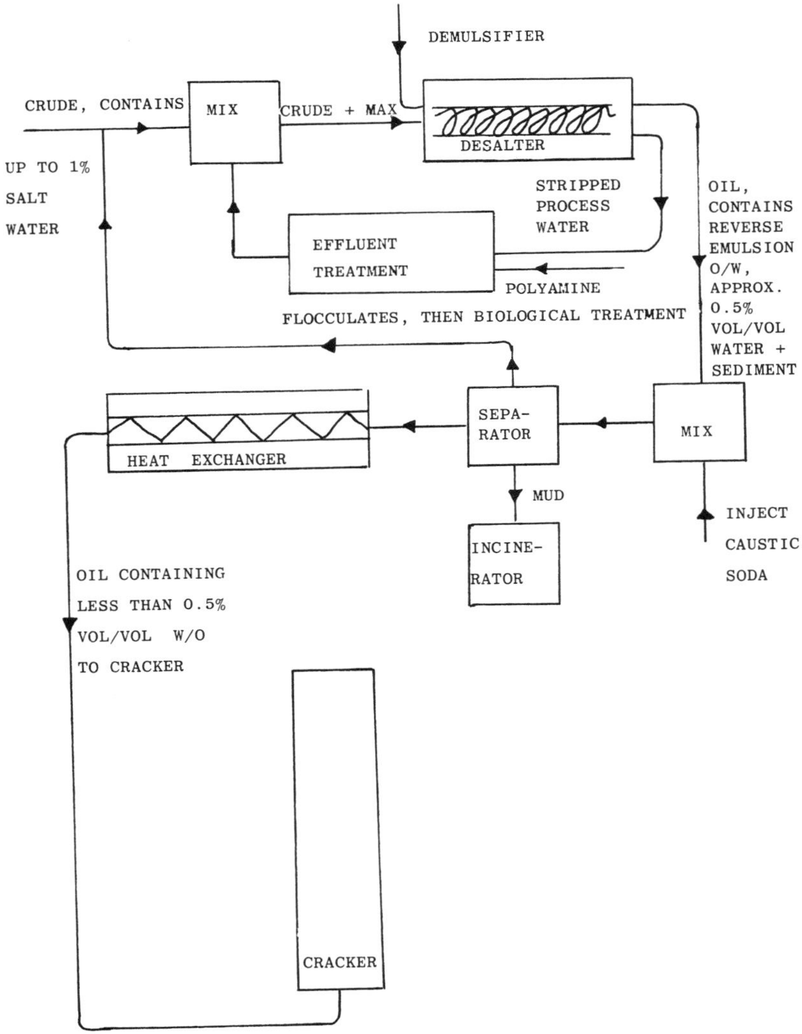

AGROCHEMICAL FORMULATING

Agrochemical active compounds such as herbicides, insecticides and fungicides generally require further formulation into a physical form more suitable for use. As all such agrochemicals are spray - applied as aqueous-based formulations, techniques of emulsification and disperison are prime in the armoury of the formulator.

Two such forms are dry: wettable powders or water-dispersible granules. Both require a wetting agent to speed their wetting when immersed in water by the farmer prior to use. The incorporation of some 0.1-1% by weight of a free-flowing spray dried sodium dioctyl sulphosuccinate, ensures that the powder/granule rapidly stirs in, aids dispersion and, thirdly, acts as adjuvant in most systems. This means that not only can the application dosage be reduced (lower surface tension means improved spreading of the active over crop or soil in thinner sprayed layers, plus improved adhesion, film formation and lower capillarity) but also the 'spectrum of kill' of many actives is synergistically enhanced.

More popular forms of active product are emulsifiable concentrates and flowables. The former comprise organic solvent-soluble actives dissolved in the solvent, together with an emulsifier system (usually anionic + non-ionic matched pair). The product should spontaneously emulsify when added to an excess of water giving particles 1-5μ in diameter. The latter comprise insoluble actives dispersed, generally in water, having been wet milled down to 1-2 μ .

It is possible to combine the two: a flowable concentrate - some call it a 'suspo-emulsion'. This technique is used for the combination of, say, a solvent soluble active with another water-insoluble one. This suggests a complex performance requirement for the stabiliser system.

In effect, a wetter/dispersant/emulsifier is sought, which normally is achieved with 3 surfactant types. However, the industry has found that where sodium dioctyl sulphosuccinate (in appropriate form) is chosen as the wetter, it provides sufficient contribution towards emulsification as to permit more flexible choice of the other two components or even better, it allows the formulator to meet those two requirements with a single product.

THE FLOWABLE CONCENTRATE

OR 'SUSPO-EMULSION'

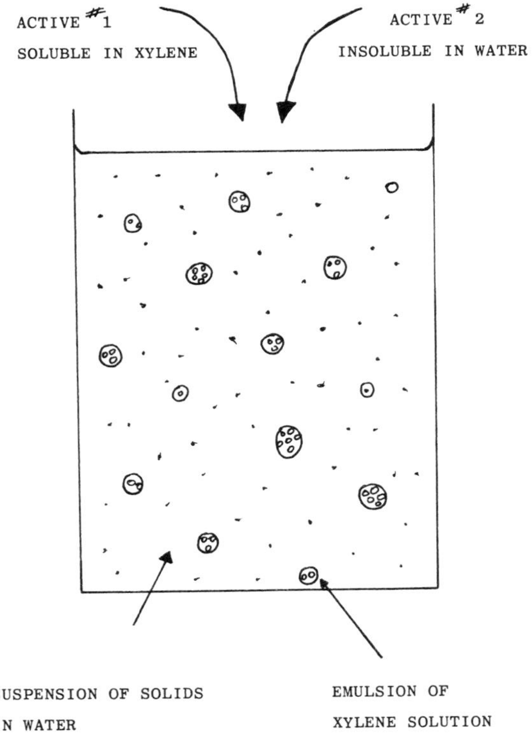

Surfactant choice is critical in control of droplet control and leaf wetting - both factors determining efficiency of pesticide usage.

Crystal size + shape can be controlled during precipitation or crystallisation, from organic media by low-level additions of long chain diester sulphosuccinates (C_8 - C_{13}). Similarly, in flushing precipitate from aqueous to organic phase, a hydrophobic sulphosuccinate has been found most effective in controlling particle size.

Phytotoxicity increases with decrease in particle size of a herbicide. Phytotoxicity of surfactants is proportional to their surface activity and to their CMC. Since in general terms some sulphosuccinates are so active, e.g., in their wetting performance and in having low CMC's, problems of phytotoxicity are more easily avoided than with other commodity types where to achieve the desired wetting effect, addition levels are raised to a level where phytotoxicity can occur.

Particle size control of emulsions is heavily dependent on surfactant type and addition level (also discussed in sector on E.P.). It has been found by Cyanamid that use of a diester sulphosuccinate having a C_{13} hydrophobe gives rather coarse emulsion particles based on its several states of micellar aggregation. It causes oil spray droplets to coalesce when sprayed on to water resulting in large single drops. A C_8 chain length in the hydrophobe gives a more hydrophilic sulphosuccinate. When added to oil it will give a finer p.s. emulsion in water and fine droplets when the oil/surfactant is sprayed in air. When these droplets are placed on the surface of water, they instantly spread out to form a thin film. A C_6 hydrophobe forms emulsion of finer p.s. still, with film spreading as above. However, the films here almost immediately break up into many minute droplets. A C_5 hydrophobe, however, gives a coarser p.s. emulsion and promotes the formation of larger, non-coalescing droplets.

The tendency to film formation is a key factor in improving cost effectiveness of pesticide sprays. This is achieved by enhanced spreading and adhesion, hence better applicability to the plant.

Where sensitivity to electrolyte is a drawback, half ester sulphosuccinates are preferred to diesters. They are also often better emulsifiers, however, their wetting performance can be enhanced by the addition of other anionics and dispersion, of course, by the incorporation of lignin sulphonates.

Non-ionic surfactants dominate this market because they are generally more soluble in commonly used solvents. This, therefore, dictates, and is borne out in our experience, that more hydrophobic diester sulphosuccinates (C_6 - C_{13}) are most in demand from this sector when sulphosuccinates are chosen.

TEXTILE PROCESSING

Classification of surfactant usage in the textile processing industry is generally into two sectors:

i) fibre manufacture
ii) textile processing

The former is speciality oriented with the relatively few producers seeking high performance in terms of lubricity, wetting of process oils, antistatic properties, and emulsification of spin finishes, for example.

DOSS is used in nylon spin finishing. The uniformity of spinning is maintained by controlling water absorbancy of the nylon. If it is too dry, it becomes brittle, if too wet, the polymer does not crystallise. This is done by reducing the water content of the spin finish emulsion. Aerosol OT as emulsifier permits low viscosity high solids o/w emulsion in which water is < 25% vol./vol. It is generally used in conjunction with nonionic (nonylphenol - and alcohol -

ethoxylates) at below 10% w.t. addition. We believe wetting at high speed of yarn production is its prime function, whilst emulsification is of secondary importance.

Turbidity at operating temperatures of 50°C arises from monoester sulphosuccinate impurity and leads to yarn breakage.

Of less"high-tec" interest is the widespread use of DOSS to wet out synthetic fibres, especially polyethylene, polypropylene and polyester, particularly when heat bonded non-wovens are treated. Applied by dip bath, also padding or spraying techniques, residual sulphosuccinate ensures not only rapid subsequent absorption of moisture where required (absorbent non-wovens) but more importantly, improved covering, spreading and adhesion of a latex coating applied in the production of textile interlinings and disposable non-woven clothing or protective garments.

In the field of textile processing, commodity surfactants are widely used to remove oil, scour, provide warp sizing, desize, foam, and assist dyeing and levelling. They also contribute to antistatic properties, wetting, softening, soil repellency, penetration of bleach, caustic etc.

Again DOSS is used in various forms for dyeing, improved scrubbing, desizing and introducing water absorbancy. It is of interest to note that in high temperature scouring, non-ionics tend to deposit on to the fibres: the surfactant in the wash bath is depleted rapidly making their use uneconomical. Anionics generally rinse out freely, and thus they are more economical. DOSS provides a useful compromise in that it is slightly substantive and partially resists rinsing - useful in providing 'built-in' lubrication and dye acceptance subsequently in the process.

PERSONAL CARE PRODUCTS

Sulphosuccinates are extensively used in mild and baby shampoos, bubble baths, bath oils, some douche scrub products, and even in liquid washing-up liquids.

Certain types are excellent foam boosters. There is the need in many of these products for stable high volume foam of a fine structure, giving it improved feel and texture. Of course, there will not be the tendency towards lowering of cloud point which effect accompanies use of alkanolamides.

Depending on the nature of the hydrophobe and therefore the HLB value, one can obtain quick flash foams which collapse rapidly, stable types of high or low volume, and large or small bubble structure.

Frequent use shampoos must be formulated with milder components which are proven less aggressive to skin and preferably non-irritant to the eyes according to the Draize rating. Sulphosuccinates, while not so mild as betaines or imidazoline products, meet the specification set by some manufacturers.

The most widely used sulphosuccinate types are monoesters based on lauryl alcohol or ethoxylated lauryl alcohol or lauric acid alkanolamide, which combine mildness with foam boosting whilst maintaining adequate biodegradeability.

Sulphosuccinates are partially substantive to hair, so offer a degree of conditioning. Monoester types have excellent electrolyte tolerance so remain effective even in hard water.

One important property of an ethoxylated nonylphenol sulphosuccinate monoester is compatibility with fatty acid ether sulphate at up to 50%. This means that up to 50% of SLES for example in a viscous or gelled product can be replaced by this type of SS without loss of structure.

In our experience, sulphosuccinates are most widely incorporated into baby bath products, baby shampoos, followed by mild foam baths and frequent use shampoos. There is no geographical preference for sulphosuccinates (as opposed to alternative mild actives) and the typical addition level is 3-4% by weight on the wet product. We recommend ~5% for optimum foam development.

One disadvantage: SS cannot be thickened by addition of salt, that is, by the commom ion effect. This is overcome by formulating in combination with, e.g., nonionic surfactants which can be thus treated.

SS can be complexed with alkylsulphates and give clear gels with amphoterics. They are excellent o/w emulsifiers (half esters) although foam can be destroyed.

Dioctyl SS is available in (spray dried) powder form for incorporation as wetter and emulsifier in powder shampoos, bleaches and bath salts for example. It improves the texture and clinging qualities of face powders and is especially effective in improving the soil-removing power of powdered shampoos.

Used in conjunction with SLS or LABS, SS's exhibit synergistic effects in that they lower the eye irritation of the SLS or LABS foams used alone.

EMULSION POLYMERISATION (E.P.)

Emulsion polymerisation is defined as the process of polymer production from monomers in the emulsified phase, generally in water. The presence of a surfactant aids emulsification of monomer and stabilises the resulting polymer latex. Such polymers are generally used as coatings in the broadest sense - paints, inks, adhesives, textile impregnations, leather, and paper coatings.

In the major polymer types - polyvinyl acetate copolymers, polyvinyl chloride copolymers and emulsion grade homopolymer, polyacrylates and modified styrene butadiene copolymers - surfactants are used for stabilisation. It is generally recognized that the surfactant system provides 3 functions;

1) To assist the emulsification of liquid monomer into water before polymerisation.
2) To provide micelles in the continuous phase which act as sites (loci) for the polymerisation.
3) To stabilise the resultant polymer in the emulsion phase by adsorption leading to both steric hindrance and electrostatic repulsion, depending on the nature of the surfactant.

There are at least 10 key performance factors which govern the success of an anionic surfactant and its acceptance for emulsion polymer systems:

1) Mechanical stability - ensures high shear conditions will not cause emulsion to 'break'.
2) Particle Size Distribution and its reproducibility.
3) Thermal stability and resistance to yellowing of latex derived films and coatings.
4) Rapid, clean, high yield polymerisation - ensures process not retarded, nor does it lead to coagulum or grits i.e.: coarse, oversize particles which have to be filtered out.

ROLE OF SURFACTANT IN EMULSION POLYMERISATION

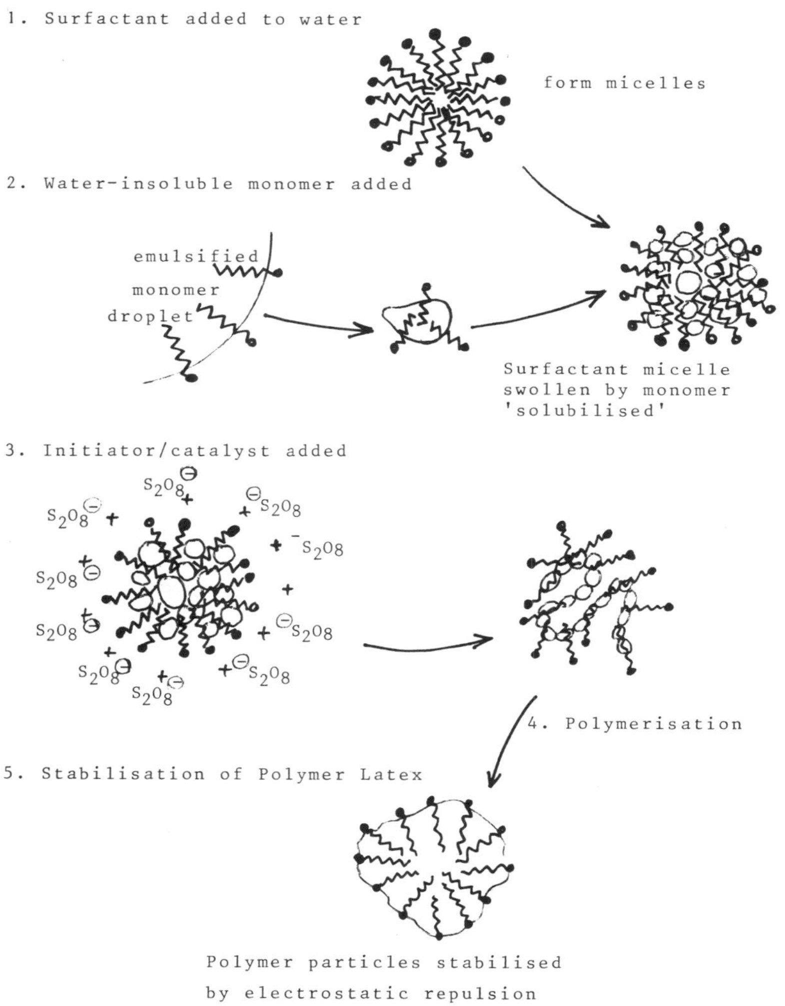

5) Viscosity stability - that even fine p.s. emulsions have comparatively low viscosity which will not significantly change on storage. Allows formulation of high solids emulsion.
6) Effective at low levels of addition. Cyanamid advises 0.5-3% real addition levels of sulphosuccinates based on monomer as a rule of thumb for starting formulations.
7) film clarity - essentially a combination of particle size distribution, surfactant compatibility and polymer Tg.
8) low foaming tendency - reduction in surface tension too low allows air to be trapped in the latex.
9) Electrolyte and pH tolerance - surfactant must not be destroyed or precipitated by, for example, hard water, other cations in the system or extremes of pH. This gives stability to the emulsion polymer when later compounded.
10) Ecological and toxicological safe handling - avoid hazard exposure of workers, users, and aids disposal of waste water without excessive treatments.

As stated above sulphosuccinates are recommended for use between 0.5 and 3% active content on monomer weight. If used in conjunction with non-ionics (used at 2-5% real weight on monomers) one can occassionally reduce the anionic content. Non-ionics enhance divalent ion stability and freeze-thaw stability. They also result in larger particle size.

Certain sulphosuccinate types have become industry standards for particular polymer types. A mixed monoester sulphosuccinate (e.g. Aerosol$^{(R)}$ 501) is a custom synthesized sole emulsifier for all-acrylic lattices for textiles and paints which gives particle size control together with non-yellowing behaviour and tolerance to cations in the system.
The ethoxylated alcohol monoester sulphosuccinate (e.g. Aerosol$^{(R)}$ A-102 Surface Active Agent) is generally recognised to be the best anionic emulsifier/stabiliser for styrene/acrylic copolymers. Its reproducibly fine particle size at relatively low levels of addition ensure more consistent end-products.

The dicyclohexyl sodium sulphosuccinate (Aerosol(R) A-196) was specifically developed for modified styrene-butadiene latex used for rug backing adhesives, upholstery backings, paper coatings etc. This surfactant has been widely accepted in this area. It offers the following properties: clean polymerisation, high surface tension latex and films and coatings of excellent water resistance. The unique structure of the cyclohexyl ester group is believed to contribute to the excellent water resistance properties observed.

Sulphosuccinates of the Aerosol(R) 22 type have a structure which comprises a long C18 chain hydrophobe with four ionisable groups (3 carboxyl, 1 sulphonate) in the hydrophile. This structure ensures a broad-based hydrophile for electrical double layer repulsion at the latex particle surface - imparts shear stability to the latex - and contributes to substrate adhesion by the latex. It also helps 'compatibilise' monomers of widely differing Tg e.g. VAC/BA, EA/MMA. It gives good vapour barrier characteristics to PVC emulsions.

Aerosol is a trademark of the American Cyanamid Co.

CONCLUSION

Sulphosuccinates and sulphosuccinamates exist in over 30 different homologues commercially available from various suppliers around the world. Their prime difference is the structure of the hydrophobe and performance differences arise from the varying 'HLB' values/solubilities which are consequent.

This means that the product for the job can be found by matching the needs one has (performance parameter under certain operating conditions) with the predicted behaviour from a particular chemical structure.

Whilst sodium dioctyl sulphosuccinate has been cited most often in the examples given, primarily for its wetting performance in water, it is occasionally non-optimal, e.g., in wetting out non-aqueous solutions on to non-metallic surfaces. Improvement here is seen when substituting either a C10 or a C6 sulphosuccinate. Hence the utility of such an extensive homologous series, within one class of surfactant should not be overlooked.

We foresee good growth prospects in E.P. and moderate prospects in textile, agricultural formulating and personal care.

There also remains scope for further development by 'bending' the sulphosuccinate molecule: Cyanamid for one, is active in screening alternative hydrophobes and modified intermediates (maleate ester chemistry) as a means to achieving custom-designed surfactants to service industry.

As performance requirements become ever more demanding, in minerals processing and industrial cleaners, to cite just two examples, we anticipate chemists in these sectors will also come to recognise the benefits of sulphosuccinate chemistry which can be harnessed to their needs.

Industrial Applications of Surfactants Derived from Naphthalene

T. Mizunuma*, M. Iizuka, and K. Izumi
WAKAYAMA RESEARCH LABORATORY, KAO CORPORATION, 1334 MINATO, WAKAYAMA 640, JAPAN

1. Introduction

Formaldehyde condensate of β-naphthalene sulphonate first appeared in the industry in 1913 by the patent issued for BASF[1], and has been widely used in many ways such as effective dispersants for dyes and pigments, as an emulsifier for synthetic-rubbers, as leather tanning agents,and etc. In the early sixties, Hattori (ex-director of Kao Corp.) and co-workers[2~5] carried out a detailed investigation on the composition and physico-chemical properties of the compound, and confirmed that it was a polymolecular mixture with properties in aqueous solutions that vary depending on the number of nuclei in a molecule. The results of further investigation revealed that high molecular weight formaldehyde condensate of β-naphthalene sulphonate was a very effective dispersant for cement particles in concrete.

Naphthalenic surfactants have rapidly proliferated over since, and recently, the annual demand in Japan has reached about 60,000 tons per year.

The properties and some examples of application of formaldehyde condensates of β-naphthalene sulphonate are reported in this paper.

Table 1 Application of Naphthalenic Surfactants

Function	Application	Structure
Dispersion	Cement, Coal-Water slurry	High condensate
	Emulsion polymerization Dyes, Pigments, Inks, Leather tanning, Agricultural chemicals	Condensate
Wetting	Fibres	Alkylnaphthalene sulphonate

A very simple relationship between the field of application and the chemical structure of naphthalenic surfactants is shown below as Table 1. The molecular weight range or the condensation degree of the naphthalenic surfactants are suitably adjusted for each application.

3. Properties of formaldehyde condensate of β-naphthalene sulphonate

As shown by Fig.1 and by the structural formula, the commercially available formaldehyde condensate of β-naphthalene sulphonate are polymolecular mixtures. Some fundamental properties of the condensates of different number of naphthalene nuclei isolated and purified from the mixture are shown in Fig.2-8.

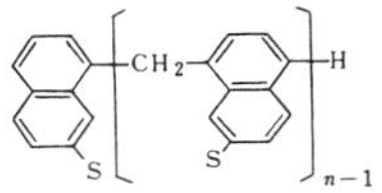

n:Number of nuclei
S:Sulfonate group

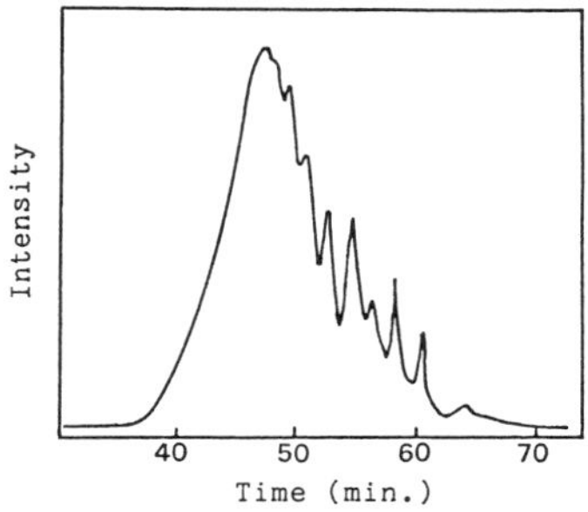

Fig. 1 GPC pattern of formaldehyde condensate of β-naphthalene sulphonate

Detector;UV spectrum
Column;G3000SW + G2000SW (Tosoh Corp.)
Mobile Phase;0.1M-Nacl/CH_3CN = 7/3

3.1 Surface tension

As shown in Fig. 2[5] and Fig. 3[5], the interfacial activity of the condensates as expressed by the reduction of surface tension is exceptionally low as compared to that of surfactants in general. Also the lack of inflexion point on the curve of concentration surface tension relationship which usually corresponds to the critical micellar concentration suggests that the condensates are poor in the capability of forming micells. With an exception of the mononuclear compound, the capability of reducing the surface tension is better for the compounds of smaller number of nuclei. The low-foaming property and the high surface tension of the compounds of larger number of nuclei is a great advantage when these compounds are used as dispersants.

3.2 Viscosity

As clearly shown by Fig.4[5], the reduced viscosity of the aqueous

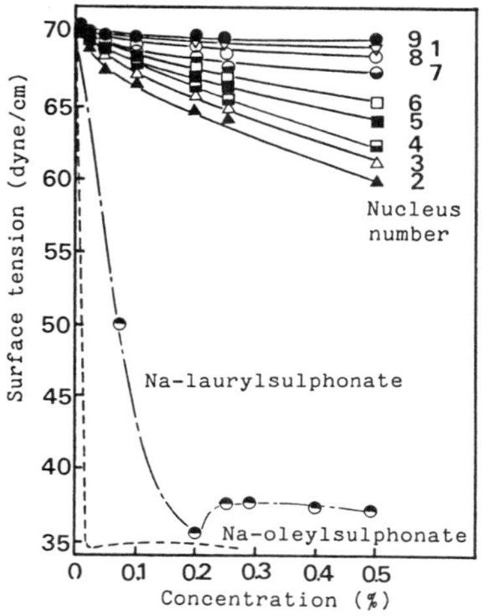

Fig.2 Surface tension of aqueous solutions of nuclear compounds (low concentration)

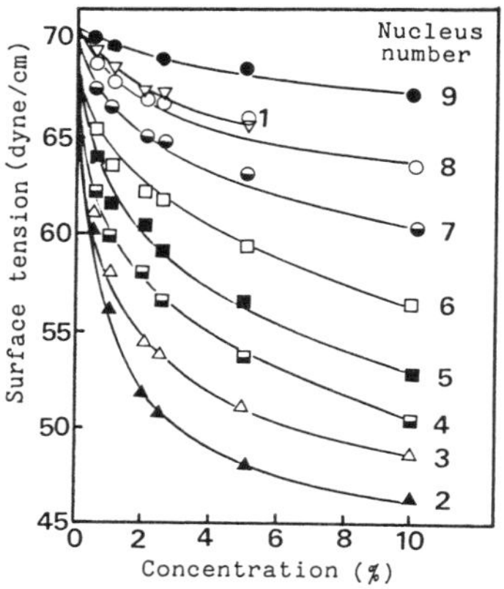

Fig.3 Surface tension of aqueous solutions of nuclear compounds (high concentration)

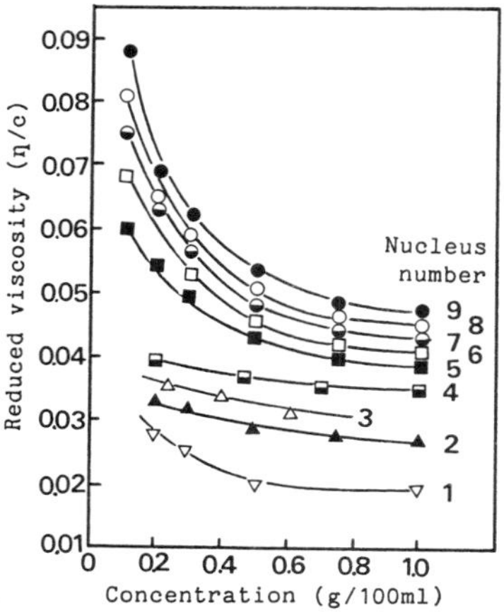

Fig.4 Concentration dependence of reduced viscosities of nuclear compounds

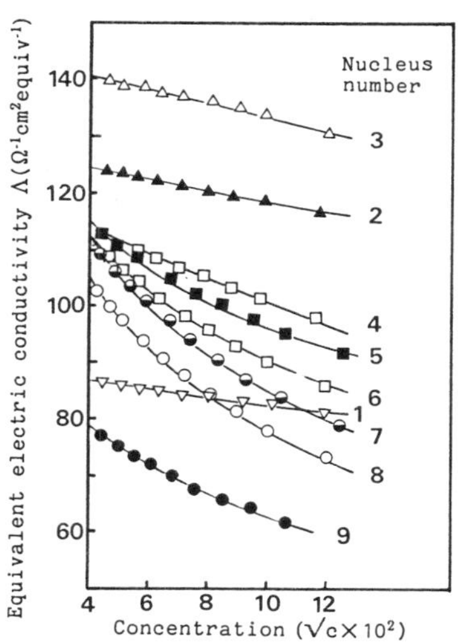

Fig.5 Concentration dependence of equivalent electric conductivities of nuclear compounds c; equivalent concentration

solutions exhibited two different trends. The concentration dependency of the reduced viscosity was very large for the compounds of 5 or larger number of nuclei in contrast to the small dependency of these of the 4 or less nuclei. Generally, the solution viscosity of polyelectrolyte tends to increase at lower concentrations owing to the spread of molecules caused by the mutual repulsion of dissociated groups in the molecules[6]. The polyelectrolyte behaviour of the compounds of 5 or more nuclei was considered to be confirmable by the viscosity data in Fig.4. The constants for the calculation of the molecular weight of the compound based on the equation proposed by Staudinger[7~8] is shown below.

$$[\eta]=A\cdot M^{a}$$

$[\eta]$; intrinsic viscosity
M ; molecular weight
$A = 2.32 \times 10^{-3}$; constants
$a = 0.36$; constants

It has been known that a for polyelectrolyte[9~10] is in the range between 0.5 and 2, 2 for rod-like molecules, 1 for random coils, and 0.5 for spherical molecules. The a value of 0.36 obtained here, which is close to 0.5, suggests a spherical shape of the molecule in water.

3.3 Electric conductivity

The relation between concentration and electric conductivity of each nuclear compound is shown in Fig.5[5]. The lack of inflexion points on the curves is considered to suggest the lack of capability of the micell formation by these compounds. From the linear relationship for the compounds of 1 to 4 nuclei which conforms to the Kohlrausch's rule, the behaviour as the low-molecular weight strong electrolytes of these compound may be suggested. The compounds of 5 or more nuclei behave as the polyelectrolytes as mentioned previously.

3.4 Solubility in water

The solubility of the condensates in water increases sharply as the number of nuclei increase from 1 to 4 as shown in Fig.6[4].

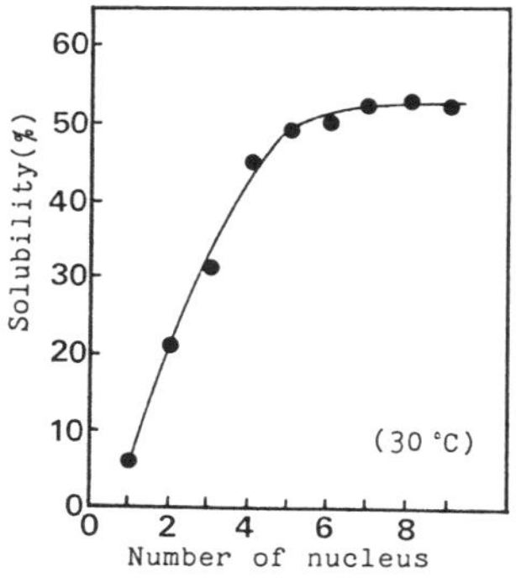

Fig.6 Solubility of nuclear compounds

The increasing trend is much less thereafter. The increasing trend of the solubility is considered to be attributable to the increase of hydration to the π-electron distribution of naphthalene nuclei based on the examination of UV spectrum.

3.5 Dispersancy

As the dispersancy is largely influenced by the amount of adsorption which is under the influence of surface condition (or the property) of the particles, the vat dye representing an organic hydrophobic particle and portland cement having the hydrophilic surface representing an inorganic particle were selected to exhibit the dispersing performance of the compounds. In both cases, the better dispersancy was observed as the number of nuclei increased. In the case for the vat dye, the improvement of dispersancy disappeared for the compound of 4 or more nuclei, but for the cement particle, the trend of improvement continued up to the 10 nuclei. From the results shown in Fig.7[5] and 8[11], the importance of selection of nuclear number depending on the kind of particles was suggested. For the formaldehyde condensates of β-naphthalene sulphonate particularly, the better dispersancy for the particles having the hydrophilic surface was expected by the compound of higher degree of condensation.

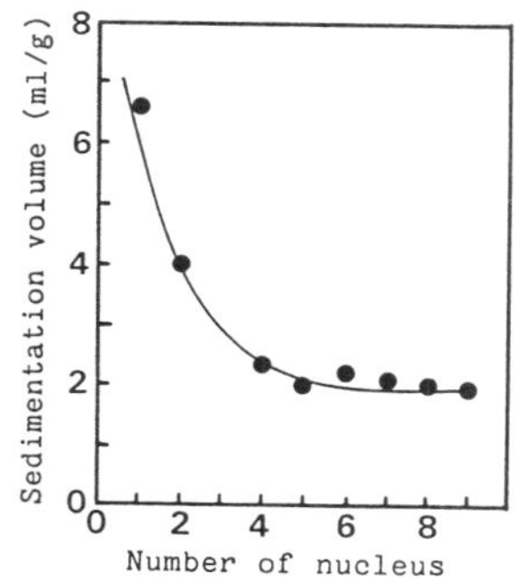

Fig.7 Influence of the nuclei number on the sedimentation volume of Nihonthrene Brilliant Green FFB

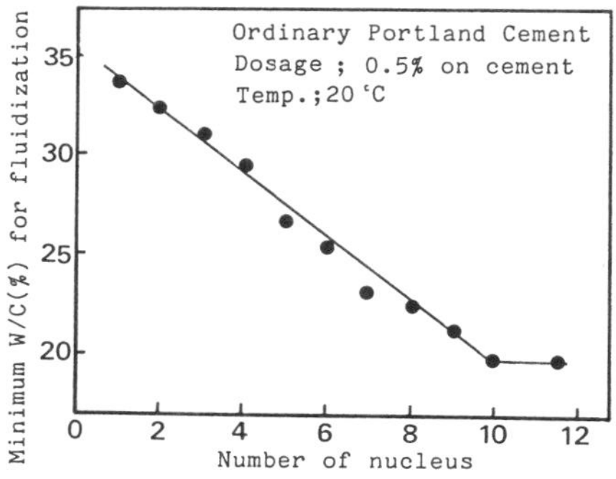

Fig.8 Relation between minimum W/C(%) for fluidization of cement pastes and neclei number

3.6 Toxicity test

An extensive test for the toxicity of the compounds were conducted[12~13]. The LD_{50} test result obtained by the Tokyo Metropolitan Health Research Institute is shown below.

LD_{50}=4880±110 mg/kg (mouse)

4. Applications as dispersants

4.1 Mechanism of dispersion

There are three main mechanisms of stabilization of dispersion known widely. One of which, the older one, is the stabilization by the electrostatic repulsion known as the coagulation rate theory[14] (stabilization by the reduction of coagulation rate) or as the DLVO (Derjaguin-Landau-Verway-Overbeek) theory. By this theory, the coagulation rate is explained by the balance between the attractive (V_A) and repulsive (V_R) potentials, or by the maximum interaction potential V_{max} which corresponds to the peak of the curve shown in Fig.9[11]. Another is the stabilization by the entropic repulsion.

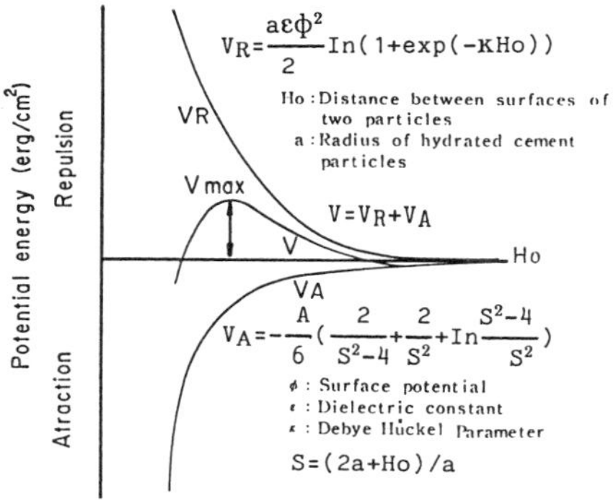

Fig.9 Curve of potential energy between particles

Although the mechanism of this stabilization process is studied quite extensively by many people, it is not easy to apply the results of these studies quantitatively to the estimation of dispersion stability in the handling practices of suspensions.

Finally, the importance of the viscosity of medium may be pointed out. In the case of using polymer solutions as the medium for dispersion, the rapid coagulation rate constant $K(=4kT/3\eta_m)$ proposed by Smoluchowski is influenced by the concentration of polymers, and the effect of stabilization may be calculable in terms of equivalent V_{max}/kT.

4.2 Cement dispersants

In 1962, the study on naphthalenic dispersants was started in the laboratory of Kao Corporation. Formaldehyde condensates of β-naphthalene sulphonate have been found to be an excellent dispersant for portland cement with less tendency of hardening retardation and air entrainment as compared to conventional chemical admixtures. With its excellent properties, the formaldehyde high condensate is used very widely in the fields related to the cement and concrete industries for improving the flowability and the strength. The annual demand in Japan is about 40,000 tons.

The flow properties of cement pastes and the mechanism of slump loss of concretes were explained by Hattori and Izumi[15~16] based on the extended version of Newton's flow equation, on the afore mentioned coagulation rate theory, and on the modified structural viscosity theory proposed by Scott Blair.[17~19]

4.2.1 Types and characteristics of cement dispersants

The structures of typical cement dispersants popularly used are illustrated in Fig.10[11]. Three major characteristics, namely, high dispersancy, low foaming, and no retardation of cement hydration are generally required for cement dispersants. In Fig.11[20] the comparative data of the surface tension of the concrete admixtures are shown. As mentioned previously, the higher surface tension is related to the lower foaming tendency, and the naphthalenic and melanin based dispersants are excellent in both the surface tension measurements and actual air content in concrete[21].

Fig.12[11] shows the relationship between the dosage level of dispersants and the flow values of cement pastes. The naphthalenic dispersant is far better than others in this regard. Fig.13[20] shows the retardation of hardening by the dispersants.

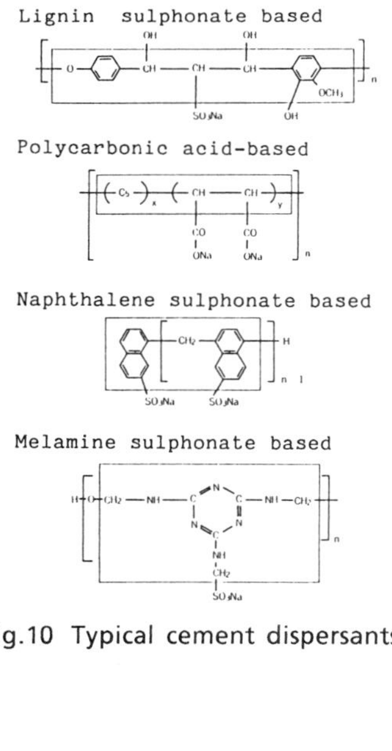

Fig.10 Typical cement dispersants

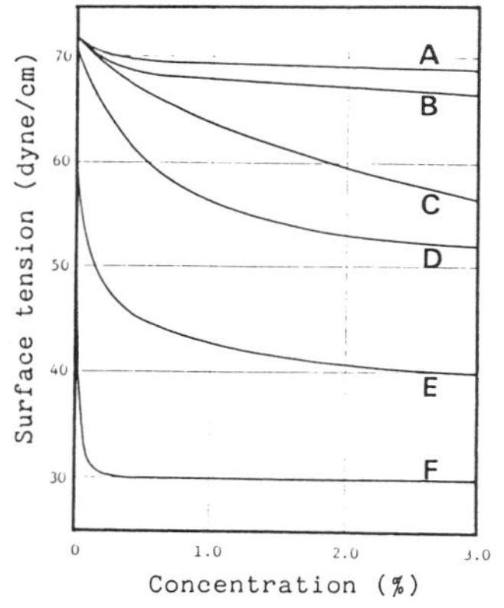

Fig.11 Surface tension of cement dispersants
A;Melamine sulphonate based
B;Naphthalene sulphonate based
C;Carbonic acid based
D;Lignin sulphonate based
E;Rosin soap
F;Nonyl phenyl ether (EO)

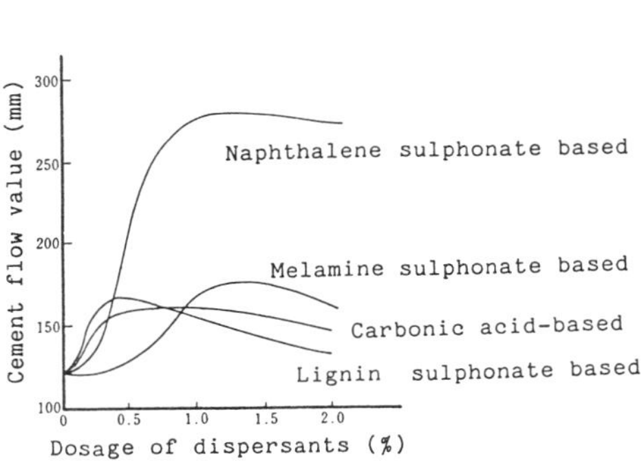

Fig.12 Dispersancy of cement dispersants
W/C ratio = 25 %

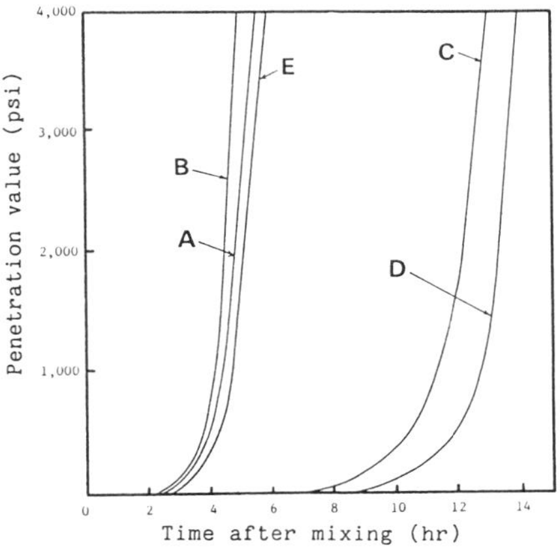

Fig.13 Effect of dispersants on the hardening time of mortar

A; Naphthalene sulphonate based
B; Melamine sulphonate based
C; Carbonic acid based
D; Lignin sulphonate based
E; Plain
※ Measured based on ASTM C 403
Cement/Sand = 1/2 , Dosage of dispersants = 1%

By the results shown above, a very favorable properties of naphthalenic dispersant as the concrete admixture were confirmed.

4.2.2 Compressive strength of concrete containing naphthalenic dispersant

The strength of concrete was known to be closely related to the water/cement ratio. Fig.14[20] shows the relationship between the compressive strength at 28 days after the placement and the W/C ratio. The minimum W/C ratio which is about 25 % to promote the hardening is attainable by the excellent dispersancy of naphthalenic dispersant, but may not be attainable by other admixtures due to the excessive retardation and to the saturation of dispersancy at the high dosage level.

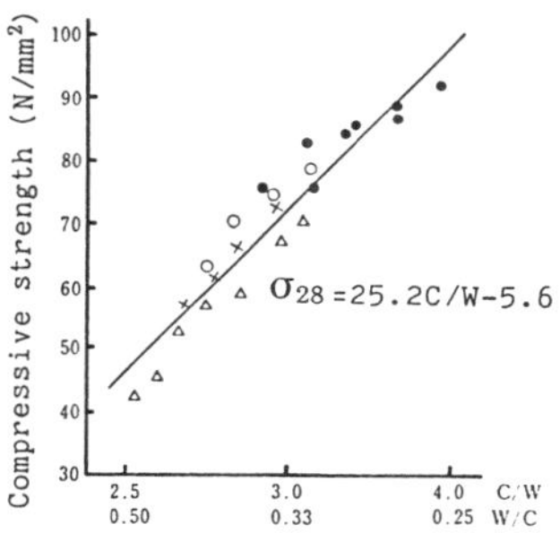

Fig.14 Relationship between W/C ratio and compressive strength of concretes

By the very high strength attained by the use of naphthalenic dispersant, concrete products such as piles, beams, poles, sleepers for railway and etc, came into realization.

4.3 Dispersants for coal-water slurry (CWS)

After a so called "oil shock", the use of coal in place of fuel oils became a large concern. To solve the problems associated with the use of coal, the use in the forms of suspension or

Table 2 Types of Dispersants and their Properties

Dispersant	Dispersancy	Stability	Dosage	Cost
Naphthalenic	○	○	○	◎
Polystyrene sulphonate	◎	○	○	×
Nonionic type (POEO)	◎	◎	×	△

slurry was considered.

As compared to the cement paste and concrete, the coal water slurry (CWS) requires a higher stability due to the longer storage time after the preparation, and the higher solid content to maintain a thermal efficiency at a satisfactory level, and these requirements are satisfied by the use of proper dispersants. By experiences, naphthalenic, sulphonates of styrene, and nonionic type dispersants shown in Table 2[22] are known to be suitable for this purpose.

4.4 Dye dispersant

In the dye industry, the dispersants are used in many process such as the milling to prepare finer particles, mixing process with soluble dyes, and dyeing process. Particularly in the milling process naphthalenic or lignin type dispersants are popularly used. To compensate the degradation of dispersancy during the high temperature dyeing process (at 120-130 °C), phenolic dispersants are frequently added after the milling process. The types and the properties of dispersants popularly used fo this purpose are listed in Table 3[22].

Table 3 Types of Dispersants and their Properties

Dispersant	Milling effect	Dispersancy at high temp.	Foaming property	Staining property
Naphthalene sulphonate formaldehyde condensate	Effective on versatile dyes	low	low	Low
Alkylnaphthalene sulphonate formaldehyde condensate	Effective on versatile dyes	Medium	Medium ~High	Low
Aryl sulphonate formaldehyde condensate	Rarely used in milling	High	High	High
Carboxylate type polymeric surfactant	Effective on fluorescent dye or special dyes	low	High	Nil

4.5 Pigment dispersants

Performance requirements for dispersants used with pigments depend largely on the surface condition of pigment particles. As the hydrophobic particles are generally poor in wettability, the dispersants with wetting property is suitable to be used with hydrophobic pigments. As the surface of hydrophilic particles are wettable, and as the electrostatic repulsion performs a main role in stabilizing the dispersion, substances which are capable of generating a larger electrostatic charge onto the surface of particles are preferable.

For the pigments of large polarity, inorganic pigment for example, substances possessing the polycarbonic acid structure are, in many cases, good dispersants. Since the dispersancy of naphthalenic surfactants is not sensitive to the pH nor to the hardness of water, these surfactants are suitable as versatile dispersants, but due to the weather instability which causes the discoloration, these are not suitable for emulsion paints and paper coatings.

4.6 Dispersants for emulsion polymerization

The surfactants to be used in the processes of emulsion polymerization have to be capable of solubilizing the monomer and of dispersing the polymer formed after the reaction. In the manufacturing processes of synthetic rubber and latex, anionic surfactants such as fatty acid soaps, rosin soap, alkylarylsulphonates, esters of alkylarylsulphonates, and formaldehyde high condensates of β-naphthalene sulphonate are popularly used. In the production processes of polymer emulsions, poly(20-80)ethoxylated alkylphenols, sulphates of polyethoxylated alkylphenols, and the mixtures thereof are frequently used. These surfactants are usually added before starting the reaction, but in the cases of latex production with anionic surfactants, nonionic surfactants may be added at the end of polymerization reaction to compensate the loss of anionic surfactants during the reaction due to the chemical instability. Kinds of polymers and the surfactants to be used for the synthesis thereof are listed in table 4[22].

4.7 Miscellaneous uses

As introduced previously in this paper, the naphthalenic surfactants are being used in the variety of industries. Other than the usage already mentioned, they are utilized in gypsum industry, agricultural chemicals, fire brick manufacturing, new ceramics, and so forth.

5. Summary

The structures, physical and chemical properties, and typical fields of application of naphthalenic surfactans are briefly reviewed. Although in the history of surfactants, formaldehyde condensates of naphthalene sulphonates belong to a relatively old generation, with their excellent performance characteristics and with possible modifications based on the new technology, they may find new fields of application in future.

Table 4 Surfactants applicable to synthetic rubber and resin industries

Emulsifiers	PVC	PVyC	PE	PP	AS	PS	ABS	SBR	CR	NBR	PVac	PVacryl
Fatty acid soaps					○	○		○	○	○		
Asymmetric rosin acid soaps					○		○	○	○	○		
Alkylbenzene sulphonates	○	○			○	○	○	○		○		
Alkyl sulphonate	○	○			○		○	○	○		○	○
Alkylnaphthalene sulphonates		○										
Dialkyl sulphosuccinates					○	○	○	○			○	○
Dicyclohexyl sulphosuccinates							○					
Polyoxyethylenealkylether sulphonates							○				○	○
Polyoxyethylenealkylaryl sulphonates							○				○	○
Polyoxyethylenealkyl phosphates							○				○	○
Polyoxyethylenealkylaryl phosphates							○				○	○
β-naphthalene sulphonate formaldehyde condensates							○	○	○	○		
Alkane sulphonate	○	○					○					
Glycerin fatty acid esters	○	○										
Sorbitan fatty acid esters	○	○										
Polyglycerin fatty acid esters	○	○										
Polyoxyethylene(20)-glycerin fatty acid esters	○	○									○	○
Polyoxyethylene(20)-sorbitan fatty acid esters	○	○									○	○
Polyethyleneglycol fatty acid esters	○	○					○				○	○
Polyoxyethylenealkyl esters	○	○			○	○	○	○	○	○	○	○
Polyoxyethylenealkylaryl esters	○	○			○	○	○	○	○	○	○	○
Polyoxyethylene-polyoxypropylene block polymers					○	○	○				○	○

Note : PVacryl; Acrylate-vinyl acetate copolymer
PVac ; Polyvinyl acetate
NBR ; Acrylonitrile-butadiene rubber
CR ; Chloroprene rubber
SBR ; Styrene butadiene rubber
ABS ; Acrylonitrile butadiene styrene copolymer
PS ; Polystyrene
AS ; Acrylonitrile styrene copolymer
PP ; Polypropylene
PE ; Polyethylene
PVyC ; Polyvinylidene chloride
PVC ; Polyvinyl chloride

6. Reference

1)BASF, D.R.P.,292531(1913)
2)K. Hattori and Y. Tanino, Bull. Chem. Soc. Japan, 1963, 66, 55
3)K. Hattori and I. Konishi, ibid., 1963, 66, 59
4)K. Hattori and Y. Tanino, ibid., 1963, 66, 65
5)K. Hattori and Y. Tanino, ibid., 1964, 67, 1576
6)R.M.Fuoss, U.P.Strauss, J. Polymer Sci., 1948, 3, 602
7)H.Staudinger, W.Heuer, Berichte, 1930, 63, 222
8)H.Staudinger, H.Scholz, ibid, 1934, 67, 84
9)A.Katchalsky, H.Eisenberg, J. Polymer Sci., 1951, 6, 145
10)W.Kuhn, Helv. Chem. Acta, 1943, 26, 1394
11)K. Hattori, Concrete Tech.(Japan), 1976, 14, 12
12)T.Futami, Y.Samata, T.Tsuda, F.Masuda, Pharmacometrics (Japan), 1976, 13, 135
13)T.Futami, Y.Samata, T.Tsuda, F.Masuda, Pharmacometrics (Japan), 1976, 13, 147
14)E.J.W.Verway, J.Th.G.Overbeek, "Theory of The Stability of Lyophobic Colloids", Elsevier Publishing Co. Inc., 1948, p139
15)K.Hattori and K.Izumi, Proc. Symposium M, Annual Meeting of Materials Res. Soc., 1982, p.14, Edited by J.P.Skalney:Nov.1-4, Boston Mass. U.S.A.
16)K.Hattori and K.Izumi, Second International Conference on the Use of Fly Ash, Silica Fume, Slag, and Natural Pozzolans in Concrete, p.21, Apr. (1986), Madrid
17)G.W. Scott Blair, Rheol. Acta, 1965, 4, 53
18)G.W. Scott Blair, Rheol. Acta, 1965, 4, 152
19)G.W. Scott Blair, Rheol. Acta, 1967, 6, 201
20)T.Yamakawa, Industry and Products Japan, 1973, 55, 103
21)K. Hattori, T. Yamakawa, A. Tsuji, M. Akashi, CAJ Review of the 23th General Meeting, 1964, P.105
22)"Surfactants - A Comprehensive guide -", Published and edited in Japan by Kao Corp., 1983

Preparation and Industrial Applications of Phosphate Esters

S. L. Paul
GAF EUROPE, 40, ALAN TURING ROAD, SURREY RESEARCH PARK, GUILDFORD, SURREY GU2 5YF, UK

Generally speaking, we all use phosphate esters. They are a part of our lives. Phosphate esters are so versatile that mother nature could not overlook their usefulness as biomolecules. There are some very familiar biomolecules that fall into this class of chemicals, for example:-

DNA and RNA

These two biomolecules are constructed from nucleotides which are the phosphate esters of nucleosides.

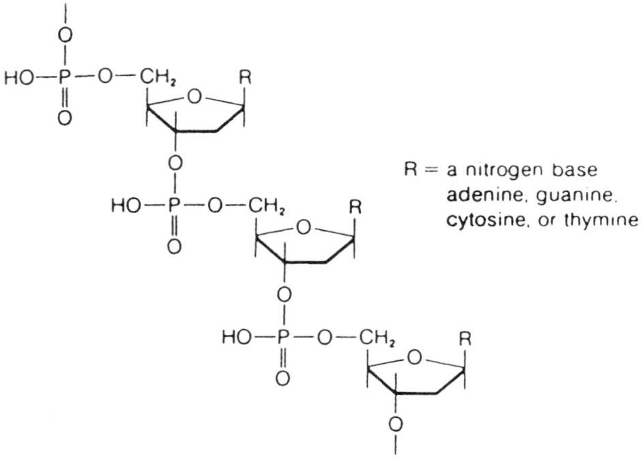

Fig. 1 Structure of the strands of deoxyribonucleic acid (DNA)

The DNA consists of two long chain polymers made up of alternating phosphate and sugar residues carrying nitrogen bases (Fig. 1.), intertwined with proteins called histones. The sugar units in DNA are D-2-deoxyribo-furanose.[1]

RNA usually consists of a single long chain polymer similar in structure to that of DNA. The sugar units in RNA are D-ribofuranose.

DNA is the genetic material in most living organisms. RNA serves the function of carrying the information contained in DNA and is essential for protein synthesis.

There are other not so familiar biomolecules that are also phosphate esters. For example ADP and GDP are used in the energy transfer cycles in biological systems. They are an important part of all biological processes requiring energy. Both molecules can be phosphorylated to form the high energy ATP and GTP. The energy thus stored in these molecules can be utilised in any energy requiring reaction.

NAD^+ and $NADP^+$ are biomolecules used in the electron transfer in biological oxidation - reduction cycles. They are an important part of oxidation in animals and photosynthesis in plants. Here these two molecules can accept a hydride from a donor molecule to form NADH and NADPH.

PHOSPHOLIPIDS are very important biological surfactants that function as barriers between polar and non-polar regions of the cell.[2]

These are just some examples of the phosphate esters in biological systems. In fact about 3% of all biological cell material is accounted for by phosphate esters.

Outside the living organism, however, we have found further and more diverse applications of phosphate esters.

In this paper I shall deal with the preparation and the industrial applications of phosphate esters. I shall describe their chemistry, their properties and then their applications in particular industries.

Phosphate esters that I will describe now are anionic surfactants. Anionic surfactants are the biggest class of surfactants in use today both by volume and by value. Over 5 million tonnes of surfactants are used every year and just over half of these are anionics.[3]

The anionic surfactants in W. Europe alone are valued at around $1 billion.[4]

Phosphate esters represent some 2 - 3% of the anionic surfactants market.

CHEMISTRY OF PHOSPHATE ESTERS

The chemistry of phosphate esters is quite complex.

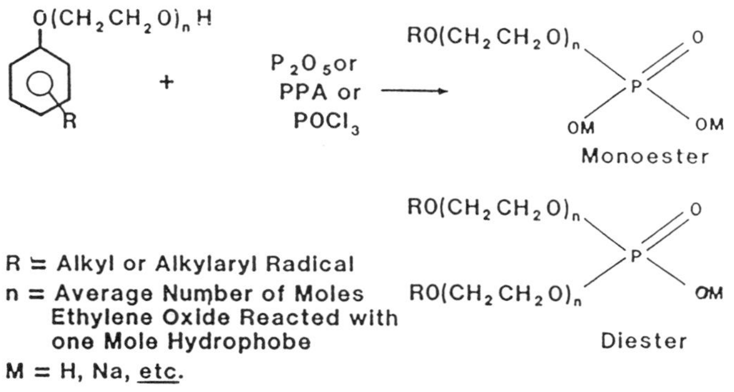

Fig. 2. Generalised Preparation of Phosphate Esters.

Fig. 2. shows that there are many variables in the preparation of phosphate esters. These are:

1. Hydrophobe (R-Group)
2. Degree of Ethoxylation (n)
3. The Phosphorylating Agent
4. Mono to Diester Ratio
5. Neutralisation of the Phosphate Ester.

The base hydrophobe is usually ethoxylated to render it water soluble or dispersible. It is then phosphorylated.

There are three main phosphorylating agents:

1. Phosphoric acid,
2. Phosphorus pentoxide,
3. Phosphorus oxychloride.

Phosphoric acid and phosphorus pentoxide are the most commonly used reagents. Phosphorus oxychloride is not so well used because it gives off hydrogen chloride as a by-product.

1. <u>Phosphoric Acid Route</u>

Phosphoric acid exists in three forms.

Orthophosphoric Acid

Pyrophosphoric Acid

Metaphosphoric Acid

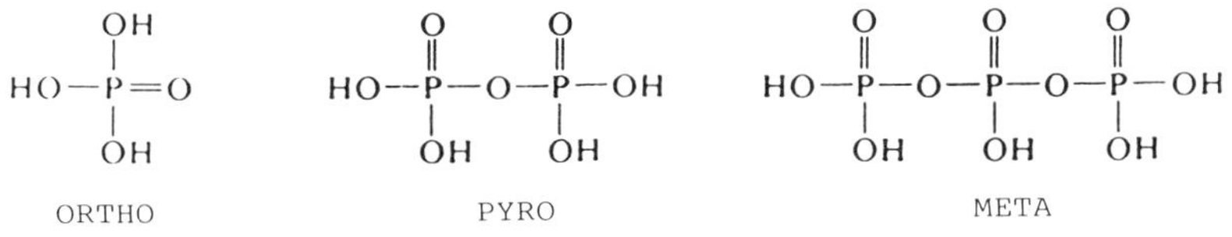

Fig. 3. The Different Forms of Phosphoric Acid.[1]

Phosphoric acid also forms higher complexes (anhydrides) when mixed with phosphorus pentoxide.

A mixture of ortho, pyro, meta and higher complexes is commercially available and is known as polyphosphoric acid (PPA).[5] It is available in 105% and 115% grades. The % is the equivalent orthophosphoric acid concentration. A typical composition of polyphosphoric acid is given in Table 1.

Table 1. Composition of Polyphosphoric Acids

	105% acid	115% acid
Orthophosphoric acid (H_3PO_4)	49–54	3–5
Pyrophosphoric acid (H_4PO_7)	41–42	9–16
Triphosphoric acid ($H_5P_3O_{10}$)	5–8	10–17
Tetraphosphoric acid ($H_6P_4O_{13}$)	0–1	11–16
Higher acids	—	46–67
Total P_2O_5%	76	83·5–84·5
Density g/cm^3	1·92	2·06
Viscosity at 25°C Pa s	0·8	30–60
at 100°C Pa s	0·035	0·5–1·7

Polyphosphoric acid (115%) is the most commonly used form of phosphoric acid. When polyphosphoric acid (115%) is used to phosphorylate, the mono ester is predominantly produced. Some orthophosphoric acid is also produced. This process is described in Fig. 4.

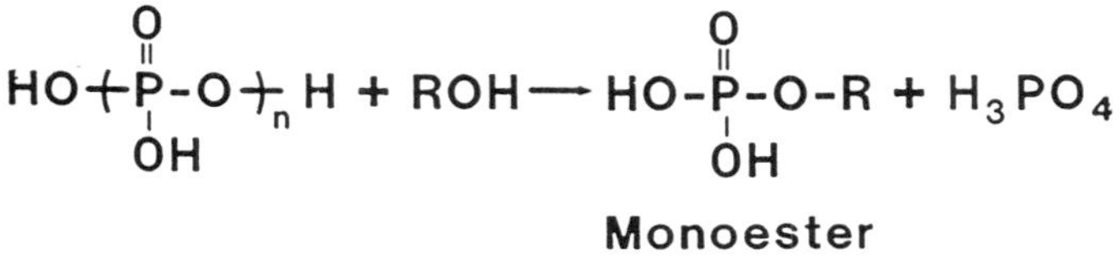

Fig. 4. Phosphorylation by Polyphosphoric Acid

2. Phosphorus Pentoxide Route

Phosphorus pentoxide (P_2O_5) is also commonly used as the phosphorylating agent. When P_2O_5 is used, it yields equimolar amounts of mono and diesters and only a very small amount of orthophosphoric acid is produced. This reaction is described in Fig. 5.

$$O=\underset{O}{\overset{O}{P}}-O-\underset{}{\overset{}{P}}=O + ROH \longrightarrow O=\underset{O}{\overset{OH}{P}}-O-\underset{}{\overset{OR}{P}}=O$$

$$O=\underset{O}{\overset{HO}{P}}-O-\underset{}{\overset{OR}{P}}=O + ROH \longrightarrow O=\underset{OR}{\overset{OH}{P}}-O-\underset{OH}{\overset{OR}{P}}=O$$

$$O=\underset{OR}{\overset{OH}{P}}-O-\underset{OH}{\overset{OR}{P}}=O + ROH \longrightarrow O=\underset{OR}{\overset{OH}{P}}-OR + HO-\underset{OH}{\overset{OR}{P}}=O$$

$$\qquad\qquad\qquad\qquad\qquad\qquad\text{Diester}\qquad\text{Monoester}$$

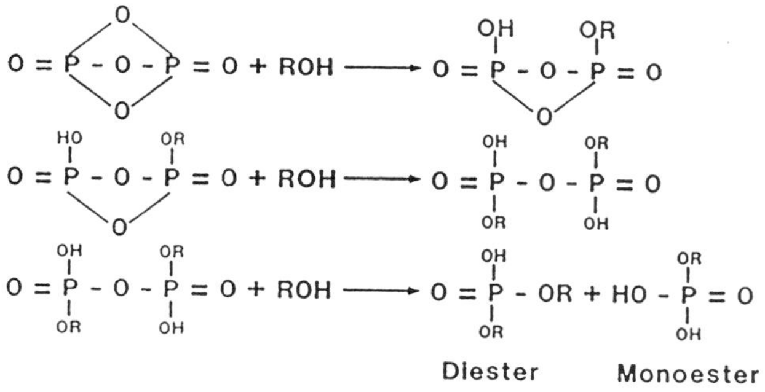

Fig. 5.   Phosphorylation by Phosphorus Pentoxide

3. Phosphorus Oxychloride Route

Phosphorus oxychloride ($POCl_3$) is used to a lesser extent to form phosphate esters. This route produces di-and triesters predominantly. One of the drawbacks of this route is the production of gaseous hydrogen chloride (HCl). This route of phosphorylation is therefore not preferred. The reaction is described in Fig. 6.

$$R-OH + Cl-\underset{\underset{Cl}{|}}{\overset{\overset{O}{\|}}{P}}-Cl \longrightarrow R-O-\underset{\underset{OR}{|}}{\overset{\overset{O}{\|}}{P}}-Cl + 2HCl\uparrow \xrightarrow{H_2O} R-O-\underset{\underset{OR}{|}}{\overset{\overset{O}{\|}}{P}}-OH + HCl\uparrow$$

$$R-O-\underset{\underset{OR}{|}}{\overset{\overset{O}{\|}}{P}}-OH + NaOH \longrightarrow R-O-\underset{\underset{OR}{|}}{\overset{\overset{O}{\|}}{P}}-ONa + H_2O$$

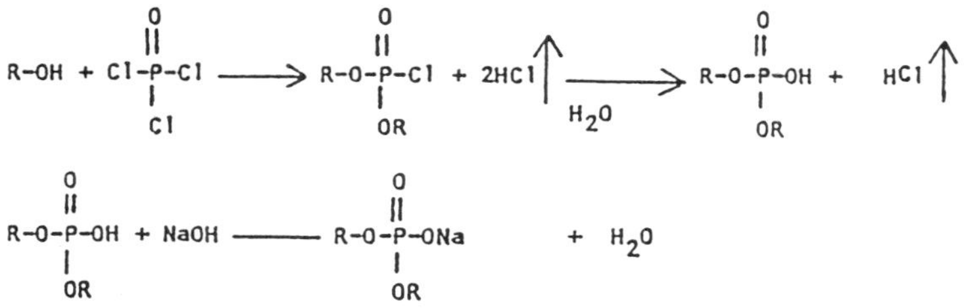

Fig. 6.   Phosphorylation by Phosphorus Oxychloride

PROPERTIES OF PHOSPHATE ESTERS

Being anionic surfactants, phosphate esters naturally possess the following usual properties:-

Emulsifying
Detergency
Wetting
Solubilising
Foaming
Dispersing.

Table 2. shows some general surface properties of phosphate esters.

Table 2. SURFACE PROPERTIES OF PHOSPHATE ESTERS

	PH 7		
PRODUCT	FOAM(MM) (0.1%)	WOT(SEC) (0.1%)	SURFACE TENSION (0.1%) (DYNES/CM)
PHOSPHATE ESTERS			
POE butoxyethyl ether phosphate	44/0	>60	63
POE phenyl ether phosphate	20/2	>60	56
POE decyl ether phosphate	124/117	>60	28
POE nonylphenyl ether phosphate	128/110	>60	29
POE octylphenyl ether phosphate	140/100	>60	32
POE nonyl ether phosphate	155/90	14	32

POE = Polyoxyethylene

In addition some phosphate esters exhibit the following special properties:-

Lubricity

High electrolyte tolerance

Hydrotropicity

Chemical stability

Corrosion inhibition

Low irritancy (neutralised esters)

Biodegradability

Table 3 shows that phosphate esters are effective hydrotropes at very low concentrations.

Table 3. HYDROTROPIC ACTIVITY OF PHOSPHATE ESTERS

Basic Formulation:	A
TKPP	10%
Nonylphenyl ethoxylate	5%
Water	85%
Hydrotrope (Active Basis)	**Hydrotrope Concentration Necessary for Formulation Compatibility 10-50°C % WT/WT. in A above**
POE nonyl ether phosphate	2.1
POE nonyl phenyl ether phosphate	2.5
POE phenyl ether phosphate	3.4
Sodium dodecyl diphenylether disulfonate	4.0
Sodium Xylene Sulfonate	4.6

Fig. 7 shows the alkaline solubility of phosphate esters.

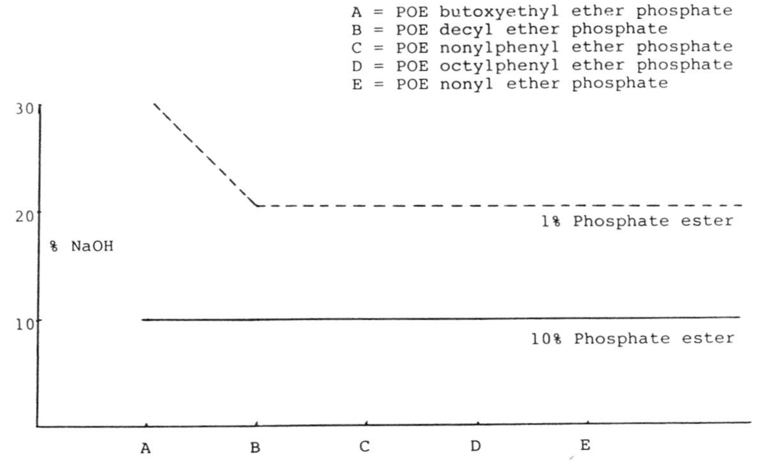

Fig. 7 Solubility Limit of Phosphate Esters

The properties of phosphate esters are dependent on:-

1. The Hydrophobe
2. The Synthetic Route
3. Neutralisation

1. <u>The Hydrophobe</u>

The hydrophobe can be an aliphatic or aromatic alcohol, but is usually an aliphatic or aromatic alcohol ethoxylate or an alkyl aryl ethoxylate. The degree of ethoxylation can be varied to give the desired solubility characteristics.

Typical hydrophobes used in manufacturing of phosphate esters are:-

Butyl ethoxylate,	Lauryl ethoxylate,
Butane diol ethoxylate,	Tridecyl ethoxylate,
Isoamyl ethoxylate,	Phenyl ethoxylate,
Hexyl ethoxylate,	Nonylphenyl ethoxylate,
Decyl ethoxylate,	Dinonylphenyl ethoxylate,
Stearyl ethoxylate,	Nonyl ethoxylate.

The base hydrophobe dictates many of the properties of the finished phosphate ester, for example, its <u>solubility</u>, <u>lubricity</u>, <u>eye and skin irritation</u> and its <u>biodegradability</u>.

Generally speaking:
The longer the carbon-chain of the hydrophobe the greater the solubility in oil.

The higher the ethylene oxide content the greater the solubility in water.

The degree of ethoxylation also has an influence on the load-carrying and rust inhibition properties of phosphate esters. Phosphate ester acids of nonylphenol and tridecyl alcohol with 45% ethylene oxide content were found to have excellent lubricant properties.[6]

The diesters are more oil soluble than the monoesters.

Monoalkyl phosphates have a lower skin irritancy than commonly used alkyl sulphates.[7]

Dialkyl phosphates have been studied in dermatological formulations and found to be non-irritant to the skin and eyes.[8]

Phosphate esters derived from aliphatic hydrophobes tend to be biodegradable.

The following phosphate esters have been found to be more than 80% biodegradable by the semi-continuous activated sludge test.

POE oleyl ether phosphate
POE lauryl ether phosphate
POE decyl ether phosphate
POE stearyl ether phosphate
POE octadecyl ether phosphate

2. The Synthetic Route

The synthetic route used can also influence the final properties of the phosphate ester. For example:

P_2O_5 route gives:-	PPA route gives:-
Superior detergency,	Poor detergency
Wetting,	Superior alkali solubility
Emulsification,	Excellent hydrotropicity
Hydrotropicity.	

3. Neutralisation

There are many phosphate esters now available as alkali metal salts.
Alkali metal salts are more hydrophilic than the free acid forms.
Alkali metal salts foam better.
Neutralised phosphate esters have a lower skin irritancy.

PHOSPHATE ESTER APPLICATIONS

Phosphate esters offer such flexibility of manufacture that a wide range is available to suit any of a number of jobs. Their usual surfactant attributes and their additional special properties make them very versatile in use.

They can be used in many different industries for example:-

Household, Industrial and Institutional (HI&I)

Metal Working

Agricultural Chemicals (Agchem)

Paper and Paint

Plastics

Textiles

Cosmetics and Toiletries

Table 4 gives a more detailed account of the types of phosphate esters used in each of the above industries.

The choice of phosphate ester in a particular application depends on the functions it must perform.

Table 4. Phosphate Ester Applications.

APPLICATION	PROPERTIES REQUIRED	PHOSPHATE ESTERS TYPES USED
HI&I		
Household and Industrial cleaners and detergents Hard surface cleaners Laundry & dry-cleaning products	Good detergency Chemical and heat stability Hydrotropicity Solvent solubility Biodegradability Bleach and alkali compatibility Hard water tolerance	Polyoxyethylene (POE) butyl, phenyl, decyl and lauryl ether phosphates (MONOESTERS + DIESTERS)
METAL WORKING		
Lubricants, high pressure additives and rust inhibitors	Emulsification, water and oil solubility Antiwear and rust inhibition	POE long chain alkyl or alkylaryl ether phosphates (MONO ESTERS, DIESTERS + TRIESTERS).

APPLICATION	PROPERTIES REQUIRED	PHOSPHATE ESTERS TYPES USED
AGCHEM		
Emulsifiers Wetting and suspending agents	Emulsification Compatibility with various insecticide and herbicide actives Wetting, suspending and penetrating properties	POE alkylaryl ether phosphates (MONO, DI + TRIESTERS)
TEXTILES		
Lubricants, caustic-level-off, sizing & scouring products	Detergency, wetting, emulsification, lubricity, antistatic, high temperature stability, chemical stability and good rinseability.	POE alkyl or alkylaryl ether phosphates (MONO, DI + TRIESTERS)
PAPER AND PAINT		
Emulsifiers Deresinators	Wetting, emulsifying, solubilising & levelling	POE alkylaryl ether phosphates

APPLICATION	PROPERTIES REQUIRED	PHOSPHATE ESTERS TYPES USED
PLASTICS		
Antistats, emulsion polymerisation and mould release products	Compatibility with all plastics and polymer emulsions. Chemical, mechanical and high temperature stability Low colour and non toxic. Emulsification and non-stick properties	POE alkyl or alkylaryl ether phosphates (MONOESTERS)
COSMETICS AND TOILETRIES		
Creams, lotions, hair grooming products, sun preparations and antiperspirants Bubble-baths and gels	Emulsification (esp. mineral oil) Low skin and eye irritation Emolliency Conditioning	POE alkyl ether phosphates (MONO AND DIESTERS)

CONCLUSIONS

Phosphate esters are very versatile in use due to their general surfactant properties and their additional special properties. A very wide range of products is available because of the flexibility in raw materials and in the synthetic routes.

Biodegradable types are available for todays more stringent environmental requirements.

REFERENCES

1. Roberts Caserio. Basic Principles of Organic Chemistry, 2nd Edn. Pub. W.A. Benjamin Inc.

2. Pine, Hendrickson, Cram & Hammond. Organic Chemistry. 4th Edition. Pub. McGraw-Hill.

3. European Chemical News, 10 April 1989. p15.

4. Frost & Sullivan. The Surfactants Market in W. Europe. August 1987.

5. Davidsohn and Milwidsky. Synthetic Detergents. 7th Edn. p178.

6. John P. G. Beiswanger, William Katzenstein and Fred Krupin. Phosphate Ester Acids as Load-Carrying Additives and Rust Inhibitors for Metal-working Fluids. ASLE Transaction 7,398-405 (1964).

7. Genji Imokawa. Comparative Study on the mechanism of irritation by sulphate and phosphate type of anionic surfactants. J. Soc. Cosmet. Chem. 31,45-66. (Mar/Apr 1980).

8. Carreras etal. Phosphate diesters for dermatological use. International Journal of Cosmetic Science 6 p159-166 (1984).

The Preparation and Applications of Alkanolamides and their Derivatives

B. Shelmerdine, Y. Garner, and P. Nelson
HARCROS CHEMICAL GROUP, SPECIALITY CHEMICALS DIVISION, MANCHESTER M30 0BH, UK

The condensation of fatty acids or esters with an alkanolamine e.g. diethanolamine, monoethanolamine, monoisopropanolamine yields nonionic surfactants of wide utility and economic importance. Diethanolamine condensates account for approximately 60% to 70% of the total production of this class of surfactant.

The reaction between a carboxylic acid and a mono or di-alkanolamine is complicated by the competing reactivity of the several functional groups present. Consequently the composition of the products can vary considerably depending on the mole ratio's of the reactants and the conditions employed during the reaction.

KRITCHEVSKY DETERGENT

Until 1937 such condensation products were mainly regarded as intermediates for alkoxylation or sulphation. It had been found that the reaction of equimolar quantities of diethanolamine with a fatty acid led to water insoluble products which contained a high level of ester amide. However in 1937 Wolff Kritchevsky noted that the reaction between 2 moles of diethanolamine and 1 mole of fatty acid led to a water soluble product which had good surfactant properties.

The chemistry of this reaction is shown in Figure 1.

FIGURE 1

$$\underset{\text{1 MOLE}}{RCO_2H} + \underset{\text{2 MOLES}}{HN(CH_2CH_2OH)_2} \longrightarrow RCON(CH_2CH_2OH)_2 + H_2O$$

R = alkyl, typically C_{12}-C_{18}

The main reaction product as shown in the above equation is fatty acid diethanolamide. In a typical reaction 2 moles of diethanolamine would be heated with 1 mole of lauric acid at between 150-180°C and water of condensation removed until the free fatty acid content is reduced to less than 5%. Studies have shown that optimum conversion to the lauric diethanolamide is achieved in 90 minutes at 180°C and over 4 hours at 160°C.[1] The resultant product is water soluble and found to contain 50 to 60% lauric diethanolamide. The product also has good detergency and foam boosting properties which are significantly superior to the products obtained by reacting equimolar quantities of diethanolamine and lauric acid.

These products were commercialised under the Ninol trade name and gradually found acceptance in a range of applications. By the early 1950's practically all of the light and heavy duty detergents, shampoo and speciality cleaning compositions contained alkanolamide. It is interesting to note that although Kritchevsky may well have synthesised the first U.S. detergent he was not actually trying to produce a surfactant but was attempting to produce oil soluble dyes. Since this time the product produced by this reaction has been known as the 'Kritchevsky Detergent'.

Kritchevsky did not know the exact chemical composition of his product but analysis has shown it to be a complex mixture of the following compounds[2].

1. Free diethanolamine
2. Fatty acid amine soap $RCOO^{-}\overset{+}{N}H_2(CH_2CH_2OH)_2$
3. Fatty amide $RCON(CH_2CH_2OH)_2$
4. Ester amide $RCONCH_2CH_2OOCR$
   $\quad\quad\quad | \; CH_2CH_2OH$
5. Diester amide $RCON(CH_2CH_2OOCR)_2$
6. Amine ester $HN\text{-}CH_2CH_2OOCR$
   $\quad\quad\quad \backslash \; CH_2CH_2OH$
7. Amine diester $HN(CH_2CH_2OOCR)_2$
8. N,N'-bis(2 hydroxyethyl)piperazine

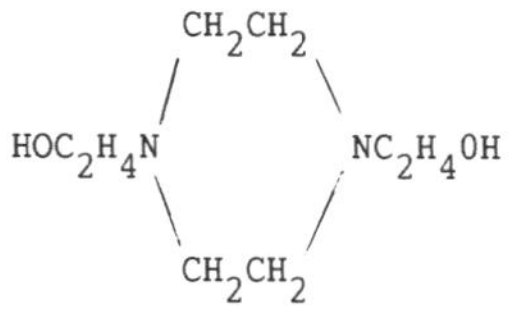

An inter-relation of these materials has been postulated by Kroll and Nadeau[3]. (Figure 2).

FIGURE 2

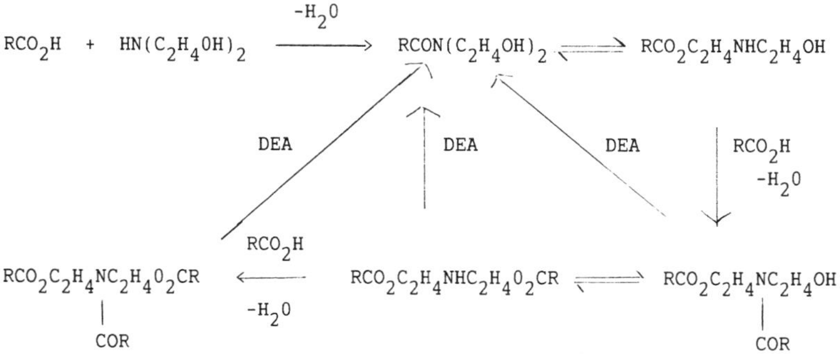

This arrangement is supported by the fact that it is possible to improve the yield of diethanolamide by aging the Kritchevsky detergent at lower temperatures. This would appear to involve the conversion of ester amine and amide to the diethanolamide by rearrangement of the amine ester to ester amide and consequent reaction with excess diethanolamine.[4]

A typical composition of the Kritchevsky detergent has in fact shown to be as follows.[2]

1. Free diethanolamine - 20 - 30%
   This shows that the diethanolamine and fatty acid are reacting in a 1:1 molar ratio and that the excess diethanolamine is there to prevent the formation of undesirable by-products as shown in Figure 2.

2. Fatty amide - 50 - 60%

3. Fatty amine soap - 3 - 5%

There is 3-5% of free fatty acid left at the end of the reaction and owing to the excess amine this will be present as fatty amine soap. In fact this soap may be responsible for the excellent water solubility of the Kritchevsky detergent.

4. Ester amide - 5 - 10%

This is formed by the reaction of 2 moles of fatty acid with 1 mole of diethanolamine. This by-product is a foam depressant rather than a foam stabiliser and is kept to a minimum in the reaction by using excess diethanolamine.

5. Amine ester - 5 - 10%

This exists in thermal equilibrium with the fatty amide, higher temperatures favouring the amine ester and lower temperatures the fatty amide.

The utility of the Kritchevsky detergent is shown by its continued use even after the introduction of superamides.

SUPERAMIDES

British Patent 631,637 (1949) to E.M. Meade describes the preparation of the so called 'superamides' with a fatty alkanolamide content of up to 95%. The reaction is a two stage process involving the production of the methyl ester (Figure 3) and subsequent conversion to fatty alkanolamide (Figure 4).

FIGURE 3

$$\begin{array}{l} CH_2OOCR \\ | \\ CHOOCR \\ | \\ CH_2OOCR \end{array} + 3CH_3OH \xrightarrow[Na]{} 3RCOOCH_3 + \begin{array}{l} CH_2OH \\ | \\ CHOH \\ | \\ CH_2OH \end{array}$$

FIGURE 4

$$RCOOCH_3 + HN(CH_2CH_2OH)_2 \xrightarrow{Na} RCON(CH_2CH_2OH)_2 + CH_3OH$$

The alcoholysis of the triglyceride with methanol could be acid or alkali catalysed although the alkali process is favoured because the reaction reaches completeness faster and can be carried out at a lower temperature. A typical reaction would be carried out at 80-95°C using a molar excess of alcohol and 98% conversion could be expected in one hour. The glycerol separates out leaving an upper layer which is mainly ester with some alcohol, catalyst soap and glycerol. The ester can be washed to improve quality.

The amidation is carried out by reacting 1 mole of methyl ester with 1.1 moles of diethanolamine using approx 0.1% sodium methylate as catalyst. The reaction will take 3-5 hours at 105°C. The use of lower reaction temperatures than with the Kritchevsky detergent mean that less by-products are produced and improved colours may be obtained. Methanol is liberated in the reaction and the removal of this is an excellent way of following the progress of the reaction. The methanol may be re-used for the first stage.

Studies have been made by Russel et al. to show the effect of mole ratio, temperature and catalyst concentration on the yield of fatty alkanolamide, using methyl laurate as a model.(1) Optimum conversion to fatty amide is achieved at a mole ratio of 1:1 to 1 diethanolamine to methyl laurate. The slight excess diethanolamine inhibits ester amide formation but extra diethanolamine above this level slows down conversion. Optimum temperature for conversion is found to be 105°C, higher temperatures encouraging side reactions and poorer colour. A catalyst level of 0.15 to 0.25% sodium methylate gives optimum conversion to amide.

The superamides may also be prepared directly from fats i.e. mixed fatty acid triglyceride esters. In these cases however the glycerol will remain in the product at a level of 8-10%.

Table 1 shows a comparison of the composition of a typical superamide and a Kritchevsky alkanolamide.(1)

TABLE 1 - Typical compositions of Kritchevsky and superamides.

	Kritchevsky %	Superamides %
Fatty diethanolamide	60	90
Diethanolamine	23	7
Fatty acid	5	0.5
Ester amines/amides	12	2.5

The higher fatty amide content of the superamides means that they are not as water soluble as the Kritchevsky product. However they are easily solubilised by other surfactants or by the addition of free fatty acid and diethanolamine to produce soap. The Kritchevsky detergent is a better wetter and detergent but poorer foam stabiliser and a more complex mixture. Its main applications are as a textile detergent, shampoos, emulsifier, rust inhibitor and dry cleaning soap. The superamides have excellent foam stabilising properties and show synergistic effects with other surfactants. They also have excellent corrosion-inhibition properties for steel. Their main applications are as foam stabilisers for anionics in laundry and dishwashing detergents and as thickeners for detergents and shampoos. Superamides are now the most widely used diethanolamides.

MONOALKANOLAMIDES

The most common monoalkanolamides are those based on monoethanolamine and monoisopropanolamine. These are usually produced by heating equimolar amounts of the fatty acid and amine under conditions in which water is removed. The monoalkanolamides of lauric acid are waxy solids and not very soluble in water and are used mainly in powder detergents, their use in liquids being limited by their poor solubility, although they are solubilised by other surfactants.

The monoalkanolamides have better hydrolytic stability than the corresponding dialkanolamide with the monoisopropanolamide derivative being more stable than the monoethanolamide probably due to steric hindrance.

A wide range of physical properties may be obtained by the correct choice of fatty chain length and method of preparation. This is illustrated in Table 2. As can be seen from this table more insoluble products and generally higher melting points are obtained from higher chain length fatty acids, monoalkanolamines and via the methyl ester route.

TABLE 2:[1] Physical Form and Solubility of Selected Fatty Alkanolamides.

Reactant		Amine/acid ratio	Product	
Fatty base	Alkanolamide		Physical form	Water solubility
Lauric acid	Diethanolamine	2:1	Soft paste	Soluble
Coco acid	Diethanolamine	2:1	Liquid	Soluble
Oleic acid	Diethanolamine	2:1	Liquid	Dispersible
Stearic acid	Diethanolamine	2:1	Paste	Dispersible
Lauric acid	Monoethanolamine	1:1	Wax or cryst. solid.	Insoluble
Lauric acid	Monoisopropanolamine	1:1	Wax or cryst. solid.	Insoluble
Myristic acid	Monoethanolamine	1:1	Wax or cryst. solid.	Insoluble
Stearic acid	Monoethanolamine	1:1	Hard cryst. wax.	Insoluble
Methyl laurate	Diethanolamine	1:1	Waxy cryst. solid.	Limited solubility
Coco methyl ester	Diethanolamine	1:1	Paste	Limited solubility

ETHOXYLATED ALKANOLAMIDES

There are two general methods of preparing ethoxylated alkanolamides.

FIGURE 5

$$RCONH_2 \;+\; n\; CH_2 - CH_2 \text{ (ring through O)} \xrightarrow{Na} RCONH(CH_2CH_2O)_nH$$

Fatty amide

FIGURE 6

$$RCONHCH_2CH_2OH \;+\; n\; CH_2 - CH_2 \text{ (ring through O)} \xrightarrow{Na} RCONH(CH_2CH_2O)_{n+1}H$$

Fatty monoalkanolamide

In Figure 5 a fatty acid amide is reacted directly with ethylene oxide in the presence of a basic catalyst. Figure 6 shows the reaction of a fatty monoalkanolamide with ethylene oxide, where the ethylene oxide adds to the free hydroxyl group and not to the hydrogen adjacent to the nitrogen. Both routes give identical reaction products.

It has been shown(5) that in both cases the ethylene oxide reacts with the terminal hydroxyl group rather than the secondary amide group. The evidence for this is by hydroxyl value, infra red spectra, and saponification resistance.

1. The hydroxyl value of lauric monoethanolamide which has been condensed with one mole of ethylene oxide is very close to that obtained by reacting lauric acid with 2(2-aminoethoxy)ethanol and indicates the presence of one hydroxy group per molecule.

   It is different from the hydroxyl value of lauric diethanolamide which would not have been the case if the ethylene oxide had reacted with the amide hydrogen.

2. Lauric diethanolamide has no absorption peak in the 6.75 micron region but the lauric monoethanolamide and its ethoxylate do show this characteristic peak.

3. Lauric monoethanolamide is more stable to alkali hydrolysis than the corresponding diethanolamide. Addition of 1 mole of ethylene oxide to the monoethanolamide increases this hydrolytic stability. The fact that this occurs points to the addition onto the terminal hydroxyl group.

   The hydrolytic stability of alkanolamides is shown in Table 3.

TABLE 3: Hydrolytic Stability of Alkylolamide Derivatives
% Breakdown After 2 hours Reflux 100°C.

Amide	H_2SO_4 Hydrolysis		NaOH Saponification	
	0.1N pH 1.2	1.0N pH 0.4	0.1N pH 12.4	1.0N pH 12.5
Lauric diethanolamide	3.2	88.2	33.6	94.1
Lauric isopropanolamide	4.3	83.7	0.6	1.8
Lauric monoethanolamide	9.5	71.9	2.7	22.2
Lauric monoethanolamide + 2 EO	2.8	22.5	0.0	6.2
Lauric monoethanolamide + 5 EO	0.0	6.8	0.0	0.0

The table shows clearly that the hydrolytic stability is in decreasing order of lauric monoethanolamide ethoxylates > lauric monoisopropanolamide > lauric monoethanolamide > lauric diethanolamide. The 5 mole ethoxylate of lauric monoethanolamide is extremely stable to acid and alkaline hydrolysis at 100°C. This hydrolytic stability makes the ethoxylated derivatives extremely useful in heavy duty alkaline systems.

Addition of ethylene oxide to the monoalkanolamide also improves other properties of the product as can be seen in Table 4.

TABLE 4: Properties of ethoxylated alkanolamides[5]

Amide	Physical form	Melting point °C	Ross & Miles[a] foam heights in m.m.		Draves[a] Wetting time in secs
			0 min	5 mins	
Lauric monoethanolamide (LMEOA)	Solid	81	-	-	-
LMEOA + 1 EO	Solid	59	-	-	-
LMEOA + 2 EO	Solid	32	-	-	-
LMEOA + 3 EO	Solid	29	116	113	5.7
LMEOA + 4 EO	Paste	25	146	144	5.0
LMEOA + 5 EO	Liquid	23	145	143	12
LMEOA + 10 EO	Liquid	29	125	115	52

(a) Carried out using 0.1% active solution in distilled water at 25°C.

The addition of ethylene oxide gives a gradual reduction of melting point showing a maximum at 5-7 moles. The basic lauric monoethanolamide is not very water soluble but the 3 mole adduct is clearly water soluble. The table shows that the optimum foaming properties are attained for the 5 mole adduct and maximum wetting for the 3-4 mole ethoxylate. The results for a coconut monoethanolamide show a similar trend.

PROPERTIES OF ALKANOLAMIDES AND DERIVATIVES

1. Excellent foam boosting/stabilisation

Soft deionised water with the absence of soil are the best conditions for generating foam with anionic surfactants. Hard water and soil, grease *etc*. reduce this foam but the addition of 1-5% coconut or lauric diethanolamide improves this dramatically. This effect is demonstrated in Tables 5 and 6.

TABLE 5[6] Effect of diethanolamide on the foaming of various surfactants

Type of surfactant	Concentration of surfactant %	Foam height at 45°C in hard water, cm			
		100% Surfactant		80% Surfactant 20% diethanolamide	
		0 min	5 mins	0 min	5 mins
TEA lauryl sulphate	0.02	38	15	110	90
MEA lauryl sulphate	0.02	30	12	105	100
Lauryl(3 EO)sulphate	0.02	160	150	150	150
Na lauryl sulphate	0.02	55	50	135	125
Na C10-14 alkyl sulphate	0.02	20	10	40	35
Octylphenol + 10 EO	0.12	140	90	170	155
Nonylphenol + 10 EO	0.12	135	125	155	145
Octylphenol + 14 EO	0.12	155	145	155	150
Nonylphenol + 14 EO	0.12	160	145	145	140

TABLE 6[2]

Effect of alkanolamide on foaming power of alkyl aryl sulphonate using plate washing test.*

Alkyl Aryl Sulphonate (active).	Coconut Diethanolamide (superamide)	Number of plates washed to foam end point		
		Distilled Water	10° Hard Water	30° Hard Water
100	0	0	8	11
80	20	4	11	12
70	30	5	12	15
60	40	6	12	15

*Assessed by washing plates soiled with fat in 0.25 g/l active solution at 40°C until the foam disappears.

These tables show that the effect on foam boosting is most pronounced for alkylaryl sulphonates and alcohol sulphates, not as great for alkyl ether sulphates and very small other nonionics.

2. Viscosity modification

Tables 7[6] and 8[2] show the effect of alkanolamides on the viscosity of dodecyl benzene sulphonate and triethanolamine lauryl sulphate solutions respectively.

TABLE 7: Effect of addition of alkylolamides to dodecyl benzene sulphonate solution

Alkylolamide	Viscosity at 25°C (Pa s)	Foam Height	
		Immediate	After 5 minutes
None	0.075	16	15
Coconut monoethanolamide	0.305	18	17
Coconut monoisopropanolamine	0.305	16.5	16
Coconut diglycolamide	0.960	18.5	17
Coconut diethanolamide	0.650	17.5	16
Oleic diethanolamide	0.840	17	17
Coconut di-isopropanolamine	0.360	17	16

TABLE 8: Effect of diethanolamide on viscosity of TEA lauryl sulphate.

Triethanolamine lauryl sulphate	High active dialkanolamide	Viscosity in centistokes.
-	10	22
10	90	44
20	80	73
30	70	97
40	60	153
50	50	410
60	40	1200
70	30	> 1300
80	20	1300
90	10	170
100	-	26

The superamides are more effective than the Kritchevsky detergent because of their higher amide content.

3. Detergency

The Kritchevsky product has good detergency in its own right but both it and the superamides have a synergistic effect with other surfactants. For instance scouring tests carried out on a 3:1 alkyl aryl sulphonate coconut diethanolamide active blend shows an improvement of 30% over the straight anionic on wool and cotton. Similar results can be shown for combinations of alkanolamides with other nonionic surfactants. These are shown in Tables 9[2] and 10[2], the figures quoted being brightness results.

TABLE 9:

Concentration of active material.	Sodium alkyl aryl sulphonate.		75 pts. active sodium alkyl aryl sulphonate: 25 pts. high active coconut diethanolamide	
gm/litre	Wool	Cotton	Wool	Cotton
0.75	38	39	51	39
1.0	44	42	58	43
1.25	48	45	61	51
1.5	50	47	64	52

TABLE 10:

Concentration of active material gm/litre	(1) Wool	(1) Cotton	(2) Wool	(2) Cotton	(3) Wool	(3) Cotton
0.5	34	42	50	43	54	44
0.75	40	46	53	45	54	48
1.0	40	54	55	55	60	57

1. Nonylphenyl polyglycol ether. Cloud point 50°C.
2. 70 parts nonylphenyl polyglycol ether. Cloud point 50°C.
   30 parts coconut diethanolamide.
3. 50 parts alkyl polyglycol ether. Cloud point 95°C.
   50 parts coconut diethanolamide.

4. Emulsification

Combinations of alkanolamides with other surfactants gives improved oil in water emulsions of waxes and mineral oils.

5. Wetting

The Kritchevsky has good wetting properties which are unaffected by hard water and acidic conditions. The superamides are not as good because of their poor water solubility but can have synergistic effects with other surfactants. Addition of ethylene oxide improves the wetting properties of coconut monoethanolamide with 3-4 moles being the optimum level.

6. Corrosion resistance

Alkanolamides in their own right are non-corrosive. However aqueous solutions of lauric diethanolamide will inhibit the corrosion of steel.

APPLICATIONS OF ALKANOLAMIDES AND DERIVATIVES

The properties of alkanolamides mean that they find use in a wide range of applications.

1. Liquid and powder detergents:

In these types of formulations use is made of the alkanolamides ability to boost detergency, foam stability and wetting properties. In liquid detergents lauric and coconut diethanolamides and ethoxylated monoethanolamides are widely used. The monoalkanolamides and ethoxylates find particular use in alkaline powder detergents. Typical formulations are shown below.

Light-duty Household Liquid Detergent

	% w/w
Dodecyl benzene sulphonic acid	10
Triethanolamine	2
Caustic soda (45% solution)	1.7
Sodium hypochlorite (10% solution)	0.6
Lauric acid diethanolamide	1
Sodium sulphate	1
Water	83.7

Spray-dried General-purpose Powder

	% w/w
Active matter as alkyl benzene sulphonic acid	25 neutralised with caustic acid.
Sodium toluene sulphonate	2.5
Lauric monoethanolamide	3
Sodium tripolyphosphate	30
Sodium silicate (1:2 ratio)anhydrous	10
CMC (100% basis)	2
Optical brightening agent	0.2
Sodium sulphate	27.3

2. Hard Surface Cleaners

Lauric and coconut diethanolamides are used in tile and wall cleaners to boost detergency and improve wetting.

3. Shampoo/bath products

In shampoos and bubble baths alkanolamides are used to modify viscosity and when added to sulphate based shampoos they increase the lather volume, speed of foam generation and stability and help lime soap dispersion.

The sodium sulphosuccinate mono-esters of lauric monoethanolamide and its ethoxylates have been used as non-irritating surfactants for shampoos. In addition lauryl sarcosine diethanolamide is an efficient foam booster for baby shampoos based on amphoteric surfactants. A typical shampoo formulation is shown below.

Typical shampoo

	% w/w
Triethanolamine lauryl sulphate	47.6
Coconut diethanolamide	3.0
Preservative	0.15
dyestuff, perfume	as required
Ammonium chloride	0.3
Water	to 100%

4. Textile Processing

The main application of alkanolamides outside detergents and shampoos is in the area of textile auxiliaries. Coconut diethanolamide is used as an anti-migration agent in dyeing polyesters with disperse dyes and as a settling agent in dyeing polyamides and cellulosics. Alkanolamides as emulsifiers for mineral oils are used in coning oils, a typical example of which is given below.

Mineral based coning oil

	% w/w
200 sec. mineral oil	90
Alcohol ethoxylate	4
PEG oleate	4
Diethanolamide	2

The diethanolamide improves the shock emulsion and scourability of the formulation and imparts some antistatic properties. Another example of their use would be in an overspray for fibre processing.

Overspray for fibre processing

	% w/w
Coconut diethanolamide	5-10
PEG 400 monolaurate	90-95

The diethanolamide here helps wetting and spreading as well as giving antistatic protection.

In an oxidised polyethylene wax emulsion for textiles a Kritchvesky type coco-diethanolamide is used not only to emulsify the wax but the free diethanolamine will also saponify the oxidised wax.

Oxidised Polyethylene Wax Emulsion

	% w/w
Oxidised PE wax	30
2:1 Cocodiethanolamide	5
Water	65

5. Antistatic for plastics

Coconut diethanolamide is used widely as an antistatic for plastics *e.g.* in polyethylene film for food packaging. It has been employed along with metallic salts as an antistat for polystyrene and in impact resistant rubber/polystyrene blends.

6. Cosmetic and pharmaceutical emulsions

Alkanolamides are used to form oil in water emulsions in this area and to prepare aqueous, transparent ringing gels of white oils in cosmetic formulations.

7. Metal working fluids

Alkanolamides are used in this application as emulsifiers for mineral oils but they also provide lubricant properties and good corrosion resistance. Groundnut based diethanolamides are particularly effective in these mineral oil systems.

Other areas where alkanolamides find application are as follows:

Antiblock agents for polyethylene - coconut diethanolamide
Mineral flotation - Kritchevsky type
Dry cleaning systems - lauric triethanolamide
Pigment dispersion
Coascervant in silk screen printing

CONCLUSIONS

Alkanolamides are a well established group of surfactants whose basic chemistry is well known and understood. However they are complex mixtures whose physical properties can vary widely depending on choice of raw materials, and reaction conditions. Because of this and their useful properties they will continue to find use in a wide variety of industrial applications.

REFERENCES

1. M J Shick, Nonionic Surfactants, 1966 Chapter 12 published Edward Arnold

2. A T Pugh, Mfg Chemist, 28 557 (1957)

3. M J Shick, Nonionic Surfactants, 1966, Chapter 8 published Edward Arnold

4. H Kroll and H Nadeau, J.Am. Oil Chemists Soc., 34 323 (1957)

5. E A Knaggs, Soap Chem Specialities, 40 (12), 79 (1964)

6. A Davidsohn, B M Milwidsky, Synthetic Detergents 6th Edition published George Godwin Limited.

Acetylenic Glycols and Derivatives

H. P. Kleintjes* and J. Schwartz
AIR PRODUCTS, CHEMICALS DIVISION EUROPE, P.O. BOX 85075, 3508 AB UTRECHT, THE NETHERLANDS

One group of additives which has been especially formulated for water-based systems is a unique line of acetylenic glycol-based, nonionic surfactants such as that typified by the structure represented in Figure 1.

STRUCTURE OF SURFYNOL 104

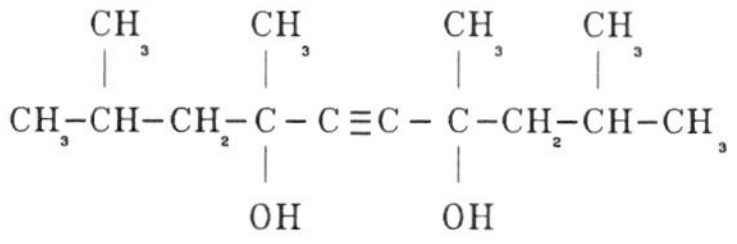

2,4,7,9-TETRAMETHYL-5-DECYNE-4,7-DIOL

Figure 1 Air Products

The ability of these acetylenic glycol surfactants to provide multifunctional benefits in water-based systems can be related to the structure of the molecules.

The combination of the triple bond and two symmetrical hydroxyl groups yields a domain of high electron density, making that portion of the molecule polar and hydrophilic.

The highly branched alkyl chains supply the hydrophobic property, creating an excellent wetting agent or surface tension reducer. This property is of particular significance for water based systems because of the high surface tension of water of 72 dynes/cm as compared to the 20 - 30 dynes/cm for the commonly used solvents. In general, foam problems are more severe in a water-based system and, therefore, the acetylenic glycols were designed to be non-foamers and/or defoamers by careful engineering of the hydrophile - hydrophobe ratio.
The six methyl groups in Surfynol 104 tend to reduce the molecular attraction between adjacent molecules and thereby minimise the tendency for local crystallization which is widely observed with chains of methyl groups. Thus, Surfynol 104 exhibits a lower melting temperature than similar straight chain molecular weight hydrocarbons (see Table 1) and it does not tend to form micelles in aqueous solutions whereas surfactants with long methyl group chains do form micelles readily.

MELTING POINTS OF VARIOUS ALCOHOLS AND GLYCOLS

NAME	STRUCTURE	# C	# CH3	# OH	MP C
1-TRIDECANOL	C13H27OH	13	1	1	30.5
TRI(N-BUTYL)CARBINOL	C13H27OH	13	3	1	20.0
1,13-TRIDECANEDIOL	C13H26(OH)2	13	0	2	76.5
1,12-TRIDECANEDIOL	C13H26(OH)2	13	1	2	60.5
1-TETRADECANOL	C14H29OH	14	1	1	40.0
1,14-TETRADECANEDIOL	C14H28(OH)2	14	0	2	84.5
2,4,7,9,-TETRAMETHYL 5-DECYNE-4,7-DIOL	C14H24(OH)2	14	6	2	32.0

Table 1 Air Products

Surfactant Mechanism

A surfactant can be defined as any substance which will greatly reduce the surface tension of a solvent at a very low concentration. The ability of a surfactant to reduce surface tension results from the combined hydrophilic as well as hydrophobic part in its structure, which makes it tend to migrate to the liquid-air interface when added to a liquid. The presence of the surfactant molecule at this liquid-air interface results in a compressive force acting on the surface - and this force reduces the surface energy or surface tension.

Many physical phenomena play a role in foam stabilisation.

- cohesive strength / surface viscosity
- charge repulsion
- steric hindrance
- surface transport / Plateau-Marangoni-Gibbs effect

The Surfynol surfactant molecule orients horizontally due to the centrally located polar group. The presence of substituted methyl groups tends to minimise intermolecular attractions thus reducing the surface viscosity. Foam stability is thereby minimised (figure 2).

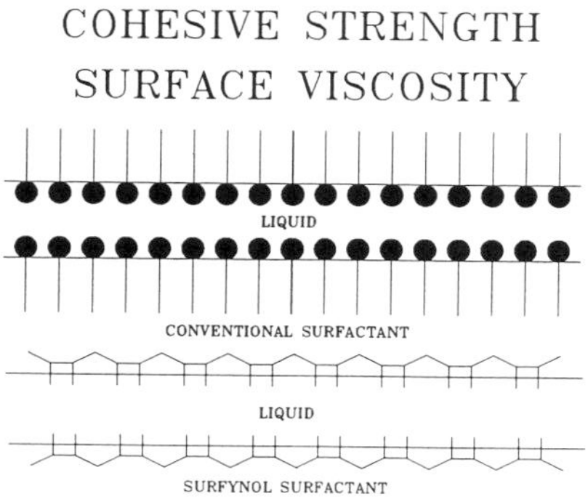

Figure 2 Air Products

In figure 3 the effect of charge repulsion is presented. Drainage of the lamellae is inhibited due to the repulsion of ionic groups.

CHARGE REPULSION

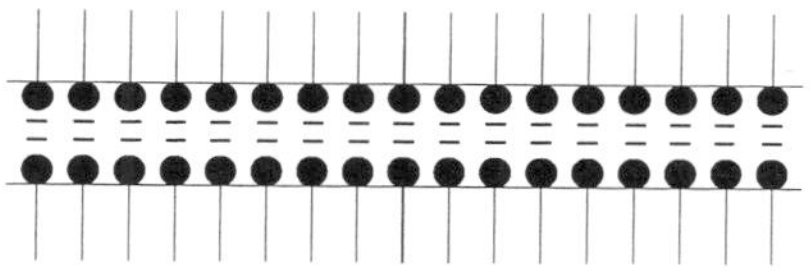

Figure 3 Air Products

In case of coalescence of foam bubbles into large unstable bubbles, the pendant hydrophilic groups prevent the close approach of the two lamellae (figure 4).

STERIC HINDRANCE

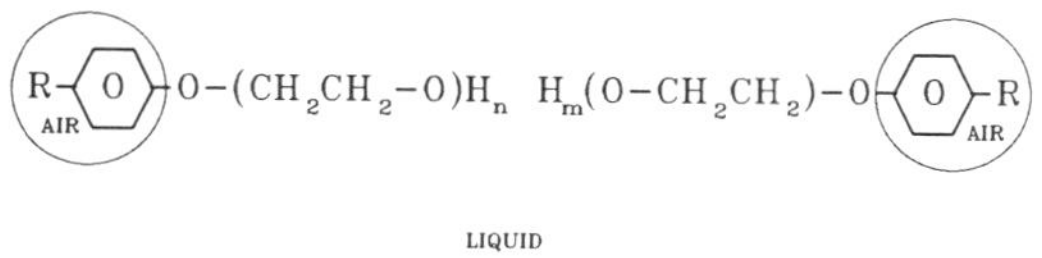

Figure 4 Air Products

Surface transport, referred to as the Plateau-Marangoni-Gibbs effect, takes place when thinning of the lamellae causes a surfactant concentration gradient. Therefore also a surface tension gradient is established. Surface transport of surfactant molecules will eliminate these gradient, but will also drag water molecules in the process, thus thickening, restoring, and stabilising the film (figure 5).

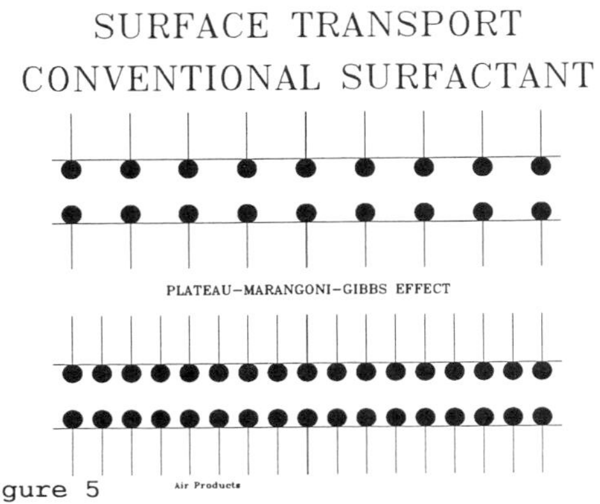

Figure 5

However, Surfynol molecules have the ability to decompress. This will eliminate the surfactant concentration gradient and not contribute toward the restoration of the lamellae. Liquid drainage continues and therefore foam is minimised (figure 6).

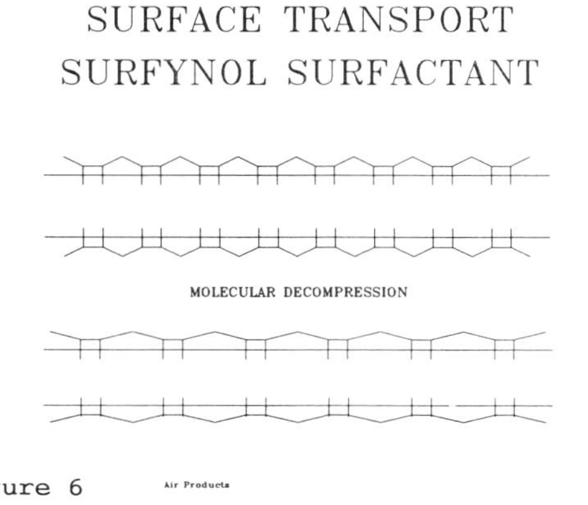

Figure 6

To summarise, the unique properties of acetylenic glycol surfactants are a result of their structures which incorporate three centrally located polar groups and many symmetrically distributed methyl groups. The centrally located polar groups cause the molecule to orient horizontally in the surface. The methyl groups minimise intermolecular attractions and tend to interface with and displace materials which form structured solid films both at air/water and oil/water interfaces.

As a result of these structural features the acetylenic glycol surfactants have defoaming properties together with excellent surface tension reduction.

Applications

Surfynol surfactants are used in many applications where wetting and/or foam problems arise, particularly in water based systems.

Coatings and inks benefit from the wetting properties of various Surfynol grades with low foaming. Improperly cleaned surfaces and difficult to wet substrates like polyolefins, are easily wetted with the aid of acetylenic glycol surfactants.

Another area is pressure sensitive adhesives (PSA), where the silicone releasepaper has to be covered with a continuous film, Surfynol surfactants improve wetting with no retraction which could lead to craters and pinholes.

Among various other applications, latex dipping is an interesting field for surfactants. In the next chapter a more detailed description is given.

Surfynol Surfactants for Latex Dipping

The latex dipping process is used to produce a wide range of products including: rubber gloves (household, medical, surgical, and industrial), balloons (toy, meteorological), condoms, finger cots, rubber tubing, and bladders. Natural rubber latex is used almost exclusively since it forms smooth, continuous films and is very strong and elastic when vulcanised. Occasionally, neoprene or nitrile lattices are blended with the natural rubber latex to enhance strength and/or solvent resistance.

Dipping Operations

There are three basic variations of the dipping process: straight, coagulant, and heat sensitive. The dipping processes are fundamentally similar; they differ only in the method used to coat the former with latex. A summary of these dipping processes follow.

Straight

Straight dipping is the simplest process. The former is immersed in the latex and withdrawn slowly to give a uniform coating. One straight dip gives a film thickness of approximately 0.05mm, depending on the viscosity and total solids of the latex. A thicker deposit can be obtained by partially drying the first layer and reimmersing the coated former in latex. These steps can be repeated until desired film thickness is obtained; then the latex film can be dried and vulcanised (figure 7).

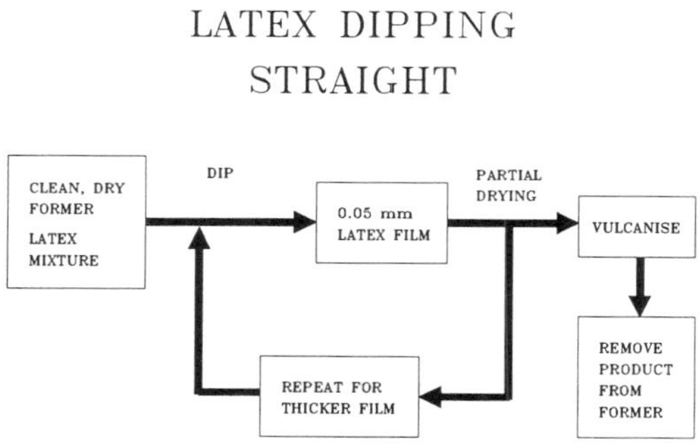

Figure 7 Air Products

Coagulant

Coagulant dipping involves an additional step. The former is immersed in a coagulant solution and allowed to partially dry. A typical coagulant solution contains 15-30 parts calcium nitrate, 2 parts dispersed talc, 1 part Surfynol TG, and the balance water.

The former is then immersed in the latex and slowly withdrawn. The latex leaves a gelled film on the former of approximately 0.2-0.8mm after one dip. Thickness depends upon the activity of the coagulant, retention time of the former in the latex, and total solids and viscosity of the latex. If a thicker film is desired, the latex on the former can be partially dried and immersed with or without reapplying coagulant solution. This step can be repeated until the required film thickness is obtained, and then the latex deposit can be dried and vulcanised (figure 8).

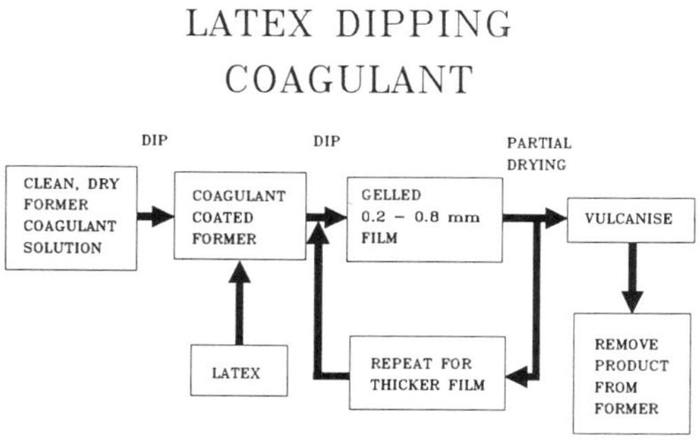

Figure 8 Air Products

<u>Heat Sensitive</u>

Heat sensitive dipping requires that the former is heated to 50-80°C before dipping into the latex. When withdrawn, the former is coated with a latex film of approximately 4mm. Film thickness is dependent on temperature and heat capacity of the former, retention time in the latex, and mixing in the latex tank.

Complications can arise when heat transferred from the former destabilises the latex mixture. Therefore, a cooling system must be employed to keep the latex in the stable temperature range. Despite the added precautions, heat sensitive dipping is the most effective process for manufacturing thick-walled products with a single dip (figure 9).

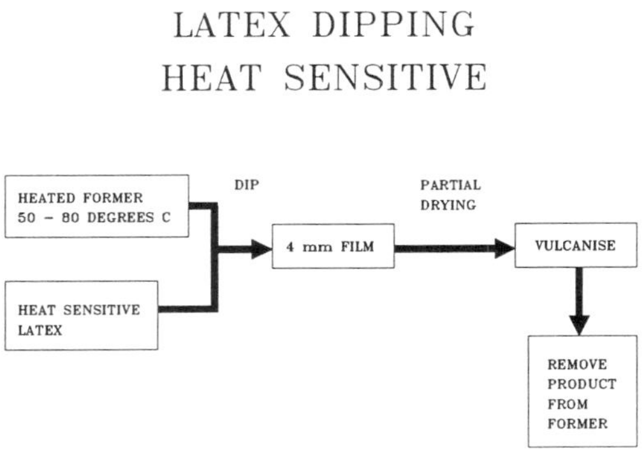

Figure 9 Air Products

Basic Post-Dip Processes

After the former has been coated with latex, the film is dried and vulcanised. This task is performed in a two-stage drying process using hot air ovens. The deposit is partially dried in an 80-90°C oven, then vulcanised in a 100-140°C oven.

The dried product is manually "stripped" (removed) from the former with compressed air. Often water (or a water/talc mixture is used to facilitate the stripping process.

Performance Processes

There are several treatments that are necessary to give the latex product desired performance properties. They include leaching, flocking, chlorination, lubrication, and solvent roughening.

Leaching (washing with water) removes all water-soluble impurities from latex-dipped products. Removal of these materials is important when producing electric-resistant or high clarity products.

Many gloves are lined with cotton to provide comfortable wear. The cotton or "flock" lining is applied using a multi-stage flocking process. The glove body is made using coagulant dipping. When the body is partially dry, a second straight dip is made into an adhesive latex. The wet former is passed through a flock cloud to give a uniform covering of flock fibers. The glove is turned inside-out when it is stripped to give a flocklined product.

Chlorination is used to reduce surface drag on latex goods. The product is immersed in dilute, aqueous chlorine solution. When removed, the chlorinated rubber surface has a lower coefficient of friction. Lubrication of the latex product with talc, mica, starch, or silicone oil is an alternate means of reducing surface drag. The preferred method for reducing the surface drag of a latex product depends on the end-use. Lubrication is used for disposable products such as surgeon's gloves or condoms, while chlorination is preferred for reusable products such as household gloves.

Solvent roughening produces a pattern on dipped gloves for both aesthetic and functional purposes. After two coagulant dips, the wet latex gel is immersed in a rubber swelling solvent. The enlarged particles leave an attractive surface finish which also provides grip to the gloves.

Surfynol Surfactants' Role

Dipping formers are typically made of porcelain, glass or highly polished metal surfaces. To keep a low rejection rate, it is important to cover the former fully with coagulant solution and/or latex mixture. Wetting agents and defoamers are added to these liquids to ensure film integrity. Surfynol surfactants' ability to wet rapidly-generated surface area makes them a natural choice for latex dipping operations. This characteristic is demonstrated in the following dynamic surface tension data in table 2, generated on a sensadyne 5000 surface tensiometer using a 0.1% aqueous surfactant solution.

DYNAMIC SURFACE TENSION OF VARIOUS SURFACTANTS

SURFACTANT	%	DYNAMIC SURFACE TENSION BUBBLE RATE	
		1 BUBBLE/SECOND	6 BUBBLES/SECOND
SURFYNOL 104	0.1	32.5	36.5
PLURONIC 25R2	0.1	39.8	43.8
TRITON X-100	0.1	33.4	44.6
NEODOL 25-7	0.1	32.4	55.8
IGEPAL CO-730	0.1	40.8	50.4
TERGITOL NP-40	0.1	50.9	54.2

Table 2 Air Products

If the coagulant solution gives poor coverage of the former, it follows that a low quality latex film is produced, causing holes and weak spots in the product. Surfynol TG surfactant is used in the coagulant solution to disperse and deairentrain the talc. Also, Surfynol TG surfactant controls foam and enhances wetting to eliminate fisheyes and prevent retraction.

Surfynol 104E surfactant can be added to an adhesives latex to provide continuous coverage over uncured latex. This is especially important when producing flocked latex products as described earlier.

Unlike many silicone defoamers, Surfynol surfactants will not cause fisheyes or weak spots in the finished product. It should be noted, however, that order of addition is important to guard against destabilisation of the latex mixture. Therefore, this fact should be considered when evaluating Surfynol surfactants in a formulation. Where greater solubility is required, the ethoxylated Surfynol 400 series surfactants can be employed.

Dewebbers

New are the XF-C41-08 and XF-C41-34 dewebbing/defoaming agents for latex dipping. These products also have surfactant properties which made them useful for wetting of dipping forms. Thin films produced using these dewebbing additives do not exhibit film inconsistencies. XF-C41-08 and XF-C41-34 are stable, 100% active organic products that can be added "neat" to latex. In addition, XF-C41-34 can also be predispersed with water.

In general, formulated systems containing waterbased latex emulsions, will benefit from the effective defoaming and wetting properties as well as the latex compatibility that these exciting new products provide.

To draw a parallel in coatings applications, these dewebbers can be used to prevent webs when dip coating sheet metal with predrilled holes. If the web dries, and afterwards a bolt or screw is used for assembly, the coating will crack and a weak spot is introduced for <u>e.g.</u> corrosion.

Quaternary Salts as Phase Transfer Catalysts

Charles M. Starks
CIMARRON TECHNICAL ASSOCIATES, TULSA, OKLAHOMA 74101-3223, USA

Introduction

Use of phase transfer catalysis (PTC) for industrial and commercial purposes is increasing rapidly and substantially throughout the world. Although the PTC technique may have been in use commercially for many decades, its enormous recent growth has developed from a clear understanding of its mechanism, its scope of application, and a predictable set of guidelines on how to use it.[1-6] It is estimated by this author that between 5 and 25 million pounds of phase transfer catalysts of all types are being used annually to catalyze an estimated 150 to 200 different commercial chemical processes. From a value perspective, it has been estimated[7] that all chemicals produced using at least one major PTC process amounts to between two billion (conservatively) and ten billion dollars annually.

Quaternary ammonium salts, and amines that are easily convertible to quats, account for much more than half of the catalysts used. Their use for this purpose in commercial applications is the subject of this review, although it will be recognized that many other types of phase transfer agents, particularly polyethylene glycols and ethers are also used, sometimes interchangeably. Industrial applications of PTC have been reviewed by Freedman.[8]

The Phase Transfer Catalysis Concept

The situation is often encountered when a chemical reaction between two chemicals cannot undergo reasonably rapid reaction with each other because they reside in different phases of the reaction mixture. Although this problem may usually be overcome by use of an appropriate mutual solvent, the problem frequently may be solved just as easily and less expensively by use of **phase transfer catalysis** (PTC). This technique works because, in principle, one can select a phase transfer agent which, added in small quantities, has the ability to repeatedly transfer one of the reagents into the natural phase of the other reagent so that reaction can occur.[1]

A simple example of PTC is the reaction of neat 1-chlorooctane with aqueous sodium cyanide, as represented by Equation 1.

$$\begin{array}{llcccccc} \text{ORG PHASE} & 1\text{-}C_8H_{17}Cl & + & Q^+CN^- & \xrightarrow{90^\circ C} & 1\text{-}C_8H_{17}CN & + & Q^+Cl^- \\ & & & \uparrow\downarrow & & & & \downarrow\uparrow \\ \text{AQUEOUS PHASE} & Na^+Cl^- & + & Q^+CN^- & \rightleftharpoons & Na^+CN^- & + & Q^+Cl^- \end{array} \qquad (1)$$

In the absence of quaternary salt, Q^+X^-, no reaction occurs. In the presence of Q^+X^- (typically $R_4N^+Cl^-$, where R is hexyl or octyl), added at a level of about 1 percent of the organic phase, reaction proceeds rapidly, and is complete in about two or three hours. Quat cyanide is easily and rapidly extracted into the organic phase where it undergoes reaction with 1-chlorooctane. Concurrently, the produced quat chloride, after extraction into the aqueous phase or at the interface, exchanges anions to regenerate quat cyanide to start a second cycle.

Although the above general mechanism greatly oversimplifies the complex kinetic behavior that occurs during PTC, it does illustrate the process quite well. It is well to recognize that there are two basic steps involved: (a) the organic-phase reaction, and (b) the sequence of steps which delivers cyanide to the organic phase. When the second step is very fast compared to the first, then the process said to be an "extraction-type" phase transfer catalysis[7a], and works best when the quat can easily extract the anion into the organic phase. When the first step is very fast compared to the second, the process is said to be an "interfacial-type" PTC, since the rate depends on how rapidly the quat can deliver anion to the organic phase where reaction occurs so fast that it is presumed to occur at or just inside the interface.

The distinction between extraction-type and interface-type PTC is important in catalyst selection.[7a] For extraction-type PCT reactions, such as cyanide displacement on 1-chlorooctane, the best catalysts are highly oleophilic quats such as tetrahexylammonium, tetraoctylammonium, or commercial quats such as ALIQUAT® 336 or ADOGEN® 464 quaternaryammonium chlorides. Use of larger quats, such as tetradodecylammonium chloride, does not improve the rate over tetrahexyl or tetraoctyl salts, because extraction is already essentially complete. For an interfacial-type PTC reaction, such as cyanide displacement with neat benzyl chloride, the best catalysts are much more water-soluble or interface-soluble, having structures such as benzyltriethylammonium, tributylmethylammonium, or surfactant quats such as hexadecyltrimethylammonium or hexadecylpyridinium. Use of completely water-soluble quats such as tetramethylammonium usually produces no catalytic effect, unless reaction occurs mostly in the water phase.

General Applicability

To recognize the great breadth of applicability of phase transfer catalysis one need only scan through the hundreds of PTC reactions in a review such as the Dehmlows' book.[6] The vast majority of reported PTC systems involve transfer of anions from either an aqueous or solid phase, liquid-liquid, or liquid-solid PTC,[8a] but examples also have been published on gas-solid, gas-liquid-liquid and gas-liquid-solid systems. Examples of inverse phase-transfer, involving transport of an organic-soluble reactant into an aqueous phase for reaction,[9] transfer of cations by anionic PTC agents,[10,11] transfer of HCl from aqueous hydrochloric acid[12], transfer of formaldehyde, from water,[13] transfer of oxygen anion radical[14], and transfer of hydrogen peroxide.

Phase transfer systems are probably abundant in nature. Clearly the ability of hemoglobin in the blood to transfer and activate molecular oxygen to different sites in the body is an essential PTC system. The phosphortransferase system of certain bacteria are trimeric phase transfer agents for the phosphate anion.[15] Substantial progress in the problem of transfer of drug molecules into the brain from the bloodstream has been made by linking a common fat-soluble carrier molecule with a drug molecule to enable the drug to be transported through the blood-brain barrier.[16]

In the search for commercially safe low-temperature bleaching systems for detergents, techniques have been developed to use water-soluble but solid and stable perborate or percarbonate salts in combination with activators. When added to water the activators deliver the active peroxide bleaching function in a low-temperature active form to the fabric surface during washing.[17]

Commercially Available Quaternary Salts and Amines for PTC

Many different quaternary ammonium salts have been reported in the literature for use as phase transfer catalysts. Indeed, for fine tuning PTC processes to achieve optimal performance, lowest cost, easiest catalyst removal, greatest catalyst stability, and optimal physical handling for commercial processes, it is always best to develop and test a variety of different catalyst structures. For most screening purposes, and often for commercial use one of the available quaternary salts listed in Table 1 will usually work very well. If one of the reagents in the reaction to be catalyzed is an inexpensive alkyl halide, use of a trialkyl amine with in situ formation of a quat is often the most cost-effective option.

Table 1

Some Commercially Available Quaternary Salts and Amines Commonly Used in the U.S.A. as Phase Transfer Catalysts

	Producer
Extraction-Type Quaternary Salts	
ALIQUAT 336	Henkel Corp., Extraction Technology Div.
ADOGEN 464	Sherex Chem. Corp.
Tetrahexylammonium Halides	R.S.A. Corp.
Interfacial-Type Quaternary Salts	
Benzyltriethylammonium Chloride	Eastman Kodak Co. Hexcel Corp. Lindan Chemicals, Inc.
Tetrabutylammonium Bromide	Chemical Dynamics Corp. Eastman Kodak Corp. Hexcel Corp. R.S.A.
Tetrabutylammonium Hydrogen Sulfate	Nobel Chemicals, Inc.
Tributylmethylammonium Chloride	Ethyl Corp.
Tetrabutylphosphonium Salts	R.S.A. Corp. Aldrich Chem. Corp.
Cetyltrimethylammonium Chloride	Hexcel Corp.
Cetylpyridinium Chloride	Hexcel Corp.
Trialkyl Amines for in Situ Quat Formation*	Akzo, Chemical Division AtoChem, Inc. Ethyl Corp., Chemicals Group Lonza, Inc.
Betaines*	Akzo, Chemicals Div. Akolac Co. Mazer Chemicals Scher Chemicals, Inc. Sherex Chemicals

*Contact companies for list of individual products available.

The synthesis and application of phase transfer catalysts has been reviewed,[18] and a comparison of some commercially available quaternary ammonium structures for some types of phase transfer catalyzed reactions has been reported.[19] Because quats lose activity at higher temperatures depending on their structure and under strongly alkaline conditions, it is well to be aware of the stability factors.[20]

A large amount of work has been devoted quaternary salts bound to insoluble resins,[21-25] and to inorganic solids.[26] These catalysts are highly attractive because of their easy removal from batch reactions or because of the possibility for using them as fixed bed catalysts in continuous reactions. Also known as "triphase catalysts"[27] these solid-bound catalysts are useful for many applications, but they also may have several severe limitations including irreversible catalyst deactivation,substantially higher cost, and physical degradation.

Commercial Applications of Phase Transfer Catalysis

General

Phase-transfer catalysis is useful in industrial applications when it offers a better overall method for dealing with the myriad of problems that must be solved in a successful commercial process. Its advantages are:

- PTC frequently allows use of lower cost reagents. For example, many alkylations can be done with potassium carbonate or aqueous sodium hydroxide, rather than much more expensive bases, such as sodium or potassium alkoxides.
- PTC frequently allows avoidance of the use of a solvent, or at least allows one to use less expensive, less toxic, easily recoverable and removable solvents. Elimination, reduction or change of solvent may allow greater throughput from a given processing plant, or it may allow debottlenecking of solvent recovery units. In a new plant is may allow reduction of capital expenditures. Use of less solvent lowers energy consumption.
- PTC provides unique options in many reactions. For example, it provides completely different opportunities in production of condensation polymers; it provides the opportunity to safely use many inorganic oxidants such as sodium hypochlorite, hydrogen peroxide, sodium persulfate, nitric acid, and electrochemical oxidations in commercial processes; it allows the industrial chemist to consider economical use of dichlorocarbene as a reagent.
- PTC may lead to process simplification, as for example in continuous counter current reactions between two liquid phases to ensure complete consumption of one or both reagents. Often it is possible to do two or more tandem PTC reactions without isolation of intermediate products.
- PTC allows excellent methods of reaction control by adjustments of catalyst concentration, agitation rate, programmed rate of addition of reagents, and adjustable rate of initiation of free radical reactions.
- PTC may allow the operator to reduce wear and tear on equipment. For example in use of strong caustic in glass lined equipment, glass erosion and damage has been minimized by switching to a PTC reaction and by addition of caustic only as rapidly as it is consumed.
- Increased product specificity. Many examples have been cited,[28] where use of PTC decreases side reactions. PTC, when carefully developed, allows production of chiral products from non-chiral starting materials.

The principle disadvantages for use of PTC are:

- The cost of the catalyst may be significant. Use of PTC requires that this cost be offset by advantages of PTC use.
- Recovery, separation, re-use or disposal of the catalyst is required. This step adds to the cost of using PTC, and must also be offset by advantages. Many methods of catalyst removal have been described, but the one of choice is usually an extraction of some sort, where the catalyst is especially selected for ease of removal.
- Substantial R&D work is often required to fine tune a PTC system, perhaps to eliminate emulsions, to select catalysts with the right degree of stability, the best reaction conditions and procedures.

Because of the complex multi-factor considerations that are normally encountered in industrial chemical processing, it is not possible here to make generalizations about the utility of PTC in commercial applications. The best path here appears to be to provide examples of how PTC is, or may be, useful in commercial practice, and to let these examples show how the advantages of PTC are beneficially used.

Commercial Uses

From the author's collection of patents and publications on phase transfer catalysis, 563 items were selected as being especially relevant to commercial and applied chemical practice, as compared to approximately 3000 other items which are more relevant to the scientific aspects of PTC. The 563 items were further subdivided into the categories shown in Figure 1.

Figure 1

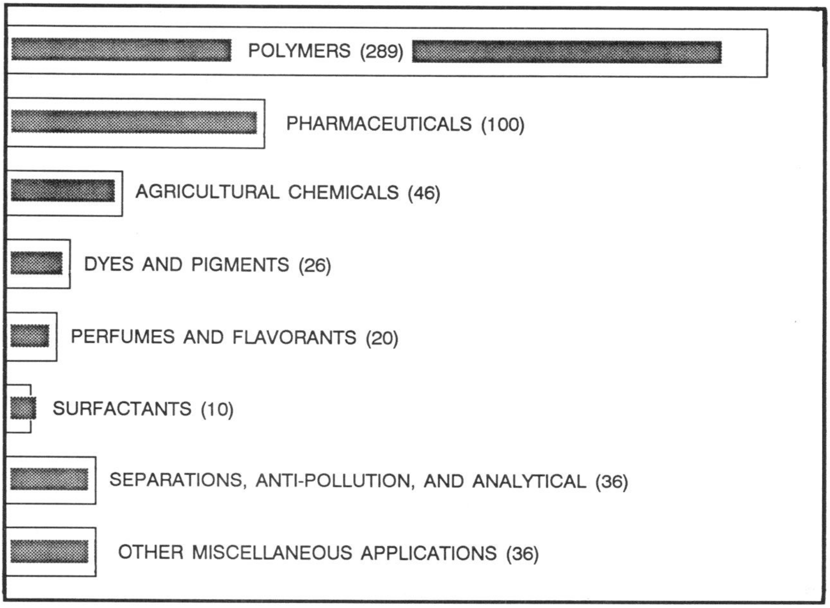

It has been the author's experience that approximately ten to twenty applied research projects take place in industrial laboratories for each one that results in a patent or publication. When applied to phase transfer catalysis, this ratio suggests that much of the industrial research work on phase transfer catalysis has not been publicized, and this is especially true for highly competitive products, or products and processes which cannot be patented in a useful manner. Indeed the author is aware of a number of commercial PTC applications have not been published.

Polymer Chemistry

Since polymers and related products constitute about 80 percent of the volume of organic chemicals sold, it is not surprising that this area would also be one of the major applications area for phase transfer catalysis. Several reviews [29-34] relating to this topic have been published. Polymer applications of PTC are sub-divided here into five categories dealing with (1) production of monomers, (2) polymerization, (3) chemistry on already formed polymers, (4) chemistry at the surface of polymers or solids, and (5) miscellaneous polymer-related chemicals.

(1) Production of Monomer

Vinyl Monomers

A large amount of work has been published on dehydrohalogenation of appropriate starting materials to prepare chloroprene[35-40], vinyl chloride,[35, 41, 42] haloacrylic acid,[43] propargyl derivatives,[44] ring halogenated styrenes for reduced flammability polymers,[45-47] and vinyl ethers of diols and esters.[48-51] Use of PTC for selective dehydrohalogenation has been reviewed.[52]

Other vinyl monomers prepared by PTC etherification and esterification reactions include alkyl ethers,[53,54] esters of methacrylic acid,[55-59] various glycidyl and epoxy monomers,[60-63] silane monomers,[64,65] and bis-analog vinyl ethers from reaction of Bisphenol A with 2-chloroethyl vinyl ethers.[66]

Arylate Monomers

Use of thermoplastic polyarylates which have high melting points, excellent mechanical properties, and excellent thermal stabilities, has grown sharply during the last decade. Polyarylates are generally defined as polymers having no aliphatic groups other than tetrasubstituted groups in the main polymer chain. Two examples of polyarylates illustrate the general nature of these materials:

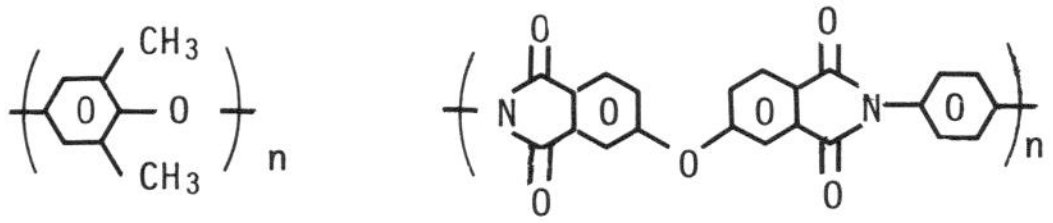

Although thousands of such polymer structures could be imagined, only a few are in actual commercial practice on a multi-million pound-per-year scale, and these are often limited in use because of their relatively high price compared to other materials. One factor which contributes significantly to their cost is the relatively high cost of producing appropriate monomers.

Much work has been done, particularly at the General Electric Co., towards successful development of new synthetic options for arylate monomers from low-cost starting materials. Many of the patented processes use phase transfer catalysis methods, as for example the reaction of Bisphenol A with N-methyl-4-nitrophthalimide.[67,68]

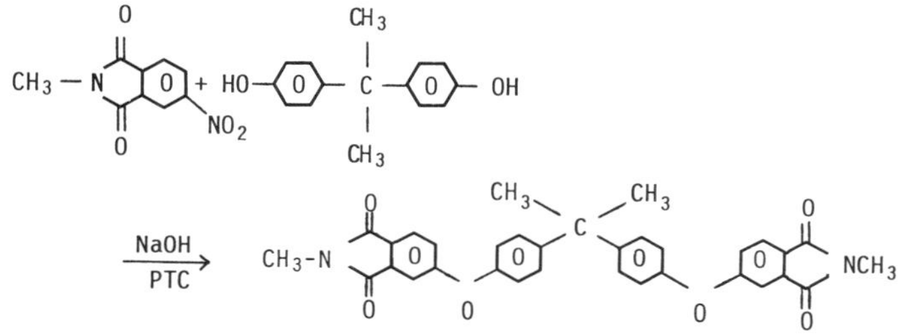

Other doubly-activated aryl chlorides, such as dinitrochlorobenzenes, also react rapidly under PTC conditions to give substitution products in high yields,[69-73], although these products are not generally suitable as monomer precursors.

Singly activated aryl chlorides, such as p-nitrochlorobenzene, are much more suitable, but their reactivity is too low for many desireable PTC reactions. If temperature is increased to obtain acceptable reaction rates (e.g., to 130-160°C), then ordinary tetraalkyl ammonium or phosphonium salts decompose too rapidly to be useful. Other types of phase transfer catalysts, particularly polyethylene glycols, may be used, although relatively large quantities are required and difficulties in their removal from products are encountered.[74] To solve this problem Brunelle and co-workers at G.E.[75-79] developed quat structures based on 4-(N,N-dialkyl)pyridinium salts which are vastly more stable than simple tetraalkyl ammonium salts.

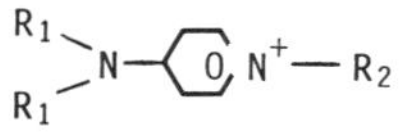

R_1 = alkyl or morphilino group
R_2 = branched alkyl group such as 2-ethylhexyl or neopentyl

For example, in the model reaction of p-nitrochlorobenzene with sodium phenolate,

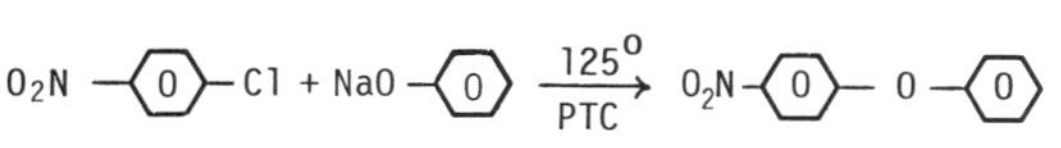

the product ether is obtained in 95 percent yield after 30 minutes with the pyridinium catalysts, but only in 12 percent after 8 hours with tetrabutyl ammonium bromide.

In a test for catalyst stability using disodium Bisphenolate A in toluene at 140° the pyridinium salts showed their clearly greater stability,[75] as listed in Table 2.

Table 2

Test of Quaternary Salts for Stability at 140°C in the Presence of Disodium Bisphenolate A[75]

Quat Cation	Time to 50% Quat Decomposition
$(C_4H_9)_4N$	8 Minutes
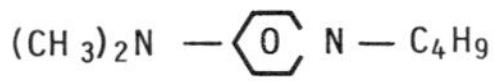 $(CH_3)_2N-\langle O \rangle-N-C_4H_9$	2 Hours
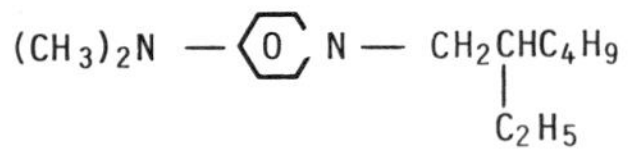 $(CH_3)_2N-\langle O \rangle-N-CH_2CH(C_2H_5)C_4H_9$	8 Hours
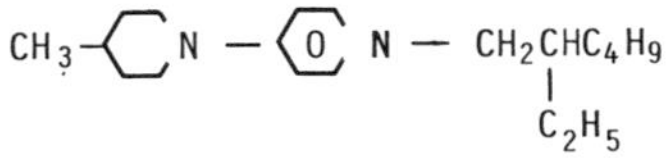 $CH_3-\langle N \rangle-\langle O \rangle-N-CH_2CH(C_2H_5)C_4H_9$	11 Hours
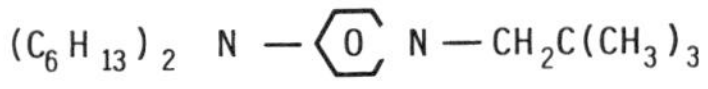 $(C_6H_{13})_2N-\langle O \rangle-N-CH_2C(CH_3)_3$	12 Hours

When disodium Bisphenolate A is used as nucleophile the reaction rate is slowed, compared to sodium phenolate, because two moles of quat are required to make the bisphenolate soluble in the organic phase. The concentration of diquaternary bisphenolate, and therefore, the reaction rate in the organic phase is proportional to the square root of catalyst concentration.[74,80] This problem could be resolved[74] by using a catalyst containing two quaternary ammonium functions, separated by an appropriate spacer group, as illustrated:

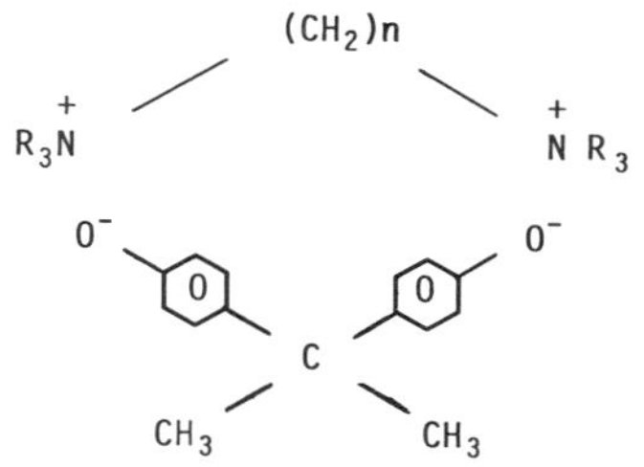

When used as catalyst these diquats greatly increased the reaction rate, and changed the kinetics to approximately first order in catalyst.

Preparations of a variety of other arylate monomers have been reported.[81-85]

(2) Polymerization via Phase Transfer Catalyzed Processes

At first glance it would seem that phase transfer catalysis should be especially useful for condensation polymerizations between reactive nucleophiles such as dihaloalkanes, reacting with difunctional electrophiles such as diols, diphenols, dithiols, etc. This is clearly true, as has been extensively demonstrated. What was not predicted was that PTC condensation polymerizations would differ so markedly in character, as is outlined in part (a) below. Neither was it so obvious that PTC would be useful in free radical polymerization, as outlined in part (b).

Beyond the above, PTC techniques have been used to prepare unique polymers such as poly (benzoin) condensation products from terephthaldehyde;[86] NYLON 4, a notoriously difficult polymer to obtain in high molecular weight,[87] and polymers from polyalkylation of phenylacetonitrile with activated dichlorides .[88,89] A normal and important feature of successful polymerization processes is avoidance of side reactions which result in chain transfer, chain termination, or unwanted cross-linking. It has been noted that phase transfer frequently increases the selectivity of many reactions to produce high molecular weight polymers, whereas with conventional techniques only low yields of low molecular weight polymers were obtained.[90-92]

(a) PTC Catalyzed Condensation Polymerization

A great deal of work has been done on phase transfer catalyzed condensation polymerization, as represented by the sample listed in Table 3. Percec and his coworkers,[93] have systematically and extensively investigated this area with regard to the chemical principles and with regard to the applied area of liquid crystals. After investigating a large number of polyether syntheses several substantial differences between conventional solution condensation polymerizations and phase transfer polymerizations were found. For example, in reactions such as 1,4-dichlorobutene-2 with disodium Bisphenolate A:

- In conventional step-wise condensation polymerization, exactly stoichometric amounts of the two monomers are required to obtain high molecular weight polymers; whereas, in PTC polymerization high molecular weight polymer is obtained even if the amounts of monomer are not balanced.

- Products from conventional step-wise condensation polymerization follow statistical behavior, giving high molecular weight polymer only at very high conversion, greater than 99.5 percent. In contrast, PTC condensation polymerizations produce high molecular weight polymers even at low degrees of conversion, meaning that the polymer chain ends react much more rapidly than monomer. Although the reason for this is not yet known, it is probably because most of the polymerization takes place very near the interface where, once polymerization has started, the produced polymer molecules remain for many reaction steps before moving back into the bulk organic phase.

Table 3

Some Phase Transfer Catalyzed Condensation Polymerizations

Dihalo Compound	Co-Reactant	Product	Reference
Phosgene	Bisphenol A	Polycarbonate	94,95
Thiophosgene	Diphenols	Polythiocarbonates	96-98
$POCl_3$	Polyphenylene oxide plus thiol-terminated polystyrenes	PPO-Polystyrene block copolymers	99
$RPOCl_2$ or $ArPOCl_2$	Diphenols, Dicarboxylic acid salts	Polyphosphonates	100-102
$SbOCl_3$	Diamines	Poly(antimonoamides)	103
CH_2Cl_2	Bisphenol A	Polyformals	104
$Br(CH_2)nBr$	Diphenols	Polyethers	105,106
n = 3 to 10	Trithiocarbonate salts	Thiopolymers for rubber stabilization	120
$ClCH_2CH = CHCH_2Cl$	Diphenols	Polyethers	107,108
α,α^1-Dihalo-p-xylene	Diphenols	Polyethers	108
	Diamines	Polyamines	109
	Sodium Sulfide	Polysulfides	110
	Benzene Sulfonamide	Polysulfonamide	111
	$(p\text{-}NaO_2SC_6H_4)_2O$	Polysulfone	112

Table 3 (Continued)

Some Phase Transfer Catalyzed Condensation Polymerizations

Dihalo Compound	Co-Reactant	Product	Reference
$(ClCH_2O)_2CO$	($\underline{p}$-$NaO_2O_{22}SC_6H_4)_2O$	Polysulfone	113
$Cl(CH_2CH_2O)nCH_2CH_2Cl$	Diphenol	Polyether	114
$ClCH_2O(CH_2)_4OCH_2Cl$	Bisphenol A	Polyether	115
Hexafluorobenzene	Diols, Dithiols	-(C_6F_4-diol)n- polymer	116
Carboxylic Acid dichlorides	Diphenols, amino-phenols, dialcohols	Polyesters and poly (amide-ethers)	117-119

- For conventional statistical condensation polymerizations the theoretical (and observed) polydispersibility is equal to 2. However, in many cases of PTC condensation polymerizations the polydispersity is less than 1.3, even at 100 percent conversion, again indicating the nonstatistical nature of the polymerization.

- In conventional polymerization with a 1:1 mole ratio of the two monomers the polymer chain ends will have a statistical distribution of functional groups. However, in the PTC system, since only a small amount of the water-soluble monomer is transferred at any time, and because this is extremely reactive, the growing chain ends only have electrophilic end groups. This result has a highly practical aspect, since these polymers can easily be used for block copolymerization or alternating block copolymerizations. Prior to the use of PTC such block copolymerizations required careful and often extensive care and effort.

- A final difference is that many conventional condensation polymerizations are run under conditions which can lead to scrambling of structures or to side reactions which lead to lower average molecular weights and broader molecular weight distributions. The extent to which these problems occur is much lower in PTC polymerizations.[121,122]

An intensively applied area of PTC condensation polymer chemistry is for preparation of liquid crystal polymers. Percec and his co-workers have done much work with polyethers,[123-131] while Keller,[132-133] Wang[134] and Chen[135] and their co-workers have produced liquid crystalline polymers by PTC modification of other polymers. Although the volume of liquid crystal polymers produced and sold is relatively small, their value is high, with great emphasis placed on being able to carefully and predictably modify their structures to obtain the desired thermal transition properties.

(b) PTC Free Radical Polymerization

Rasmussen and Smith[136] first reported examples of free radical initiated polymerizations using aqueous potassium persulfate with quaternary salts or crown ethers as phase transfer agents. They found the solution polymerization of acrylic monomers to be much more facile than when common organic-soluble initiators were used. This technique is believed to involve transfer of the divalent persulfate anion into the organic monomer phase where it slowly decomposes to generate radicals,[137-139] although under some conditions radical ions formed in the aqueous phase are believe to be the principal species transferred.[140] Adjustment of the kind and amount of PTC catalyst and of persulfate allows one to precisely control the rate of initiation at any given point during polymerization, rather than be at the mercy of conventional initiator decomposition kinetics.

Organic initiators are frequently the second most costly ingredient charged to a free radical polymerization, so that substitution of low-cost persulfate may represent substantial cost savings. Use of PTC radical initiation allow production of polymers with higher molecular weights and more uniform molecular weight distributions.[141]

Many of the practical features of PTC polymerization have been extensively researched.[142-149] Future work in this field may allow precise rates of free radical generation through electrochemical reactions in the aqueous phase, followed by transfer of the radicals or radical ions into the monomer phase for initiation of polymerization. Aside from the benefits of precise-control of initiation, this technique would also allow the temperature to be independently controlled to improve the quality of the polymers.

(3) Chemistry on Polymers via Phase Transfer Catalysis

Since polymer solutions are difficult to handle, particularly when trying to conduct reactions with inorganic species, the PTC technique offers exceptional promise as a means to achieve chemical modifications of polymers. For example, a number of different kinds of reaction involving nucleophilic displacement, oxidation, reduction, grafting, cross-linking and chain scission can be expected. In fact, a substantial number of such PTC processes have been reported, including such reactions as capping of polyphenylene oxide for stabilization,[150] use of PTC reaction of NaOH to de-vulcanize scrap rubber,[151]
use of PTC chemistry to add dichlorocarbene to the double bond of polybutadiene to obtain stiffer products,[152,153] use of PTC for cross linking,[156] and modification of halo-polymers to adjust glass-transition temperatures, crystallinity, strength, impact, and stiffness properties.[157] The area of polymer modification using PTC has been reviewed.[158]

PTC-PVC Chemistry

Polyvinyl chloride (PVC) which is produced in enormous quantities throughout the world would be an ideal candidate for use in phase transfer catalyzed reactions. However, because of alternating secondary chloro-atom positioning in PVC, eliminations rather than displacement is favored. Moreover, once elimination has started, further elimination to yield long strings additional eliminations occur (the "zipper effect") to yield conjugated polyenes in such profusion as to cause the polymer to become black, cross-linked, and intractable. This elimination process has been carefully studied.[159,160]

In spite of the great tendency of PVC toward elimination, some successful PTC displacements on PVC have been disclosed. Treatment with sodium acetate under mild conditions cause the most reactive chlorine atoms to be substituted by acetate groups with consequent stabilization of the PVC. This reaction also provides an analytical method to determine the amount of labile chlorine present in the PVC.[161] Similarly, certain regions of PVC, notably the isotactic triads with GTTG conformations, undergo facile displacement with sodium thiophenolate under PTC conditions.[162] Treatment of PVC in THF or THF-water mixtures with sodium phenoxide gives both elimination and substitution products: the more homogeneous the reaction mixture the less substitution and more elimination.[163] Grafting of a variety of functional groups onto solid PVC in an aqueous slurry-suspension can be accomplished using PTC, with reaction rates depending on diffusion through the reacted layer,[164] and accelerated if small amounts of PVC swelling agents are present.[165]

PTC Reaction with α-Chloromethylstyrene Polymers

Commercially available α-chloromethylstyrene polymerizes or co-polymerizes with a variety of other vinyl monomers to products having the highly reactive benzyl chloride function. For example:

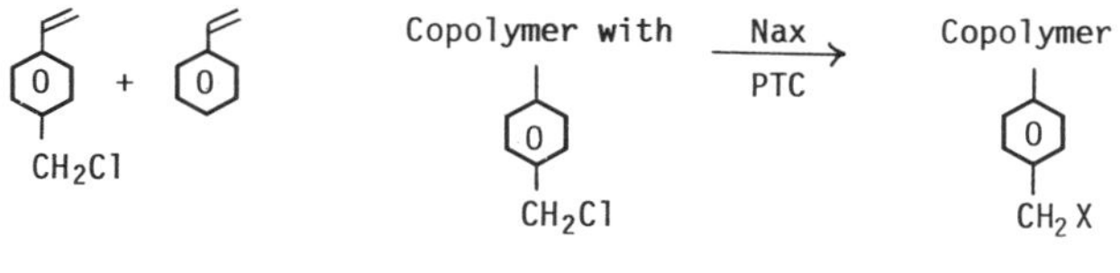

x may be thiols, hydroquinones, fatty acid groups, other polymers, etc.

A large amount of work on this and related polymer systems has been reported for preparation of specialty polymers.[166-180] In fact a phase transfer catalyzed method for chlorination of methyl groups of methyl vinyl aromatic polymers using aqueous sodium hypochlorite has been patented.[181]

Other Chlorinated Polymers

Polyepichlorohydrin,[182-184] poly(vinylchloroformate),[185-188] and poly(2-chloroethylvinyl ether)[189-190] have also been explored for their ability to participate in phase transfer catalyzed reactions to produce specialty polymer products.

PTC for Cellulose Modification

Use of phase transfer catalysis is highly effective in bleaching colored wood specks by aqueous sodium hypochlorite in pulping of tropical woods.[191] More fundamental modifications of cellulose is accomplished by etherification with benzyl chloride, ethyl chloride, chloroacetic acid, and acrylonitrile.[192] For these reactions it was shown that tetramethylammonium chloride was an especially good catalyst, indicating that the slow step of the process probably takes place in a largely aqueous environment. Modified starch derivatives exhibiting reversible gelation properties are prepared by reaction of an amylose starch with an etherification or esterification agent which introduces long alkyl chains onto the amylose.[193] Ethylation of carboxymethylcellulose under PTC conditions gives carboxymethyl ethylcellulose with good physical and pharmaceutical properties.[194] Many other cellulose modifications have been reviewed[195] including use of milk by-products to make coating materials.[196]

(4) PTC-Polymer Surface Modification

Use of phase transfer catalysis permits a rather unique method for performing chemical modifications to surfaces, particularly polymer surfaces. Such treatments are valuable to promote adhesion of a polymer with other materials, to prepare a surface for painting or coating, to alter the appearance or physical properties of the surface. For example, PTC has been used in hide liming,(197) in dyeing of wool,(198) and to make the surfaces of polymers electrically conductive.(199)

Much of the surface modification chemistry using PTC has been done on reaction of poly(vinylidene fluoride) surfaces with aqueous sodium hydroxide.(200-204) This treatment causes the polymer to be dehydrofluorinated to a depth of about 10Å, leaving the surface more wetable by water, stable against air oxidation, and chemically resistant. However, the surface could be chlorinated to become adhesively compatible with other polymers. The same PTC caustic treatment of fluoroelastomers gave points of unsaturation within the polymer which could be used for cross-linking to make a thermoset rubber from an otherwise chemically inert material.(205) Treatment of vinylidene chloride-vinyl chloride copolymer film with NaOH in the presence of phase transfer catalysts gives a conjugated polyene surface structure which, after doping with iodine vapor, becomes electrically conductive.(206)

(5) Miscellaneous Products Related to Polymers

Phase transfer catalyzed methods have been used in the production of plasticizers,(207-209) flame retardants for polymers,(210) preparation of Ziegler-Natta catalysts,(211) photopolymerization initiators,(212-213) delayed vulcanization accelerators,(214) as stabilizers and antioxidants,(215-220) and to control polymerization inhibition.(221)

Use of PTC in Manufacture of Pharmaceuticals

Phase transfer catalysis techniques have been widely explored in pharmaceutical chemistry, with more than 100 patents and papers directly concerned with the application of PTC. These involve a wide variety of chemical reactions which are not easily summarized into neat categories. However, this section will briefly outline three particular applications which illustrate three different advantages in use of phase transfer techniques.

The first application is the large-scale esterification of benzylpenicillin using α-chloroethyl ethyl carbonate.(222)

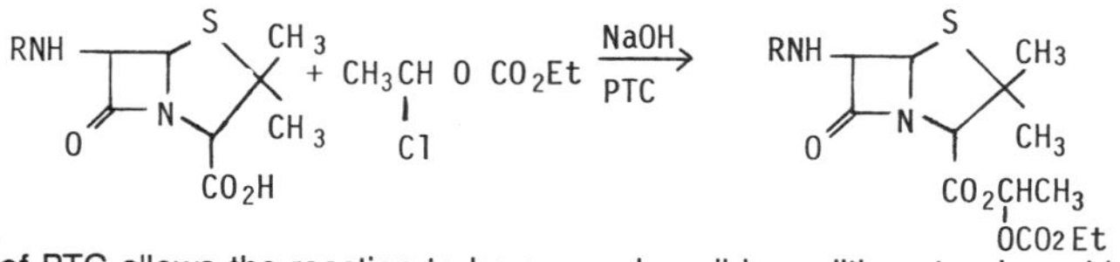

Use of PTC allows the reaction to be run under mild conditions to give a high yield of ester. In the absence of quaternary salt catalyst, the α-chloroethyl ethyl carbonate is rapidly hydrolized.

The second application is the alkylation of phenylacetonitrile, phenyl acetone and related derivatives. Makosza and his co-workers(223) have demonstrated that a number of derivatives having the partial structure

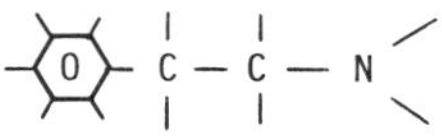

can be produced using a series of low-cost phase transfer reactions. The part structure above has been found to be pharmaceutically active in a variety of applications, and therefore, the synthetic method is widely useful. Also, Makosza's alkylation method allows use of aqueous sodium hydroxide instead of sodium ethoxide, avoiding the expense and manufacturing problems associated with this reagent.

The third application illustrates the ability to use phase transfer catalysis as a method to obtain chiral products from achiral starting materials. Although it was recognized at an early date that chiral phase transfer catalysts offered the possibility of chiral selectivity, the large amount of early work was, at best, only moderately successful in producing optically active products. As this work proceeded it was recognized that the chiral part of the catalyst structure needed to reversibly bind, but strongly orient the substrate molecule prior to its reaction at the cation-anion reaction center. When this was done with carefully selected catalysts and substrates, then enantiomeric excesses (ee) of 30 to 60 percent could be obtained. While this indicated that the premise of chiral direction could be realized, it was far from the >90 percent ee required in practical manufacturing processes.

Dolling, Davis, and Grabowski[224-226] were the first to clearly demonstrate that a PTC technique can be used to prepare chiral products in high enantiomeric excess. Using N-p-trifluoromethylbenzylcinchonium bromide as phase transfer catalyst,

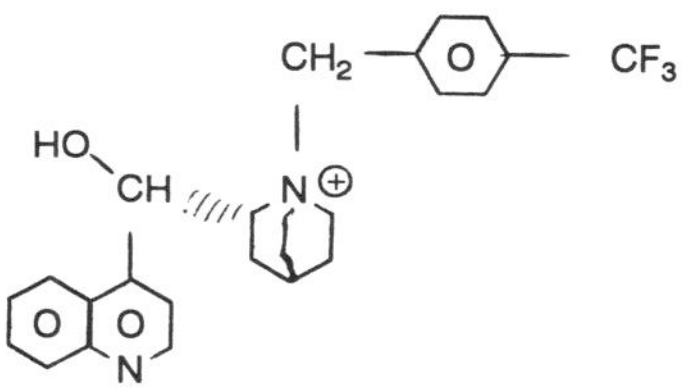

ee values as high as 97 percent and chemical yields of 95 percent of the desired product were obtained, as represented by the equation:

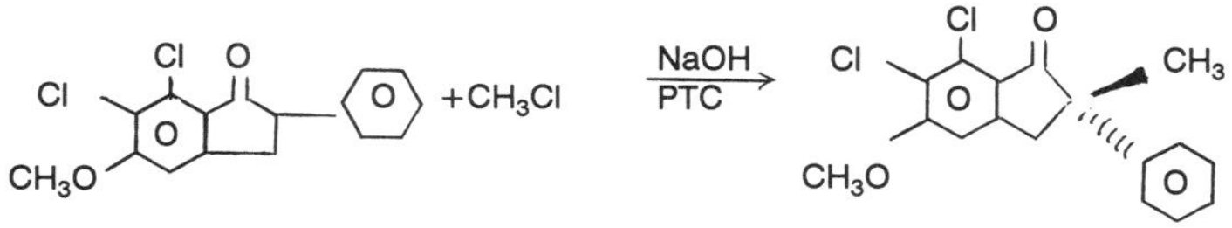

The product is an intermediate in their commercial synthesis of the pharmaceutical S-(+)indacrinone. Like the results obtained by many other research groups, their initial trials with N-benzylcinchoninium chloride gave ee's of 20 to 30 percent. However, by a series of changes in solvent, catalyst structure, reagents, and reaction conditions the ee was improved step-by-step until the desired results were obtained. The key to success during this improvement process was the clear recognition that tight binding of the intermediate carbanion from the ketone starting material needed to be firmly and stereospecifically established before alkylation occurred.

PTC in Production of Agricultural Chemicals

Although it is known that phase transfer catalysis is extensively and widely used to manufacture agricultural chemicals, this field is so competitive that few details about the actual chemical processes are publicly known. The count of patents and publications relating to agricultural chemical numbered to about 50, although these are believed to represent only a small fraction of the processes in use. Processes have been described for production of herbicides,[227-236] insecticides and pesticides,[237-246] insect pheromone intermediates,[247-248] plant growth regulators,[249-252] nematocides,[253] grass control agents,[254] fungicides for plants[255-256] and "agricultural chemicals and intermediates".[257-270] Freedman[8] lists other pesticides which are commercially produced using PTC methods, including two synthetic pyrethroids:

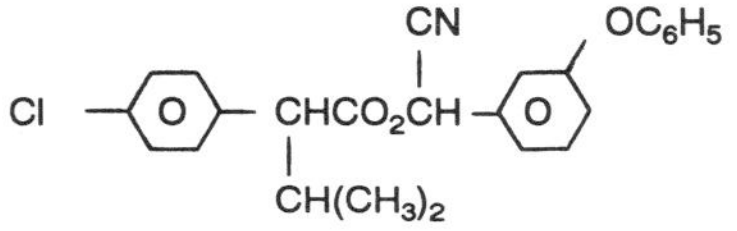

Fenvalerate

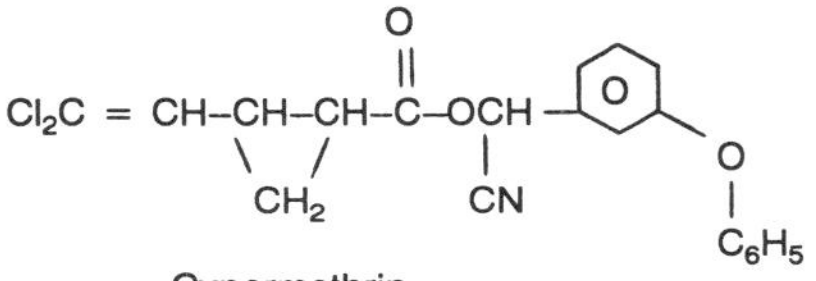

Cypermethrin

PTC in Dyes and Pigments

Between 1980 and 1988 at least twenty-six patents and papers have been published in which phase transfer catalysis is described as a useful method for synthesis or improvement in synthesis of dyes and pigments.[271-296]

PTC in Production of Flavorants, Perfumes and Cosmetics

PTC is used to prepare many perfumes, flavorants, and their intermediates, as reviewed by Vernin[297] and by Gebauer.[198] These processes frequently take advantage of the selectivity and mild reaction conditions afforded by phase transfer techniques. A particularly innovative application is use of PTC to simulate high dilution reaction in preparation of macrocyclic rings.[299] Use of PTC in preparation of UV absorbing agents has been patented.[300,301]

Surfactant Chemicals

A surprisingly small amount of work has been reported on the use of PTC in preparation of surface chemicals, other than perfumes and odorants for detergents, probably because only a few truly "specialty" surfactants are in commercial use. PTC is helpful in preparation of sulfonates by sulfite or bisulfite displacement reactions with alkyl halides, although this route to anionic surfactants is only rarely used because of its greater expense than other routes.

PTC methods have been described for production of high purity glycerol monostearate,[302] and for fluorocarbon surfactants.[303-305] Use of PTC in saponification of fats and oils to make soap has been reported[306,307] and also in production of optical brighteners.[308]

Two practical methods for preparation of long-chain olefin epoxides have been described using either aqueous hydrogen peroxide[309] or sodium hypochlorite[310] as oxidants:

$$RCH=CH_2 \quad + \quad \begin{matrix} H_2O_2 \\ \text{or} \\ NaOCl \end{matrix} \quad \xrightarrow{PTC} \quad R-\overset{\overbrace{\quad O \quad}}{CH-CH_2}$$

Many patents on the use of epoxides of this type for specialty surfactant applications appeared in the period from 1960 to 1980, but they were not used commercially, mostly because no low-cost epoxidation method was available. The two methods above give epoxides in high yields and selectivity, and this author expects these versatile epoxides to appear as commercial products in the future.

Oxidation Processes

Although not specific to any type of commercial products, the use of phase transfer catalysis for commercial oxidation processes deserves special mention. Oxidation of organic chemicals with inorganic reagents prior to the advent of PTC required the use of special solvents, special oxidizing agents, and very careful operations to safely obtain selected products in high yields. The major difficulties arose because the least expensive oxidizing agents are either insoluble in most organic solvents, or they attack the solvent as well as the substrate to be oxidized, often violently. With PTC it is possible to transfer oxidant into the organic phase in a highly controlled manner, and to use solvents which are relatively inert. Some oxidants can be transferred directly, such as hypochlorite,[(311)] permanganate[(312)] or chromate,[(317)] or oxygen from air.[(318,319)] More commonly, best results are usually obtained by use of a variable-valent metal in anionic form as the transferred oxidant, in conjunction with an inorganic oxidizing agent which regenerates the metal anion to its oxidant form.

Metal mediators, or co-catalysts, include anionic forms of cerium, silver, gold, manganese, chromium, nickel, molybdenum, tungsten and others. Oxidants includes hypochlorite,[(313-319)] nitric acid,[(320-321)] persulfate,[(322)] or hydrogen peroxide.[(309,323-329)] Even more promising from a low-cost standpoint is the potential for using either oxidation or reduction systems which employ electricity, via electrochemical reaction in the aqueous phase.[(330-342)] Another remarkable use of PTC is in photochemical oxidations by coupling the phase transfer catalyst with a singlet oxygen photosensitizer, such as Rose Bengal, so that solvents of low polarity can be used, for example to oxidize olefinic and aromatic hydrocarbons.[(343)] Other PTC-photochemical reactions offer insights into commercially interesting chemistry.[(344-347)]

Separation, Purification, Antipollution, and Analytical Uses

General uses of phase transfer techniques for separation and purification of organic materials was pioneered by Brandstrom.[(348)] Extensions to involve PTC methods for removal of impurities has been demonstrated: to make dioxane pure enough for polymerization to high molecular weight poly(formaldehyde),[(349)] for removal of trace amounts of phenolics from waste water,[(350)] for detoxification of dihalocompounds,[(351)] removal of PCB's,[(352)] removal of carbonyl sulfide from hydrocarbons,[(353)] removal of aldehydes from ethylene,[(354)] for purification of phenoxybenzaldehyde,[(355)] epoxy resins,[(356)] m-dinitrobenzene,[(357)] 2-phenylbenzotriazole,[(358)] and for separation of hindered acids.[(359)]

Phase transfer catalysis techniques have been useful in analysis of several commercial products such as: direct titration of unsaturated fats and oils with permanganate;[(360)] direct determination of porphyrins;[361] determination of trace amounts of chloride, bromide, iodide, cyanide, thiocyanate and nitrate in waste water,[362-366] for methylation of carboxylic acids in complex mixtures, such as the suberin fraction of birch bark,[367] for determination of partial structures in coal[368] and coal oxidation products.[369]

References

[1]C. M. Starks, J. Am. Chem. Soc., 93, 195 (1983); C. M. Starks and C. L. Liotta, "Phase Transfer Catalysis, Principles and Techniques," Academic Press, New York 1978.

[2]M. Makosza, Russian Chem. Revs., 46, 1151 (1977); in "Survey of Progress in Organic Chemistry," Academic Press, New York, 1979, Vol. IX.

[3]A. Brandstrom, "Principles of Phase Transfer Catalysis by Quaternary Ammonium Salts," Adv. Phys. Org. Chem. 15, 267 (1977).

[4]W. P. Weber and G. W. Gokel, "Phase Transfer Catalysis in Organic Synthesis," Springer Verlag, New York, 1977.

[5]W. E. Keller, "Compedium in Phase Transfer Reactions and Related Synthetic Methods," Fluka, Switzerland, 1979.

[6]E. A. Dehmlow and S. S. Dehmlow, "Phase Transfer for Catalysis," Chem. Verlag, 1983.

[7]M. Halpern, Dow Chemical Co., Midland, Michigan, private communication, October, 1988.

[7a]M. Rabinovitz, Y. Cohen, and M. Halpern, Angew. Chem., 98, 958 (1986).

[8]H. H. Freedman, Pure & Appl. Chem. 58, 875 (1986).

[8a]G. Bram, A. Loupy, and J. Sansoulet, Israel J. Chem., 26, 291 (1985); H. A. Yee, H. J. Palmer and S. H. Chen, Chem. Eng. Prog., 83, 33 (1987).

[9]L. J. Mathias and R. A. Vaidya, J. Am. Chem. Soc., 108, 1093 (1986).

[10]G. Lipiner, I. Wilner, and Z. Aizenshtat, Nouv. J. Chim., 10, 91 (1986) [C.A. 106 195791].

[11]V. Rod, Z. Sir, and A. Gruberova, Chem. Eng. Res. Dev., 63, 96 (1985).

[12]R. Vladea, D. E. Oltean, T. L. Simadan, L. M. Rusnac, and C. Vladea, Rom. Pat RO 89, 387 (1986) [C.A. 107 6763].

[13]V. Z. Sharf, K. A. Kasymova, and E. F. Litvin, Izv. Akad. Nauk. SSSR, Ser. Khim, 1013 (1986) [C.A. 105 171750].

[14]A. D. Grebenyuk and L. V. Kosaveva, Khim. Prir. Soedin, 515 (1982) [C.A. 98 53344].

[15]J. Deutscher, K. Beyreuther, H. Sobek, K. Stueber, and W. Hengstenberg, Biochemistry 21, 4867 (1982).

[16]G. Bylinsky, Fortune, March 28, 1988, p. 115.

[17]A. Smith, E. Smith, and I. Mackirdy, J. Am. Oil. Chem. Soc.,66, 177 (1988).

[18]J. V. D'Souza and N. Sridhar, J. Sci. Ind. Res. 42, 564 (1983) [C.A. 100 50657].

[19]F. E. Friedi, T. L. Vetter and M. J. Bursik, J. Am. Oil Chem. Soc., 62, 1058 (1985).

[20]D. Landini, A. Maia, and A. Rampoldi, J. Org. Chem., 51, 3187 (1986).

[21]W. T. Ford, Chemtech, 14, 436 (1984).

[22]W. T. Ford and M. Tomoi, Adv. Polym. Sci., 55, 49 (1984).

[23]F. Montanari, D. Landini, A. Maia, S. Quici, and P. L. Anelli, ACS Symposium Series 326, 54 (1987).

[24]M. J. Pugia, A. Czech, B. P. Czech, and R. A. Bartsch, J. Org. Chem, 51, 2945 (1986).

[25]F. Svec, J. Kahovec, and J. Hradil, Chem. Listy, 81, 183 (1987) [C.A. 106 139343].

[26]R. A. Sawicki, to Texaco Inc., U.S. Patent 4,421,675 (1983); ACS Symposium Ser. 326, 143 (1987).

[27]S. L. Regen, J. Am. Chem. Soc., 97, 5956 (1975).

[28]E. D'Incan, P. Viout and R. Gallo, Israel J. Chem., 26, 277 (1985).

[29]F. L. Cook and R. W. Brooker, Polym. Preps. (ACS, Div. Polym. Chem.), 23, 149 (1982) [C.A. 100 7174].

[30]L. J. Mathias and C. E. Carraher, eds., "Crown Ethers and Phase Transfer Catalysis in Polymer Science," Polym. Sci. and Technol., Vol. 24, Plenum Press, New York, 1984.

[31]A. Jayakrishnan and D. O. Shah, J. Polym. Sci., Polym Chem. Ed., 21, 3201 (1983); J. Appl. Polym. Sci., 29, 2937 (1984).

[32]Y. Imai, Yuki Gosei Kagaku Kyokaishi, 42, 1095 (1984); Kobunshi Kako, 33, 581 (1984) [C.A. 102 185538].

[33]T. Nishikubo, Kobunshi, 35, 132 (1986) [C.A. 104 186865].
[34]V. Percec and T. D. Shatter, in "Advances in Polymer Synthesis," J. E. McGrath and B. M. Culbertson, Eds., Plenum Press, New York, 1985, p. 133.
[35]G. A. Chukhadzhyan, R. G. Karapetyan and K. N. Babayan, Arm. Khim. Zh., 35(12) 766-80 (1982) [C.A. 98 106743].
[36]J. Heinrich, R. Casper and M. Beck, Ger. Offen DE 3,208,796 (1983) to Bayer A.-G. [C.A. 100 008204].
[37]E. M. Asatryan, G. S. Grigoryan, A. Ts. Malkhasyan and G. T. Martiroysyan, Arm. Khim. Zh. 36(10), 644 (1983) [C.A. 100.035554].
[38]E. M. Asatryan, V. O. Kirakosyan, A.Ts. Malkhasyan, and G. T. Martirosyan, Arm. Khim. Zh. 39(1), 32-8 (1986) [C.A. 100.034321].
[39]L. A. Khachatryan, K. B. Emirzyan, R. A. Kazaryan, A. Malkhasyan, Arm. Khim. Zh. 40(1), 36-40 (1987) [C.A. 106.215283].
[40]E. M. Asatryan, G. S. Grigoryan, A. Ts. Malkhasyan and G. T. Martirosyan, Arm. Khim. Zh. 36(8), 527-30 (1983) [C.A. 100.023398].
[41]H. J. Pettelkau, Ger. Offen. 3,007,634 (1981), to Fed. Rep. Ger. [C.A. 96.007208].
[42]I. M. Rostomyan, A. G. Israelyan, V. A. Matosyan and G. A. Chukhadzhyan, Arm. Khim. Zh. 37(11), 719 (1984) [C.A. 102.149822].
[43]T. Kondo, T. Matsuda and Y. Funae, Japan Kokai Tokkyo Koho JP 61/36245 (1986) [C.A. 105.061055].
[44]A. E. Kalaidzhyan, S. G. Akopyan, and K. A. Kurginyan, Arm. Khim. Zh. 39(4), 237-42 (1986) [C.A. 106.195843].
[45]S.L.J. Daren, D. Vofsi and M. Asscher, U.S. Patent 4,292,453 (1981) to Makhteshim Chemical Works [C.A. 96.006345].
[46]P. F. Jackish, U.S. Patent 4,423,262 (1983), to Ethyl Corp. [C.A. 100.102912].
[47]C. H. Kolich, U.S. Patent 4,633,026 (1986) to Ethyl Corp. [C.A. 107.040520].
[48]B. A. Trofimov, Z. A. Akhmedzhanova, and V. K. Stankevich, Zh. Prikl, Khim 54(8), 1913-14 (198) [C.A. 96.019615].
[49]L. J. Mathias, J. B. Canterberry and M. South, Publ. Int. Union Pure Appl. Chem., p. 212 (1982) [C.A. 99.140493].
[50]R. Bicker, Ger. Offen, DE 3,237309 (1984) to Hoechst A.-G. [C.A. 101.171917].
[51]J. Pielichomski, R. Popielarz, and R. Chrzaszcz, J. Polym. Sci., Polym. Lett. Ed., 23(7), 387-93 (1985) [C.A. 103.105321].
[52]K. A. Kurganyan, Arm. Khim. Zh., 38, 228 (1985) [C.A. 103 105321];Zh. Vses. Khim. O-va. im. D.I. Mendeleeva, 3, 164 (1986) [C.A. 105 171420].
[53]V. I. Lavrov, B. A. Trofinov, Z. A. Akhmedzhanova, and L. N. Parshina, Zh. Org. Khim., 18(8), 1610-12 (1982) [C.A. 97.162287].
[54]T. Kondo, T. Matsuda, and Y. Funae, Japan Kokai Tokkyo Koho JP 61/36245 (1986) [C.A. 105 061055].
[55]J. L. Charlton, V. A. Sayeed, and G. N. Lypka, Synth. Comm. 11(11), 931-4 (1981) [C.A. 97.163526].
[56]P. J. Shannon, Macromolecules 16(10), 1677-8 (1983) [C.A. 99.176304].
[57]T. Matsuda, H. Sugisawa, Y. Funae, T. Kondo and N. Takatani, Japan Kokai Tokkyo Koho JP 62/63541 (1987) [C.A. 107.134812].
[58]X. Xiang and X. Zheng, Faming Zhuanli Shenqing Gongkai Shuomingshu, CN 86102380 (1986) Peop. Rep. China [C.A. 107.218217].
[59]X. Sun, D. Kong, G. Lu, Y. Fu, Shiyou Huagong, 15(12), 766-9 (1986) [C.A. 107.078276].
[60]Daikin Kogyo Co., Ltd., Japan Kokai Tokkyo Koho JP 58/131976 (1983) [C.A. 100.006309].
[61]P. Panster, A. Karl, W. Buder, and P. Kleinschmit, Ger. Offen, 3,047,995 (1982), to Degussa A.-G. [C.A. 97.198370].
[62]Sanko Chemical Co., Ltd., Japan Kokai Tokkyo Koho 8231679 (1982) [C.A. 96.217680].

[63]C. E. Monnier and F. Stockinger, European Patent Application 226,543 (1987), to Ciba-Geigy A.-G. [C.A. 108.006958].
[64]P. Panster, A. Karl, W. Buder, and P. Kleinschmit, Ger. Offen. 3,047,995 (1982), to Degussa A.-G. [C.A. 97.198370].
[65]C. R. White, U.S. Patent 4,418,229 (1983) to Mallincrodt, Inc. [C.A. 100.067983].
[66]R. R. Galluchi and R. C. Going, J. Org. Chem., 48, 342 (1983).
[67]T. Evans, to General Electric Co., U.S. Patent 4,520,204 (1985).
[68]J. Verbickey and A. M. Colley, To General Electric Co., U.S. Patent 4,577,033 (1986).
[69]M. Makosza, M. Jaqusztyn - Grochowska, M. Ludwikow, and M. Jawodosiuk, Tetrahedron, 30, 3723 (1974).
[70]M. Jawodosiuk, M. Makosza, E. Malinowski, and W. Wilczynski, Pol. J. Chem., 52, 2189 (1978).
[71]A. Frimen and I. Rosenthal, Tetrahedron Lett., 2809 (1976).
[72]B. R. Cho, D. Sung and J. I. Lin, Pollimo, 7, 199 (1983) [C.A. 99 139423].
[73]T. D. Shaffer and V. Percec, Makromol. Chem., 187, 1431 (1986).
[74]D. J. Brunelle, ACS Symposium Series; 326, 38 (1987).
[75]D. J. Brunelle, to General Electric Co., U.S. Patent 4,460,778 (1984).
[76]D. J. Brunelle and D. A. Singleton, to General Electric, U.S. Patent 4,517,141 (1985).
[77]D. J. Brunelle, to General Electric Co., U.S. Patent 4,595,760 (1986).
[78]D. J. Brunelle, to General Electric Co., U.S. Patent 4,681,949 (1987)
[79]D. J. Brunelle and D. A. Singleton, Tetrahedron Lett., 25, 3383 (1984).
[80]N. Yamazki and K. Imai, Polym. J. (Tokyo), 17, 377 (1985) [C.A. 102 167269].
[81]J. E. Hallgren, to General Electric, U.S. Patent 4,361,519 (1982).
[82]C. R. White, to Mallincrodt, U.S. Patent 4,418,229 (1983).
[83]T. L. Evans, Synth. Comm., 14, 435 (1984).
[84]M. A. Beretta, G. Bausani, G. Bottaccio, S. Campolmi, and F. Montanari, European Patent Application 145,377 (1985) to Montedison S.p.A. [C.A. 104.005634].
[85]General Electric Co., Japanese Patent 60/188368 (1985).
[86]C. Nalliah, J. Polym. Mater., 3, 11 (1986) [C.A. 105 227457].
[87]R. Backskia, Polym. Sci. Technol. (Plenum) 24, 183 (1984) [C.A. 100 175378].
[88]Y. Imai, A. Kameyana, T. Q. Nquyen and M. Ueda, J. Polym. Sci., Polym. Chem. Ed., 19, 2997 (1981).
[89]Y. Imai and M. Ueda, Polym. Prepr. (ACS, Div. Polym. Chem.) 23, 164 (1982).
[90]C. E. Carraher, Jr., R. J. Linville, and H. S. Blaxall, Polym. Prep. (ACS Div. Polym. Chem), 23, 160 (1982).
[91]A. Nathansohn, J. Polym. Mater., 2, 129 (1985).
[92]T. Otsuki, H. Tsuchikawa, and S. Kimura, to Japan Synthetic Rubber Co., Ltd., Japanese Patent 61/238826 (1986) [C.A. 106 177094].
[93]V. Percec, ACS Symposium Series 326, 96 (1987), and refercnes contained therein.
[94]F. L. Keohan, R. G. Freelin, J. S. Riffle, I. Yilgor, and J. E. McGrath, J. Polym Sci., Polym. Chem. Ed., 22, 679 (1984).
[95]L. H. Tagle and F. R. Diaz, Eur. Polym. J., 23, 109 (1987).
[96]L. H. Tagle, F. R. Diaz, J. C. Vega, and P. F. Alquinta, Makromol. Chem. 186, 915 (1985).
[97]L. H. Tagle, F. R. Diaz, and N. Valdebenito, Polym Bull. (Berlin) 18, 479 (1987) [C.A. 108 113072].
[98]L. H. Tagle, F. R. Diaz and P. E. Riveros, Polym. J. (Tokyo) 18, 501 (1986) [C.A. 105 153619].
[99]D. M. White, to General Electric, U.S. Patent 4,495,333 (1985); U.S. Patent 4,356,290 (1982).
[100]C. E. Carraher and R. J. Linville, Polym. Prepr. (ACS Div. Polym. Chem.) 23, 160 (1982) also 25, 31 (1984).
[101]A. Natansohn, J. Polym. Mater. 2, 129 (1985) [C.A. 104, 187228].

[102]T. Otsuki, H. Tsuchikawa, and S. Kimura, Japanese Patent 61/238826 (1986) [C.A. 106 177094].

[103]C. E. Carraher and M. D. Naas, Polym. Preps. (ACS Div. Polym. Chem.) 23, 158 (1982).

[104]A. S. Hay, F. J. Williams, G. M. Loucks, H. M. Rolles, B. M. Boulette, P. E. Donahue and D. S. Johnson, Polym. Prepr. (ACS Div. Polym Chem) 23, 117 (1982) [C.A. 100 175452], and J. Polym. Sci., Polym. Lett. Ed. 21, 449 (1983).

[105]V. Percec, T. D. Shaffer, and H. Nava, J. Polym. Sci, Polym. Lett. Ed. 22, 637 (1984); T. D. Shatter and V. Percec, Makro. Chem. Rapid Comm., 6, 97 (1985); T. D. Shatter, M. Jamaludin and V. Percec, J. Polym. Sci., Polym. Chem. Ed., 23, 2913 (1985); 24, 15 (1986), J. Polym. Sci., Polym. Lett. 23, 185 (1985); T. D. Shaffer and V. Percec, J. Polym. Sci. Polym. Chem., 24, 451 (1986); and references contained threin.

[106]J. I. Jin and J. H. Change, Polym. Sci. Technol., 24, 91 (1984); J. I. Jin, K. S. Lee, J. H. Chang, and S. J. Kim, Pollimo, 6, 60 (1982) [C.A. 96 200227]; 7, 185 (1983) [C.A. 99 88616].

[107]T. D. Nguyen and S. Baileau, Polym. Sci. Technol. (Plenum) 24, 59 (1984); T. D. Nguyen, O. Mahamat and S. Baileau, Conv. Ital. Sci. Macromol [Atti] 6th. Vol 2, 283 (1983) [C.A. 101, 24044]; N. Yamazaki and Y. Imai, Polym. J. (Tokyo), 15, 603 (1983) [C.a. 99 140475].

[108]N. Yamazaki and Y. Imai, Kobunshi Ronbunshu, 43, 105 (1986); V. Percec and B. C. Auman, Polym. Bull. (Berlin), 12, 253 (1984) [C.A. 102 25140].

[109]A. S. Erkinov, Z. K. Irmatova and A. T. Dzhalilov, Izv Vyssh. Uchebn Zaved. Khim., Khim. Tekhnol, 30 (10), 129 (1987) [C.A. 108 54602].

[110]L. Tagle, F. R. Diaz, M. P. de La Maza, and J. C. Vega, J. Polym. Sci., Polym. Chem. 24, 495 (1986).

[111]Y. Imai and H. Kamata, Kobunshi Ronbunshu 40, 165 (1983) [C.A. 98 198832].

[112]M. Sato and M. Yokoyama, Makromol. Chem., 158, 629 (1984).

[113]M. Sato, Makromol. Chem. Rapid Commun., 5, 151 (1984).

[114]T. Shaffer and V. Percec, J. Polymer Sci., Polym. Chem., 25, 2755 (1987), and references contained therein.

[115]T. D. Shaffer, K. Antolin, and V. Percec, Makromol. Chem., 188, 1033 (1987).

[116]R. Kellman, R. F. Williams, G. Dinotsis, D. J. Gerbi and J. C. Williams, ACS Symposium Series 326, 128 (1987); D. J. Gerbi, G. Dimotsis, J. L. Morgan, R. F. Williams, and R. Kellman, J. Polym. Sci., Polym. Lett. Ed. 23, 551 (1985), and references contained therein.

[117]Y. Imai, S. Abe, and M. Neda, J. Polym. Sci., Polym. Chem. Ed., 19, 3285 (1981).

[118]Y. Li, D. Lin, H. Liu, H. Yu, and Z. Wu, Gaodeng Xuexiao Xuaxue Xuebao, 8, 473 (1987); X. Wang, H. Wu, and S. Li, Huadong Huagong Xueyuan Xuebao, 12, 285 (1986) [C.A. 106 85123].

[119]V. Aulacovschi, D. Bejenaru, and C. I. Simionescu, Mater. Plast. (Bucharest), 24, 69 (1987) [C.A. 107 199019].

[120]A.F.L.G. Devaux and P. G. Monicotte, to Monsanto Europe SA, European Patent 128,890 (1984) [C.A. 102 114962].

[121]V. Percec, to B. F. Goodrich Co., U.S. Patent 4,638,039 (1987).

[122]T. D. Nguyen and S. Boileau, Polym. Prepr. (ACS, Div. Polym. Chem.) 23, 154 (1982).

[123]V. Percec, T. D. Shaffer, and H. Nara, Polym. Prepr. (ACS, Div. Polym. Chem.) 25, 45 (1984) [C.A. 102 46358]; J. Polym. Sci., Polym. Lett. Ed. 22, 637 (1984).

[124]V. Percec, H. Nara and H. Johnson, J. Polym. Sci., Polym. Chem., 25, 1943 (1987).

[125]V. Percec and H. Nava, J. Polym. Sci., Polym. Chem., 25, 405 (1987).

[126]V. Percec, B. C. Auman and H. Nava, J. Polym. Sci., Polym. Chem., 26, 721 (1988).

[127]C. Pugh and V. Percec, Polym. Bull. (Berlin), 16, 513 (1986); 16, 521 (1986).

[128]T. Shaffer and V. Percec, J. Polym. Sci., Polym. Lett. Ed., 23, 185 (1985).

[129]T. Shaffer and V. Percec, Makromol. Chem., 187, 111 (1986), 187, 1431 (1986).

[130]T. D. Shaffer and V. Percec, J. Polym. Sci., Polym. Chem., 24, 451 (1986); 25, 2755

(1987).

[131]T. D. Shaffer, M. Jamaludin and V. Percec, J. Polym. Sci., Polym. Chem. Ed., 24 15 (1986).

[132]P. Keller, Mol. Cryst. Liq. Cryst., 155, (Part B) 37 (1988) [C.A. 108 167983].

[133]P. Keller, Macromolecules, 18, 2337 (1985); 17, 2937 (1984); 20, 462 (1987); Mol. Cryst., Liq. Cryst., Letter Sect. 2, 102 (1985).

[134]X. Wang, H. Wu, and S. Li, Huadong Huagong Xueyuan, Xuebao, 12, 1 (1986) [C.A. 106 120320].

[135]S. H. Chen and Y. F. Maa, Macromolecules, 21, 904 (1988).

[136]J. K. Rasmussen and H. K. Smith, II, J. Am. Chem. Soc., 103, 730 (1981); Makromol. Chem., 182 701 (1981).

[137]J. K. Rasmussen, S. M. Heilmann, L. R. Krepski, and H. K. Smith, II, ACS Symposium Series 326, 116 (1986).

[138]J. K. Rasmussen, to 3M Co., U.S. Patent 4,326,049 (1982).

[139]J. K. Rasmussen and H. K. Smith, II, in "Crown Ethers and Phase Transfer Catalysis in Polymeric Science," L. J. Mathias and C. E. Canahen, Jr., eds., Plenum Press, New York 1984, pp. 105-109.

[140]However, see N. N. Gosh and B. M. Mandal, Macromolecules, 19, 19 (1986).

[141]K. Choi and C. Y. Lee, Ind. Eng. Chem., Res. 26, 2079 (1987).

[142]M. Takeishi, H. Ohkawa, and S. Hayama, Makromol. Chem. Rapid Comm., 2, 457 (1981).

[143]N. Kunieda, H. Taguchi, S. Shiode, and M.Kinoshita, Makromol. Chem. Rapid Comm., 3, 385 (1982).

[144]H. Taguchi, N. Kunieda, and M. Kinoshita, Makromol. Chem. Rapid Comm., 3, 495 (1982).

[145]H. Taguchi, N. Kunida, and M. Kinoshita, Makromol. Chem., 184, 925 (1983).

[146]H. Ryoshi, N. Kunida, and M. Kinoshita, Makromol. Chem., 187, 263 (1986).

[147]N. Kunida, S. Shiode, H. Ryoshi, H. Taguchi, and M. Kinoshita, Makromol. Chem. Rapid Comm., 5, 137 (1984).

[148]A. Jayakrishnan and D. O. Shah, J. Polym. Sci., Polymer Chem. Ed. 21, 3201 (1983); J. Appl. Polym. Sci., 29, 2937.

[149]C. Simonescu, C. Mihailescu, and V. Bulacovschi, Acta Polym. 38, 502 (1987) [C.A. 107 237345].

[150]D. M. White and G. R. Loucks, ACS Symposium Series 282, 187 (1985).

[151]P. P. Nicholas, Prepr. ACS Div. Pet. Chem., 30, 421 (1985).

[152]G. S. Howard, P. P. Nicholas, and S. E. Hornes, Jr., J. Polym. Sci. Polym. Chem. Ed., 23, 2005 (1985).

[153]Ube Ind. (Japan), Japanese Patent 83/111804 (1983) [C.A. 99 195973].

[154]P. Hodge, B. J. Hunt, J. Waterhouse, and A. Wightman, Polym. Comm., 24, 70 (1983).

[155]Y. Nakamura, K. Mori, and K. Wada, Nippon Gomu Kyokaishi; 57, 561 (1984) [C.A. 102, 7996].

[156]J. E. Mark and S. J. Pan, Makromol, Chem., Rapid Comm. 3(10), 681-5 (1982) [C.A. 92.217677].

[157]E. S. Percec and G. S. Li, to Standard Oil of Ohio, U.S. Patent 4,699,634 (1985).

[158]D. C. Sherrington, Macromol. Chem. (London) 3, 303 (1984).

[159]K. T. Howang, K. Iwamoto, M. Seno, and H. Kise, Makromol. Chem., 187, 611 (1986).

[160]H. Kise and Y. Uno, to Mitshubishi Petrochemical, Japanese Patent 61/101505 (1986) [C.A. 105 115602]; J. Polym. Sci., Polym. Chem. Ed. 20, 3189 (1982); Japanese Patent 58/142905 (1983) [C.A. 100 86332].

[161]J. Lewis, M. K. Nagui, and G. S. Park, Polymer Prepr. (ACS Div. Polym. Chem), 23, 140 (1982) [C.A. 100 34949].

[162]G. Martinez, P. Terroba, C. Miyangos, and J. Millan, Rev. Plast. Mod. 49, 63 (1985).

[163]C. I. Simionescu, V. Bulacovschi, D. Macocinschi, G. Stoica, and I. I. Negulescu,

[164]A. Nikansah and G. Levin, Polym. Sci. Technol. (Plenum) 21, 109 (1983) [C.A. 100 69119].
[165]G. Levin, Org. Coat. Appl. Polym. Sci. Proc. 46, 70 (1981) [C.A. 100 7311].
[166]T. Iizawa, S. Akatsuka, and T. Nishikubo, Polymer J. (Tokyo) 19, 1413 (1987).
[167]T. Iizawa, K. Hayasgi, Y. Endo, and T. Nishikubo, J. Polym. Sci., Polym. Lett. 23, 623 (1985).
[168]M. Akashi, I. Yamashita, and N. Miyauchi, Angew. Makromol. Chem., 122, 147 (1984).
[169]S. I. Hong, K. A. Kim, and D. W. Jeon, Pollimo, 9, 545 (1985) [C.A. 104 130450].
[170]C. Ungurenasu and C. Cotzur, Polym. Bull, (Berlin) 14, 411 (1985).
[171]T. Nishikubo, T. Iizawa, K. Kobayshi, and Y. Matsuda, Macromolecules, 16, 722 (1983).
[172]H. Kise and H. Sato, Makromol. Chem. Rapid Comm., 5, 759 (1984).
[173]B. Kolarz and A. Rapak, Makromol. Chem., 185, 2511 (1984).
[174]P. Hodge, B. J. Hunt, E. Khoshdel and J. Waterhouse, Polym. Prep. (ACS, Div. Polym. Chem) 23, 147, (1982).
[175]K. T. Howang, M. Takahashi, K. Iwamoto, and M. Seno, Makromol. Chem., 188, 1383 (1987).
[176]T. D. Nguyen, J. C. Gautier, and S. Boileau, Polym. Prepr. (ACS Div. Polym. Chem.) 23, 143 (1982).
[177]T. Nishikubo, T. Iizawa, S. Akatsuka, and M. Okawara, Polym. J. (Tokyo) 15, 911 (1983).
[178]A. S. Gozdz, Polym. Bull, (Berlin), 5, 591 (1981).
[179]T. Nishikubo, A. Sahara and T. Shim Okawa, Polym. J. (Tokyo), 19, 991 (1987).
[180]T. Iizawa, T. Nishikubo, Y. Masuda, and M. Okawara, Macromolecules, 17, 992 (1984).
[181]M. Mohanraj and W. t. Ford, U.S. Patent, 4,713,423 (1986).
[182]T. Iizawa, T. Nishikubo, M. Ichikawa, Y. Sugawara, and M. Okawara, J. Polym. Sci., Polym. Chem. Ed. 23, 1893 (1985).
[183]A. S. Gozdz, Polym. Bull. (Berlin) 6, 375 (1982).
[184]T. Nishikubo, T. Iizawa, Y. Mizutani, and M. Okawara, Makromol. Chem., Rapid Comm., 3, 617 (1982).
[185]S. Boivin, P. Hemery, and S. Boileau, Polym. Prepr. (ACS Div. Polym. Chem) 27, 3 (1986); ACS Symposium Series 364 37 (1988).
[186]S. Boivin, P. Hemery, J. P. Senet, and S. Boileau, Bull. Soc. Chim. Fr. (5-6, Part 2) 201 (1984).
[187]S. Boivin, A. Chettouf, P. Hemery, Polym. Bull, (Berlin) 9, 114 (1983).
[188]G. Meunier, S. Boivin, P. Hemery, J. P. Sevet, and S. Boileau, Polym. Sci. Technol. (Plenum) 21, 293 (1983).
[189]R. R. Gallucci and R. C. Going, J. Org. Chem. 48, 342 (1983).
[190]T. Iizawa, T. Nishikubo, M. Ichidawa, and M. Okawara, Makromol. Chem. Rapid Comm., 4, 93 (1983).
[191]T. Tanaka, S. Tachibana, and M. Sumimoto, Mokuzai Gakkaishi 32, 103 (1986) [C.A. 105 45024].
[192]W. Daly, J. D. Caldwell, K. Van Phung and R. Tang, Polym. Sci. Technol. (Plenum) 24, 45 (1984) [C.A. 100 193794] and Polym. Prepr. (ACS Div. Polym. Chem) 23, 145 (1982) [C.A. 100 8801].
[193]P. T. Irzasko, M. M. Tessler and T. A. Dirscherl, European Patent Application 188,237 (186) [C.A. 105 228855].
[194]Kojin Co., Ltd., Japanese Patent 58/147,402 (1983) [C.A. 100 105396].
[195]Yu. A. Zhdanov and Yu. E. Alekeseev, Zh. Vses. Khim. O-va. Lm. D.I. Mendeleeva, 31, 188 (1986) [C.A. 105 134244].
[196]M. Guidini, R. D. Denise, and B. Bariou, Lait, 63, 463 (1983) [C.A. 101 74677].
[197]W. Nowicki, J. Swietoslawski, and A. Silowiecki, Poludniowe Zaklady Przemyslu Skorzanego "Chehmek," Pol. Patent 115,374 (1982) [C.A. 98 217,604].
[198]M. Dewath, Bor-Cipotech, 33, 8 (1983) [C.A. 98 217046].

[199]Mitsubishi Petrochemicals Co., Ltd., Japanese Patent 83/142905 (1983) [C.A. 100 86332].
[200]H. Kise and H. Ogata, J. Polym. Sci., Polym. Chem. Ed., 2, 3443 (1983).
[201]A. Dias and T. J. McCarthy, Polym Mater. Sci. Eng., 49, 547 (1983).
[202]B. Hahn and V. Percec, J. Polym. Sci., Polym. Chem., 25, 783 (1987).
[203]A. Dias and T. J. McCarthy, Macromolecules, 17, 2529 (1984).
[204]M. W. Urban and E. M. Salazar - Rojas, Macromolecules, 21, 372 (1988).
[205]G. Moggi, P.Bonardelli, G. Chiodini and S. Conti, to Ausimont Spa, European Patent Appl. EP 195,565 (1986) [C.A. 105 210240].
[206]F. F. He and H. Kise, Makromol. Chem., 186, 1395 (1985).
[207]G. M. Greene and C. A. Bunnell, European Patent Application 51,457 (1982), to Eli Lilly and Co. [C.A. 97.144590].
[208]Z. Liu and X. Chen, Hauxue Shijie 23(4), 104-6 (1982) [C.A. 100.087638].
[209]V. K. Krishnakumar and M. M. Sharma, Ind. Eng. Chem. Process Des. Dev., 24(4), 1293-7 (1985) [C.A. 103.123594].
[210]H. Kawahara, S. Yoshida and S. Narisawa, Contemp. Top. Polym. Sci., 4, 197-208 (1984) [C.A. 100.140153].
[211]C. T. Berge and M. P. Mack, U.S. Patent 4,412,065 (1983), to Conoco, Inc. [C.A. 100.068962].
[212]L. Avar and E. Kalt, Switz Patentschrift CH 642,652 (1984), to Sandoz A.-G. [C.A. 101.090768].
[213]M. Koehler, M. Roemer and C. P. Herz, Ger. Offen DE 3,512,541 (1986) to Merck Patent G.M.b.H. [C.A. 106.032575].
[214]K. Tatsuoka and K. Tobinaga, Japan Kokai, Tokkyo Koho JP 60/158161 (1985) [C.A. 104.090286].
[215]R. W. Layer, European Patent Application 113,666 (1984) to B. F. Goodrich Co. [C.A. 101.231774].
[216]A. F. L. G. Devaux and P. G. Moniotte, European Patent Application 128,890 (1984) to Monsanto Europe [C.A. 102.114962].
[217]Ricoh Co., Ltd., Japan Kokai Tokkyo Koho JP 60/84261 (1985) [C.A. 103.123145].
[218]T. Fujii, et al, European Patent Application 172,413 (1986) to Sumitomo Chem.[C.A. 104.208382].
[219]J. Ertl, Ger. Offen, DE 3,524,542 (1987) to Hoechst A.-G. [C.A. 106.139328].
[220]R. J. Paessun, M. P. Serve and W. A. Feld, Org. Coat. Plast. Chem., 42, 274-8 (1980) [C.A. 96.052241].
[221]J. Silhanek, J. Bartl, R. Mateju and M. Zbirovsky, React. Kinet. Catal. Lett 19 (1-2), 115-8 (1982) [C.A. 97.038239].
[222]L. Lindbloom and M. Elander, Pharm. Tech. 39 (1980); E. P.D'Incan, P. Viout and G. Gallo, Israel J. Chem., 26, 277 (1985).
[223]M. Makosza, Pure Appl. Chem., 43, 439 (1975) and references contained therein.
[224]U.-H. Dolling, P. Davis, and E. J. Grabowski, J. Am. Chem. Soc., 106, 446 (1984).
[225]U.-H. Dolling, D. L. Hughes, A. Bhattacharya, K. M. Ryan, S. Karady, L. M. Weinstock and E. J. J. Grabowski, ACS Symposium Series 326, 67 (1987).
[226]U.-H. Dolling, to Mereck, U.S. Patent 4,605,761 (1986).
[227]J. P. Chupp, to Monsanto, U.S. Patent 4,351,667 (1982).
[228]G. H. Alt, to Monsanto, U.S. Patent 4,451,285 (1984).
[229]M. Li, X. Zhan and J. Chen, Xibei Shifan Xueyuan Xuebao, Ziran Kexueban (2), 31 (1984) [C.A 103.070996].
[230]K. Takemoto, to Sumitomo, Japanese Patent 60/158147 (1985) [C.A. 104 33845].
[231]D. R. Maulding, to American Cyanamide, Ger. Offen 3,506,972 (1985) [C.A. 104 109625].
[232]J. J. Maul, to Occidental Chemical, European Patent appl. 180,057 (1986).
[233]T. Kawahara, S. Yamamoto, K. Hasegawa and K. Konno, Japanese Patent 61/103872

(1986) [C.A. 105.226552].

[234]G. Hamprecht, W. Rohr, and J. Varwig, Ger. Offen DE 3,514,183 (1986) to BASF A.-G. [C.A. 106.033078].

[235]Y. Li, J. Chen, and M. Li, Xibei Shifan Xueyuan Xuebao, Ziran Kexueban, (2), 47-9 23, (1986), Peoples Trp. China [C.A. 106.175898].

[236]D. R. Maulding to American Cyanamide, U.S. Patent 4,675,423 (1987).

[237]R. Lantzsch and L. Elbe, Ger. Offen, DE 3,210,725 (1983) to Bayer A.-G. [C.A. 100.034148].

[238]N. Cortese, Jr., to American Cyamide, U.S. Patent 4,521,629 (1985).

[239]Mueller, W. W. Wiersdorff, M. Thyes, W. Kirschenlohr, and H. Scherer, Ger. Offen, DE 3,341,306 (1985), BASF [C.A. 103.178161].

[240]P. Maggioni, F. Minisci and M. Correale, European Patent Application 151392 (1985) to Brichima S.p.A. [C.A. 104.019411].

[241]T. Mueller and H. G. Werchen, Ger. (East) DD 228,808 (1985) VEB Fahlbert-List [C.A. 105.133359].

[242]H. Kobayashi, et al, Japanese Patent JP 61/148138 (1986) to Nippon Kayaku Co., Ltd. [C.A. 106.032407].

[243]K. Gu, Faming Zhuanli Shenqing Gongkai shuomingshu CN 85100500 (1986) [C.A. 107 78110].

[244]H. Xu, S. Qian, C. Chen, X. Ren, D. Fu, C. Wang, W. Xong, and K. Fu, Faming Zhuanli Shenqing Gongkai Shuo. CN 85100241 (1986) [C.A. 107.115777].

[245]K. Maehara and H. Kobayashi, Japanese Patent 62/81348 (1987), Nippon Kayaku Co., Ltd. [C.A. 107.1977798].

[246]P. K. Wehrenberg, to Stauffer Chemical, U.S. Patent 4,704,467 (1987).

[247]R. Rossi, A. Carpita, M. G. Quirici and M. L. Gaudenzi, Tetrahedron, V 38 (5), p. 631, 37 (1982) [C.A. 97.038435].

[248]J. Villieras, M. Rambaud, and M. Graff, Tetrahedron Lett. 26 (1), 53-6 (1985), French [C.A. 102.184863].

[249]D. A. Wood and P. H. Briner, European Patent Application EP 161,722 (1985) to Shell Int. Research [C.A. 104.207129].

[250]S. Balint, et al, PCT Int. Appl. WO 86/904 (1986), to Nitrokemia Impart. [C.A. 105.134144].

[251]I. Liao, H. Zhang, and Q. Guo, Faming Zhuanli Shenqing Gongkai Shuomingshu CN 85102994 (1986) Peoples Rep. China [C.A. 107.077805].

[252]Z. Huang, S. He, S. Wang, M. Lin and Y. Chen, Huaxue Shiji 9 (5), 299-301 (1987), Peoples Rep. China [C.A. 108.145317].

[253]A. T. Au, to Dow Chemical, U.S. Patent 4,389,411 (1983).

[254]A. T. Au, to Dow Chemical Co., U.S. Patent 4,391,630 (1983)

[255]A. L. Hall, et al, to Nat. Distillers, U.S. Patent 4,522,950 (1985).

[256]H. Marda, et al, Japanese Patent 62/153272 (1987) [C.A. 108 55896].

[257]G. Chapelet, et al, to ELF-Union S.A., Ger. Offen. 2,844,756 (1979) [C.A. 91 107661].

[258]J. Bourdon, to Rhone-Poulenc, European Patent appl. 36,378 (1981) [C. A. 96 51982].

[259]J. T. Lai, to B. F. Goodrich, U.S. Patent 4,347,379 (1982).

[260]J. M. Renga, to Dow Chemical, U.S. Patent 4,384,115 (1983).

[261]Seitesu Kagaku, Japanese Patent 83154561 (1983) [C.A. 100 120891].

[262]A. Ginebreda, et al, Afinidad, 4, 190 (1984)[C.A. 101 54667].

[263]M. E. Brokke, et al, to Stauffer Chemical, European Patent Application 140,454 (1985) [C.A. 103 123014].

[264]R. F. Mason, et al, to Shell Int. Res., European Patent Application 169,603 (1986) [C.A. 105 60509].

[265]R. K. Singh, to Monsanto, U.S. Patent 4,584,376 (1986).

[266]T. Kondo, et al, Japanese Patent 61/36245 (1986) [C.A. 105 61055].

[267]H. Zhang, et al, Xiamen Daxue Xuebao, Ziran Kexueban, 25, 315 (1986) [C.A. 105

226459].
[268]W. Schroth, et al, East German Patent 241,412 (1985) [C.A. 107 39188].
[269]H. Maeda, et al, Japanese Patent 62/153261 (1987) [C.A. 107 197774].
[270]T. J. Adaway, to Dow Chemical, U.S. Patent 4,701,531 (1986).
[271]H. Hiller, H. Eilingsfeld, R. Heinz, H. Reinicke and F. Traub, Ger. Offen. 2,910,716 (1980) to BASF A.-G. [C.A. 94.032179].
[272]W. A. Daniels and R. K. Zawadzki, European Patent Application 24503 (1981) to American Cyanamid Co. [C.A. 95.080454].
[273]P. M. Wuan, J. S. Hunter and P. J. Duggan, British Patent 1,598,799 (1981) to Imperial Chem. Ind. Ltd. [C.A. 96.124512].
[274]P. Dopfer, J. Ostermaier, and S. Schwerin, European Patent Application 57,880 (1982) to Hoechst A.-G. [C.A. 98.036088].
[275]N. V. Monich, A.F. Vompe, E. P. Shchenlkina and A. V. Kazymov, U.S.S.R. Patent 1,002,329 (1983) [C.A. 98.199852].
[276]M. Dewath, Bor-Cipotech V.33 (1), p. 8-13 (1983) Hung. [C.A. 98.217046].
[277]P. Kniel, European Patent Application 71,576 (1983) to Ciba-Geigy A.-G. [C.A. 98.217175].
[278]L. Avar and E. Kalt, Switz, Patentschrift 634,814 (1983) to Sandoz A.-G. [C.A. 99.038480].
[279]W. Bauer and K. Kuehlein, Ger. Offen. DE 3,216,126 (1983) to Cassella A.-G. [C.A. 100.053206].
[280]W. Bauer, K. Kuehlein, and G. Nagl, Ger. Offen. DE 3,216,125 (1983), to Cassella A.-G. [C.A. 100.053207].
[281]S. Nath, A. Bhattacharyya and P. K. Sengupta, J. Indian Chem. Soc., 60 (8), 801-2 (1983) [C.A. 100.120728].
[282]M. Hattori, A. Taguma, T. Morimitsu and A. Takeshita, European Patent Application EP 105,762 (1984) to Sumitomo Chem. [C.A. 101.173037].
[283]M. Hattori, M. Nishikuri and A. Takeshita, Ger. Offen DE 3,400,747 (1984) to Sumitomo Chem. [C.A. 101.212666].
[284]L. L. Pushkina, L. V. Dubovaya, O. P. Shelyapin and S. M. Shein, Zh. Org. Khim., 20 (9), 1939-43 (1984) [C.A. 101.231927].
[285]P. Baumgartner, Switz. Patenschrift CH 644,099 (1984) to Ciba Geigy [C.A. 101.231942].
[286]H. Nishi, T. Furukaga, K. Kitahara and S. Tokita, Nippon Kagaku Kaishi (2), 255-9 (1985) [C.A. 103.055421].
[287]Yu. Shlykov, et al, Ger. (East) DD 226,294 (1985) VEB Chemickom, Bitterfield [C.A. 105.210411].
[288]M. Hattori, M. Nishikuri and Y. Ueda, Japanese Patent JP 61/28554 (1986) to Sumitomo Chem. [C.A. 106.034646].
[289]M. Takayama and T. Suzuki, Japanese Patent JP 61/271255 (1986) to Ihara Chemical Ind. [C.A. 106.213563].
[290]U. Seitz and J. Daub, Synthesis (8), 686-9 (1986) [C.A. 107.023060].
[291]G. Somio, Ger. Offen. DE 3,631,172 (1987) to Ciba-Geigy A.-G. [C.A. 107.023105].
[292]X. Huang, Faming Zhuanli Shenqing Gongkai Shuom. CN 85108951 (1986) [C.A. 107.197784].
[293]M. Hattori and S. Kajitani, Ger. Offen. DE 3,700,303 (1987), to Sumitomo Chem. [C.A. 108.023349].
[294]S. Rohr, G. Hanisch and S. Werner, Ger (East) DD 242,050 (1987) to VEB Filmfabrik Wolfen [C.A. 108.023353].
[295]J. Sanders and D. Dieterich, Ger. Offen. DE 3,609,813 (1986) to Bayer A.-G. [C.A. 108.057894].
[296]F. Seela and W. Bourgeois, Heterocycles 26(7), 1755-60 (1987) [C.A. 108.187138].
[297]G. Vernin, R. M. Zamkotsian and J. Metzger, Parfums. Cosmet. Aromes, 50, 83-6, 91-

7 (1983) French [C.A. 99.014428].

[298]H. Gebauer, Seifen, Oele, Fetta, Wachse, 112, 513 (1986) [C.A. 105 196963].

[299]Y. Kimura and S. L. Regen, J. Org. Chem., 48, 1533 (1983).

[300]A. Furangen, to Astra Pharm. Prod. AB, Soedertaelje 5-151,85 Swed. [C.A. 104 129232].

[301]T. Kanechika, H. Okamura and C. Ebina, Japanese Patent 61/161269 (1986), to Sumitomo Chem. [C.A. 106.033067].

[302]A. Aserin, et al, Ind. Eng. Chem. Prod. Res. Dev., 23, 452 (1984).

[303]I. Tabuse and N. Morioka, Japanese Patent 60/228453 (1985), to Nippon Oils and Fats Co. [C.A. 105.081150].

[304]K. Von Werner, to Hoechst, German Offen. 3,034,641 (1982) [C.A. 97 91725].

[305]M. Ikeda, M. Miura and A. Aoshima, European Patent 100,488 (1984), to Ashai Chem. [C.A. 101.072593].

[306]B. Robinson, A. Safdar, and R. S. Davidson, British Patent Application 2,170,507 (1985).

[307]J. W. Hill, et al, J. Chem. Educ., 59, 788 (1982).

[308]F. Seela and H. Steker, Liebigs Ann. Chem. (LACHOL) (9), p. 1576-87 (1983) [C.A. 100.007039].

[309]C. Venturello, E. Alneri, and G. Lana, German Offen 3,027,349 (1981) [C.A. 95 42876].

[310]E. Guilmet and B. Meunier, French Patent 2,518,545 (1983), to Produits Chim. Ugene, Kuhlmann [C.A. 99.213027].

[311]G. A. Lee and H. H. Freedman, Israel J. Chem., 26, 229 (1985), and references contained therein.

[312]D. G. Lee, E. J. Lee, and K. C. Brown, ACS Symposium Series 326, 82 (1987).

[313]F. Montanari, M. Penso, S. Quici, and P.Vigano, J. Org. Chem., 50, 4888 (1985).

[314]J. C. Dobson, W. K. Seok, and T. J. Meyer, Inorg. Chem., 25, 1513 (1986).

[315]Daikin Kogyo Co., Ltd., Japanese Patent 58/131976 (1983) [C.A. 100 6309].

[316]H. Yoon and J. C. Burrows, National ACS Meeting, Toronto, 1988, Abstract No. 151.

[317]M. Juaristi, J. M. Aizpurua, B. Lecea, and C. Palomo, Can. J. Chem., 62, 2941 (1984) [C.A. 102 24207].

[318]R. Neumann and Y. Sasson, J. Org. Chem., 49, 1282 (1984).

[319]K. Januszkiewicz and H. Alper, Tetrahodron Lett, 24, 5163 (1983).

[320]F. Gasparrini, M.Giovannali, D. Misiti, G. Natile, and G. Palmieri, Congr. Naz. Chim. Inorg. [Atti], 16th, 215 (1983) [C.A. 101 22765].

[321]G. Gasparrini, M. Giorannoli, G. Natile, and G. Palmieri, Congr. Naz. Chim. Inorg. [Atti], 15th, 178 (1982) [C.A. 101 6358].

[322]E. V. Dehmlow and J. K. Makrandi, J. Chem. Res. Synop. 32 (1986) [C.A. 104 185804].

[323]J. A. S. J. Razenberg, A. W. Van der Made, J. W. H. Smeets, and R. J. M. Nolte, J. Mol. Calal, 31, 27 (1985).

[324]L. T. McElligott, to Union Camp, European patent application 151,941 (1985) [C.A. 104 149198].

[325]FMC Corp., Japanese Patent 62/230778 (1987) [C.A. 108 96629].

[326]C. Venturello, E. Alneri, and M. Ricci, J. Org. Chem., 48, 3831 (1983).

[327]O. Bortolini, ViConte, F. diFuria, and G. Modena, J. Org. Chem, 51, 2661 (1986).

[328]Y. Salto, S. Araki, Y. Sugita and N. Kurata, to Nippon Shokubaikagaku Kogyo Co., Ltd., and to Nihon Joryu Kogyo Co., Ltd., European Patent application 193,368 (1986) [C.A. 105 190703].

[329]G. Barak and Y. Sasson, J. Chem. Soc. Chem. Comm., 1266 (1987).

[330]O. C. Trivedi, Bull. Electrochem., 2, 285 (1986) [C.A. 105 160706].

[331]S. R. Ellis, et al, Appl. Electrochem, 12, 687 (1982) [C.A. 97 225493].

[332]E. Laurent, et al, C. R. Seances Acad. Sci. Ser. 2, 295 339 (1982) [C.A. 98 88746].

[333]S. R. Ellis, et al, J. Appl. Electrochem, 12, 693 (1982) [C.A. 98 24531].

[334]Asahai Chem. Ind., Japanese Patents 8358289 (1983); 58/207383 (1983); 58/207382 (1983); [C.A. 99 95855; 100 164384; 100 164349].
[335]J. Lin-Cai and D. Pletcher, J. Electroanal. Chem. Interfacial Electrochem., 152, 157 (1983) [C. A. 99 183816].
[336]Fleischmann, C. L. K. Tennakoon, P. Gough and J. H. Steven, J. Appl. Electrochem. 13 (5), 603-10 (1983) [C.A. 99.165725].
[337]S. R. Ellis, D. Pletcher, W. N. Brooks and K. P. Healy, J. Appl. Electrochem, 13 (6), 735-41 (1983) [C.A. 100.058707].
[338]E. Laurent, G. Rauniyar and M. Thomalla, J. Appl. Electrochem, 14 (6), 741-8 (1984) [C.A. 102.069247].
[339]E. Laurent, G. Rauniyar and M. Thomalla, J. Appl. Electrochem, 15 (1), 121-7 (1985) [C.A. 102.069249].
[340]Z. Ibrisagic, D. Pletcher, W. N. Brooks, and K. P. Healy, J. Appl. Electrochem, 15 (5), 719-25 (1985) [C.A. 103.185776].
[341]J. Koryta, Chem. Listy, 81, 897 (1987) [C.A. 107 207237].
[342]S. R. Forsyth, D. Pletcher and K. P. Healy, J. Appl. Electrochem, 17 (5), 905 (1987) [C.A. 108.157990].
[343]A. Guarini and P. Tu, J. Org. Chem., 52, 3501 (1987).
[344]Z. Goren and I. Wiliner, J. Am. Chem. Soc., 105 (26), 7764 (1983) [C.A. 100.005612].
[345]K. Lein and A. C. Wee, Can. J. Chem., 60, 425 (1982).
[346]T. Kitamura and S. Koba, Chem. Lett. 1523 (1984); J. Org. Chem., 49, 4755 (1984).
[347]J. Brunet and C. Sido, J. Org. Chem., 48, 1166, 1919 (1983).
[348]A. Brandstrom, J. Mol. Catal., 20, 93 (1983), and references contained therein. See also: F. Michaelis, J. Pharm. Sci., 58, 201 (1969).
[349]J. T. Fenton, to Conoco Inc., U.S. Patents 4,423,238 (1983); 4,513,144 (1985), and 4,574,158 (1986).
[350]V. K. Krishnakumar and M. M. Sharma, Ind. Eng. Chem. Process Des. Dev. 23 (2), 10-13 (1984) [C.A. 100.108607].
[351]D. Martinetz, Z.Chem., 22, 257 (1982) [C.A. 97 215435].
[352]D. J.Brunell, to General Electric, U.S. Patent 4,410,422 (1983); Chemosphere, 12, 167 (1983).
[353]R. R. Bhave and M. M. Chen, Chem.Eng. Sci., 38, 141 (1983).
[354]Z. Wu, C. Zuo, H. Sun, Huaxue Shijie, 26, 247 (1985) [C. A. 103 179939].
[355]Toyo Soda Mfg., Co., Japanese Patent 82200323 (1982) [C.A. 99 22115].
[356]R. Tanaka and H. Goto, Japanese Patent 62/64817 (1987), to Yuka Shell Epoxy K. K. [C.A. 107.177140].
[357]G. Qi and X. Wang, Huadong Huagong Xueyuan Xuebao, 13, 294 (1987) [C.A. 108 7873].
[358]T. Kanechika, H. Okamura and C. Ebina, Japanese Patent 61/233677 (1986), to Sumitomo Chem. [C.A. 106.067322].
[359]B. Funn and J. C. Lechleiter, European Patent Application EP 138,553 (1985) to Eli Lilly and Co. [C.A. 103.141465].
[360]J. W. Hill and T. C. Boyd, J. Chem. Educ. 56 (12), 824 (1979) [C.A. 92.075324].
[361]P. I. Demyanov, N. Malo and V. S. Petrosyan, Vestn. Mosk. Univ., Ser. 2: Khim 26 (2), 223 (1985), Russ. [C.A. 103.038938].
[362]W. Faigle and D. Klockow, Fresenius' Z. Anal. Chem. 310 (1-2), 33-8 (1982) [C.A. 96.109866].
[363]K. Funazo, H. L. Wu, K. Mórita, M. Tanaka, and T. Shono, J. Chromatogr. 319 (2), 143 (1985) [C.A. 102.124764].
[364]T. Chikamoto and T. Maitani, Anal. Sci. 2 (2), 161 (1986) [C.A. 105.021156].
[365]S. H. Chen, H. L. Wu, M. Tanaka, T. Shono, and K. Funazo, J. Chromatog. 396, 129-37 (1987) [C.A. 107.108370].
[366]M. Tanaka, H. Takigawa, Y. Yasaka, T. Shono, K. Funazo, and H. L. Wu, J.

Chromatog., 404 (1), 175-82 (1987)[C.A. 107.249083].

[367]C. A. R. Skow and M. K. L. Bicking, Chromatographia 21 (3), 157-60 (1986) [C.A. 106.084094].

[368]E. S. Olson and B. W. Farnum, Prepr. Pap. - Am. Chem. Soc., Div. Fuel Chem. 26, 60 (1981) [C.A. 97.147361].

[369]R. Narayan and G. T. Tsao, Prepr. Pap. - Am. Chem. Soc., Div. Fuel Chem. 28, 261 (1983) [C.A. 100.123791].

An Overview of the Biocidal Activity of Cationics and Ampholytes

B. Davis and P. Jordan
A.B.M. CHEMICALS LIMITED, LEEDS, UK

Introduction

The necessity for biological as well as physical cleaning is becoming increasingly apparent, particularly in industries involved in food preparation and food handling. This has been highlighted by the recent Salmonella scare with regard to egg production. Keeping the kitchen hygienic means keeping the surfaces free from both dirt and infection. It is particularly necessary to remove the types of soil which can harbour and protect bacteria from biocidal action. Using biocides and surfactants together therefore makes good sense whether these are two separate chemicals or whether both functions are combined in one product. The object of this talk is to review the biocidal surfactants which are normally available, for example the quaternary ammonium compounds and the biocidal ampholytes. When used in general disinfectant cleaning formulations they have certain advantages over other types of biocide (Table 1).

Table 1. Advantages over other Biocides

Relatively low toxicity at use dilutions.
Environmentally acceptable.
Minimum tainting.
Not corrosive.
Non staining.
Reasonable hard water tolerance.
Reasonable resistance to de-activation.

For example they exhibit relatively low toxicity particularly at the level used in formulations and at the actual use dilution. They are considered to be biodegradable and generally environmentally acceptable with a lower possibility of tainting than phenols. Furthermore, they do not have the corrosive characteristics of chlorine or the staining problems of iodine or the rapid deactivation shown by both of these compounds [1]. A further virtue is the reasonable hard water tolerance and the rapid activity against both Gram-positive organisms such as Staphylococcus and Gram-negative organisms such as Escherichia and Salmonella. Reports of problems against Pseudomonas aeruginosa are not unique to quaternary ammonium compounds as this bacteria is simply hard to kill, being protected by a proteinaceous slime.

Quaternary Ammonium Compounds

Effective Structures

A typical quaternary ammonium compound structure is illustrated in Figure 1 which represents Cetrimide.

Figure 1. Chemical Structures

$$\text{Cetrimide} \quad - \quad C_{14}H_{29} - \overset{\overset{\displaystyle CH_3}{|}}{\underset{\underset{\displaystyle CH_3}{|}}{N^+}} - CH_3 \quad Br^-$$

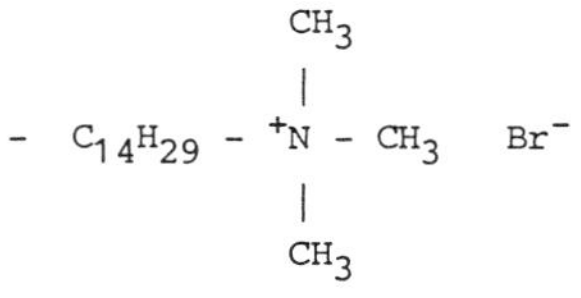

$$\text{Benzalkonium chlorides} \quad - \quad C_{14}H_{29} - \overset{\overset{\displaystyle CH_3}{|}}{\underset{\underset{\displaystyle CH_3}{|}}{N^+}} - CH_2-\!\!\!\bigcirc \quad Cl^-$$

$$- \quad R-\!\!\!\bigcirc\!\!\!-CH_2 - \overset{\overset{\displaystyle CH_3}{|}}{\underset{\underset{\displaystyle CH_3}{|}}{N^+}} - CH_3 \quad Cl^-$$

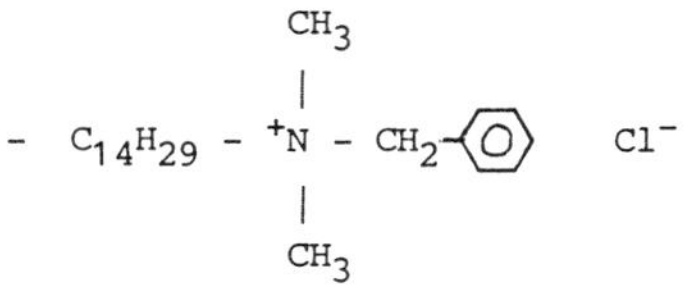

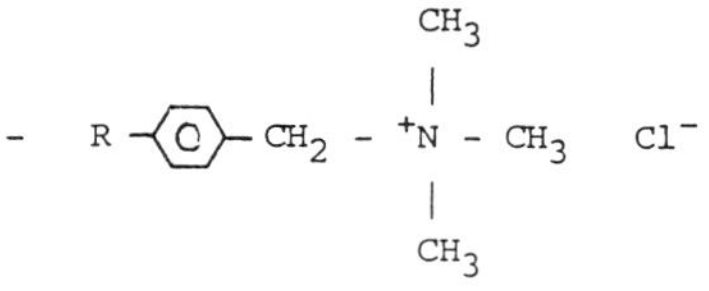

The biocidal activity can be improved by including a benzyl group and the popular Benzalkonium chlorides are widely used. In fact the benzyl group can be put in two alternative places in the molecule as shown, giving equivalent biocidal effectiveness. Dialkyl quaternaries, as shown in Figure 2, are becoming increasingly popular, perhaps because of their rapid action.

Figure 2. **Alternative Structures**

Didecyl quat.

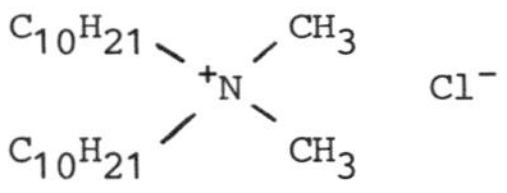

Pyridinium quat.

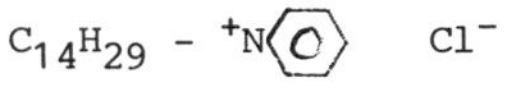

Pyridinium quaternaries such as myristyl pyridinium chloride are particularly powerful biocides but they are not generally commercially available. Some comparative bactericidal results are illustrated in Table 2.

Table 2. **Comparative bactericidal activity (BS:6471)**

	Bactericidal Dilution (100% actives)
Cetrimide	1:3750
Alkylbenzyl dimethyl amm. chloride	1:5000
Alkylbenzyl trimethyl amm. chloride	1:5000
Didecyl dimethyl amm. chloride	1:5000
Didecyl ethyl methyl ethosulphate	1:3000
Myristyl pyridinium chloride	1:6000

In this case the test method according to BS:6471 was employed and this will be discussed later. It is interesting to note that the negative ion can have a drastic effect on the effectiveness of the quaternary compound, for example the ethosulphate quaternary is very much less effective.

Investigations of Benzalkonium chlorides with varying alkyl chain lengths [2] indicate that a chain length of about 14 carbon atoms is an optimum value in terms of biocidal activity (Figure 3).

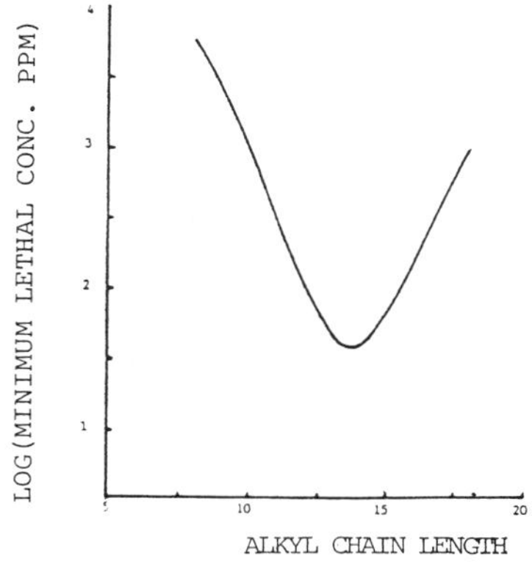

Figure 3: Effect of alkyl chain length. B.A.C. *vs*. *Ps. serguinosa*

Natural blends containing C_{12}-C_{14} mixtures, or synthetic blends containing C_{13}-C_{15} mixtures are popular commercial products.

They are effective against bacteria (Table 2) and also against many other organisms such as algae, fungi and yeasts (Table 3).

Table 3. Microbiostatic Properties
(Benzalkonium chloride)

		m.i.c. (ppm)
Fungi	*Aspergillus niger*	40-200
	Chaetomium globosum	8-40
Algae	*Chlorella vulgaris*	<1
	Stigeoclonium sp.	<1
Yeasts	*Saccharomyces cerevisiae*	40-200
	Rhodotorula rubra	8-40

The figures show the minimum inhibitory concentration (m.i.c.), that is the level of biocide in parts per million which prevents the growth of the microorganism, controlling it without completely killing it. Somewhat higher levels are necessary for complete biocidal effect.

Applications

Some typical areas of application are given in Table 4, which gives some idea of the versatility of these particular products, from skin application, e.g. with Cetrimide, to their use in swimming pools or their more general use in disinfectant surface cleaners of all types.

Table 4. Some Applications

Food contact surfaces.
Food equipment and utensils.
Poultry farms and Abattoirs.
Swimming pools.
Hospitals.

Typical formulations contain some nonionic surfactant to increase the wetting power and to improve the detergency. Figure 4 illustrates that the choice of wetting agent is of great importance because deactivation of the quaternary ammonium compound can be quite severe.

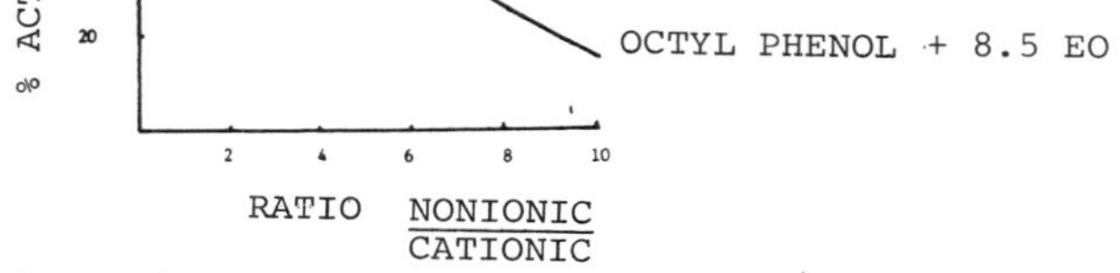

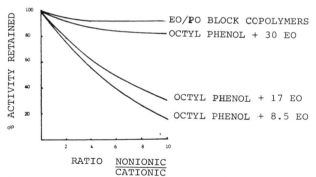

Figure 4: Effect of nonionics on the bactericidal activity of QACs

The more highly ethoxylated alkyl phenol based surfactants appear to be less deactivating and it is suggested in the literature [3] that this is probably connected with micelle formation. Above the critical micelle concentration of the nonionic surfactant it is envisaged that the quaternary compound is absorbed into these micelles being less available therefore for attack on the bacteria. The higher ethoxylates have less tendency to form such deactivating complexes.

Generally speaking, quaternary ammonium biocides are compatible with both betaines and ampholytes, but when they are mixed with anionic surfactants severe deactivation can often occur. In extreme cases large molecular weight insoluble anionic-cationic complexes will be formed completely deactivating the biocide. This is in fact the basis for the titration and analysis[4] of typical quaternary ammonium compounds using sodium lauryl sulphate as the anionic material. It is interesting to note, however, when formulating quaternary ammonium compounds, that the presence of certain types of anionic surfactants can often be tolerated. If the anionic portion of the surfactant is highly ethoxylated and sufficiently water soluble then the precipitation of an insoluble complex does not occur. A foamy formulation could in fact be achieved by using the anionic component illustrated in Figure 5.

Figure 5. Compatible Anionic

$$C_8H_{17}-C_6H_4-(O-CH_2-CH_2)_{8.5}O-SO_3^{-}\ {}^{+}Na$$

Within certain limits it is possible to develop a successful biocidal formulation without significant deactivation as illustrated in Table 5, where the cationic material is a popular alkyl benzalkonium chloride.

Table 5. Compatible Blends

Anionic : Cationic	Bactericidal Dilution*
0:1	1:10,000
1:1	1:10,000
1.5:1	1:5,000
2:1	< 1:1,000

*<u>Ps. aeruginosa</u>, distilled water, 10 min. contact at 20°C

Sometimes the production of foam is a positive advantage, for example when cleaning vertical surfaces and walls in a food processing factory. Foam, on the other hand, can be a positive disadvantage because it requires rinsing off and can interfere with an efficient detergent process. There appears to be an increasing demand for relatively low foaming quaternary ammonium compounds that are quick acting at very low concentrations so that the rinsing stage can be omitted without leaving significant amounts of the surfactant on the surface to contaminate subsequent food material. The short chain dialkyl quaternaries are popular in this respect, being based on C_{10} or C_8 dialkyl carbon chains. To reduce the level of foaming even further it may be possible to blend antifoaming agents in with the quaternary compound, for example silicone oil might be solubilised in very small quantities but in some industries this is not an acceptable solution. In this case it may be possible to solubilise a defoaming nonionic surfactant into the quaternary ammonium compound. The success of the nonionic antifoam will probably be dependent upon the temperature of the washing process as it is much easier to formulate effective antifoams which work at higher temperatures, above the cloud point of the nonionic component.

Mechanism

It is suggested[5] that the cationic biocide is attracted to the negatively charged cell wall of the bacteria which has a porous structure. The biocide must penetrate this porous structure to reach protein material on the cytoplasmic membrane where interactions can occur between suitably charged sites in the protein such as carboxyl groups. This causes disorganisation of the membrane molecules, resulting in denaturation and precipitation of protein material, therefore disrupting the normal functions of the cell. Gram-negative bacteria cells are known to be more resistant to most biocides and in particular to quaternary ammonium compounds. This is probably because they have an extra outer layer of lipoprotein and lipopolysaccharide material giving extra protection against anti-microbial agents. It is interesting to note [6] that the use of chelating agents such as EDTA or NTA in formulations will often increase susceptibility of the bacteria to the biocide as illustrated in Table 6. This may be the result of the chelating agent taking calcium and magnesium ions out of the outer protective layer, thus disrupting its structure and making it more porous to the attacking biocide.

Table 6. Effect of Chelating Agents on Benzalkonium Chloride

Agent	Bactericidal Dilution*	
(ppm)	$NTA.Na_3$ (38%)	$EDTA.Na_4$ (38%)
0	1:2400	1:2400
10	1:2400	1:2400
100	1:60,000	1:120,000
1000	1:150,000	1:600,000

*Ps. aeruginosa, distilled water, 10 min. contact at 20°C

Formulations

British Standard BS:6424 1984 designates three grades of disinfectant where the QAP number is the recommended maximum use dilution. Typical formulations are shown in Table 7.

Table 7. Typical Disinfectants (BS:6424)

	QAP30	QAP50	QAP100
Benzalkonium chloride (50%)	1.5%	2.5%	5.0%
Non-ionic surfactant	1-2%	1-2%	1-2%
Pine oil or fragrance	0.5-2%	0.5-2%	0.5-2%
Water to	100%	100%	100%

A heavy duty disinfectant cleaner might contain co-surfactants and chelating agents, as shown in Table 8.

Table 8. Heavy Duty Disinfectant (QAP100)

Benzalkonium chloride (50%)	5%
Alkyl betaine (40%)	5%
$NTA.Na_3$ (38%)	5-10%
Butyl Oxitol (optional)	2-5%
Water to	100%

Biocidal Ampholytes

This class of biocide is usually considered less irritant and less persistent that quaternary compounds, although biocidal potency may not be as great in many cases. In formulations it is often similar to quaternary ammonium compounds in compatibility to nonionics and anionics and it is also improved by the presence of chelating agents. Whereas quaternary compounds are not totally stable in strongly alkaline formulations, the ampholytes can tolerate high levels of caustic soda when they will exist in the anionic form yet still remaining biocidal. This, of course, suggests some difference in mechanism in their action on the bacteria or other organisms such as fungi, algae and yeasts.

Effective Structures

Biocidal activity seems to be enhanced as the number of nitrogen atoms increases [7] as shown in Table 9.

Table 9. Structure and Biocidal Activity

	Bactericidal action*
$C_{12}H_{25}NHCH_2COOH$	10 mins.
$C_{12}H_{25}NH(CH_2)_2NHCH_2COOH$	5 mins.
$C_{12}H_{25}NH(CH_2)_2NH(CH_2)_2NHCH_2COOH$	1 min.

* 0.05% solutions vs. *St. aureus*

Furthermore, it is likely that a synergistic effect is observed when more than one structure is involved as illustrated by the commercial blend [8] shown in Figure 6.

Figure 6. Commercial Blend

$$C_{12}H_{25}NH(CH_2)_3\ NH\ CH_2\ COOH$$

$$C_{12}H_{25}NH(CH_2)_2NH(CH_2)_2NH\ CH_2\ COOH$$

Introducing more anionic carboxyl groups to the molecule seems to reduce the biocidal effectiveness. Figure 7 shows a structure with very little biocidal activity.

Figure 7. Anionic Groups

$$C_{12}H_{25} - \underset{\displaystyle CH_2COO^-}{\underset{|}{N}} - (CH_2)_3 - \underset{\displaystyle CH_2COO^-}{\underset{|}{N}} - CH_2COO^-\ .\ 3Na^+$$

Similar structures such as betaines, shown in Figure 8, also show very little biocidal activity but occasionally a useful synergistic effect is observed with the more normal ampholyte types.

Figure 8. Betaine Structures

$$C_{12}H_{25} - \overset{\displaystyle CH_3}{\overset{|}{\underset{\displaystyle CH_3}{\underset{|}{{}^+N}}}} - CH_2\ COO^-$$

$$C_{12}H_{25} - CONH - (CH_2)_3 - \overset{\displaystyle CH_3}{\overset{|}{\underset{\displaystyle CH_3}{\underset{|}{{}^+N}}}} - CH_2\ COO^-$$

For example, the ampholyte betaine blends listed in Table 10 illustrate a real synergistic effect.

Table 10. Betaine Synergism

Ampholyte %	**Betaine %**	**Total Actives %**	**Biocidal Dilution***
0.0	32.5	32.5	<1:50
32.5	0.0	32.5	1:600
14.5	18.0	32.5	1:800
12.5	20.0	32.5	1:600
10.5	22.0	32.5	1:500
8.5	24.0	32.5	1:400
6.5	26.0	32.5	1:300

* *E. coli* in hard water, 15 mins. contact time in the presence of horse serum.

Applications

In many areas the excellent wetting and detergency of this class of compound plus stability in highly alkaline solutions makes them a preferable choice to quaternary ammonium compounds. Other factors in their favour include their greater effectiveness in the presence of protein soils [9] and their better rinse-off characteristics, which makes them a good choice for use in the dairy industry, brewery industry and other food related industries (Table 11).

Table 11. Areas of Application

Breweries	Food Areas
Dairies	Poultry Farms
Soft Drinks Factories	Abattoirs
Hand Cleaning	Hospitals

In the literature [10] there are claims that ampholytes offer good protection against bacteria and other organisms found in meat products, for example Salmonella, Beef trichophyta, Chicken aspergilli, Swine Fever Virus, Newcastle Disease Virus and several others. Therefore the bactericide can be used in the cleaning and disinfection of food processing factories, poultry farms and slaughterhouses. Their low toxicity and low skin irritancy make this type of compound ideal for incorporation into hand cleaners, skin cleansers and shampoos where some ampholytes give sufficient protection to the formulation without the addition of extra preservatives.

Ideally an ampholyte requires the potency and cost effectiveness of a quaternary biocide plus extra low foam characteristics.

Formulations

The uses of biocidal ampholytes are illustrated by the following examples. Table 12 shows a possible formulation for a biocidal hand cleaner, making use of the synergistic effect of the betaine.

Table 12. Hand Cleaner

Ampholyte actives		1.75%
Betaine actives		2.8%
Thickener		2-5%
Water	to	100%

Biocidal dilution	1:60
Expected biocidal dilution	1:28

Table 13 shows a more powerful cleaner in which both the betaine and the chelating agent are having an enhancing effect on the biocidal activity of the ampholyte.

Table 13. Biocidal Detergent

Ampholyte actives		2%
Betaine actives		3.2%
$NTA.Na_3$		15%
Blend of alcohol ethoxylates		6.5%
Hydrotrope		2%
Sodium metasilicate.$5H_2O$		4%
Water	to	100%

Biocidal figure	1:100
Expected biocidal figure	1:32

Biocide Testing

The test methods used in evaluating and comparing different biocides are of great importance as illustrated in Table 14.

Table 14. Different Test Conditions

	Bactericidal Dilution (*vs. E. coli*)		
	Test 1	Test 2	BS6471:1984
Didecyl dimethyl amm. chloride (50%)	1:80,000	1:16,000	1:2500
Benzalkonium chloride (50%)	1:35,000	1:10,000	1:2500

Test 1 - Distilled water, 20°C, 10 mins. contact.
Test 2 - Hard water (250 ppm), 20°C, 5 mins. contact.

Obviously different results can be obtained under different conditions and it is particularly important to use test methods appropriate to the chemical type of biocide being used. Eventually a British Standard test was developed for quaternary ammonium compounds, namely BS:6471 (1984). The test conditions are outlined in Table 15, showing a test against one bacterium at a fixed time and temperature in the presence of protein deactivating material.

Table 15. BS:6471 (1984)

Water Hardness	:	200 ppm
Test organism	:	*Escherichia coli* ATCC 11229
Temperature	:	22°C
Contact time	:	10 minutes
Organic matter	:	5% sterile horse serum
Inactivator	:	3% Tween 80 + 2% Soya Lecithin
Result	:	Conc. to give 99.99% kill

We have also found this test very convenient to use when evaluating ampholyte biocides.

To evaluate a biocide thoroughly, however, it is necessary to carry out more than one test. For example, you may wish to know the effectiveness of the biocide against a whole variety of organisms, various bacteria, algae, fungi and yeasts. Furthermore, the speed of action of the biocide may be a critical factor. Its performance under real conditions will also be affected by the types of soil present. Some types of soil may be particularly deactivating to some biocides. The developing of the best series of tests is therefore quite a problem. The 555

Test is an attempt to give more information about the product by testing against the five organisms shown in Table 16, with a contact time of 5 minutes to establish the dilution required to produce a 5 log 10 reduction in organisms, e.g. 10^8 down to 10^3. Again protein material is normally added as a possible deactivating agent.

Table 16. 555 Test Organisms

Pseudomonas aeruginosa
Staphylococcus aureus
Proteus mirabilis
Streptococcis faecium
Saccharomyces cerevisiae

A recent development in the field of disinfectant testing has been the introduction of the European Suspension Test (EST). This is a modified 5 5 5 test and is described as a screening method for evaluating disinfectants for food hygiene. At the time of writing (February 1989) its current status in the U.K. is as a BSI draft for development. It is currently receiving a great deal of attention primarily because modifications to the test protocol are thought to be necessary. However, there is a distinct possibility, particularly in the light of 1992, that it will be eventually adopted as a European Standard for the food industry.

REFERENCES

1. A.K. Pryor and R.S. Brown, Journal of Environmental Health, 37 (4), 326 (1975).

2. R.A. Cutter, E.B. Cimijotti, T.J. Okolowich, W.F.Wetterau. Proceedings of the 53rd Annual Meeting C.S.M.A., 102-103 (1966).

3. I.R.Schmolka, J. Soc. Cosmet. Chem. 24 (9), 577-92 (1973).

4. D.E. Hering, Lab. Pract. 11, 113-115 (1962).

5. A.D. Russell, W.B. Hugo, G.A.T. Aycliffe, Principles & Practice of Disinfection, Preservation & Sterlisation, 164-165 (1982).

6. M.R.W. Brown, R.M.E. Richards, Nature 207 1391 (1965).

7. G. Sykes, Disinfection and Sterlisation, 2nd Ed., 377-378 (1965).

8. Seymour S. Block, Disinfection, Sterlisation and Preservation, 3rd Ed., 348 (1983).

9. S. Kendereski, M. Ilic, Hrana Ishrana 10, 433-438 (1969).

10. Seymour S. Block, Disinfection, Sterilisation and Preservation, 3rd Ed., 357 (1983).

Amine Oxides and their Applications

H. Rörig and R. Stephan
AKZO CHEMICALS RESEARCH CENTER, KREUZAUER STR. 46, D-5160 DÜREN, WEST GERMANY

Historical Introduction

Actually this presentation is a birthday present. The chemical class of compounds called amine oxides is celebrating its 90th anniversary this year, first described by Dunsting and Goulding [1] in 1899 under the name oxyamines. Since those days a lot of articles have been written about amine oxides and of course a lot of patents have been filed beginning with a German patent by the former IG-Farben [2] in 1934 and a Swiss patent by Ciba [3] dating from the same year. In these publications amine oxides were praised as excellent wetting agents, foam boosters, emulsifiers and powerful cleaning agents. This paper will demonstrate how all these attractive properties can be employed in practical applications.

Structure and Chemistry

Amine oxides are made by reacting tertiary amine with H_2O_2. This synthesis is well-known and well described in the literature [4]. Therefore it will not be discussed here, but this presentation will concentrate on applicational values and related physical properties of amine oxides.

As illustrated in figure 1 the term amine oxide describes a tetrahedral coordinated nitrogen atom with one linkage to oxygen. This specific atomic bond can be formulated either as dative binding (structure a) or as a semipolar bond (structure b). The polarity of this binding is in these structures indicated by an arrow (a) resp. by plus and minus signs (b), representing a dipole moment of 4 - 5 Debye [4b].

FIGURE 1: AMINE OXIDE STRUCTURES

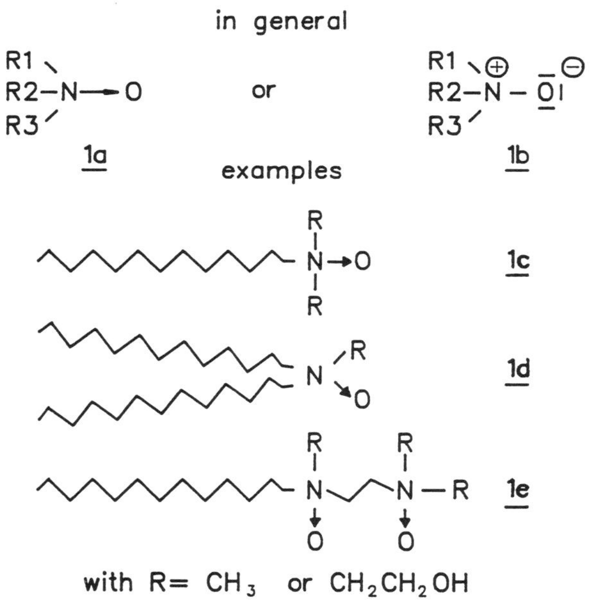

A vast range of different radicals are suitable as group R^1 to R^3, coordinating the central nitrogen atom. Some examples are given in the structures 1 c, d and e. Our considerations, however, will concentrate on fatty acid derived amine oxides of formula 1 c exhibiting alkyl chain lengths of 10 to 18 carbon atoms.

In the following chapters:

- Physical Properties
- Applicational Properties and Formulations
- Ecotoxicology
- Different Amine Oxides

we will evaluate typical physical properties and their applicational relevance, and demonstrate the application of amine oxides in some selected formulations. As a central item the influence of chemical structure on physical parameters and applicational properties will be discussed.

Physical Properties

Due to the highly polar NO-headgroup amine oxides are readily soluble in water and other polar solvents. Also due to this NO-group they exhibit an amphoteric character as illustrated in figure 2. In acidic solutions the negatively charged oxygen is protonated and a cationic quaternary ammonium species is formed.

FIGURE 2: AMINE OXIDES IN AQUEOUS SOLUTIONS

$$\sim\!\!\sim\!\!\sim\!\!\sim\!\!\underset{R}{\overset{R}{\underset{|}{N}}}\!\!\rightarrow\!\!O \;+\; H_3O^\oplus \;\rightleftharpoons\; \sim\!\!\sim\!\!\sim\!\!\sim\!\!\underset{R}{\overset{R\;\oplus}{\underset{|}{N}}}\!\!-\!OH \;+\; H_2O$$

cationic behaviour in acidic media

nonoionic behaviour in alkaline media

$$\sim\!\!\sim\!\!\sim\!\!\sim\!\!\underset{R}{\overset{R}{\underset{|}{N}}}\!\!\rightarrow\!\!O \;+\; 2\,H_2O \;\rightleftharpoons\; \sim\!\!\sim\!\!\sim\!\!\sim\!\!\underset{R}{\overset{R}{\underset{|}{N}}}\!\!\rightarrow\!\!O \begin{matrix} \cdots H-O^{\diagup H} \\ \cdots H-O_{\diagdown H} \end{matrix}$$

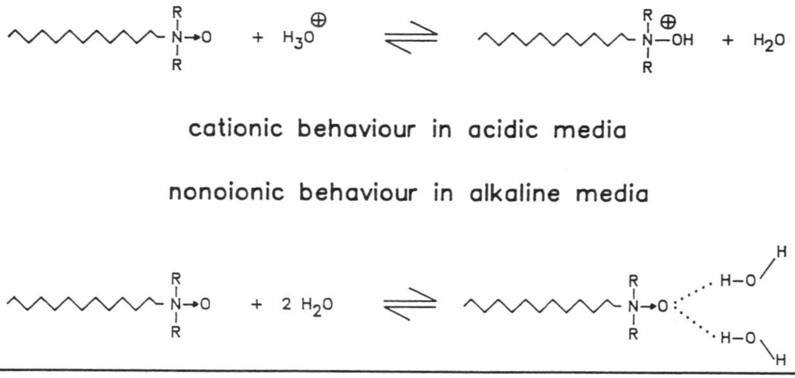

In neutral to alkaline solutions amine oxides behave like nonionic surfactants solubilized via hydrogen bondings. This pH dependent behaviour changing from cationic to nonionic is reflected in a lot of physical properties as demonstrated in the following figures.

Figure 3 for example shows a surface tension *vs.* concentration diagram of a coco-amine oxide at two different pH-values. The critical micelle concentration (CMC) of the nonionic species at pH 11 is about 0.1 g/l, whereas the cationic form at pH 3 exhibits a CMC of 0.7 g/l.

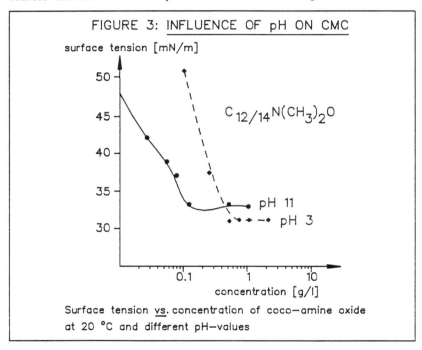

FIGURE 3: INFLUENCE OF pH ON CMC

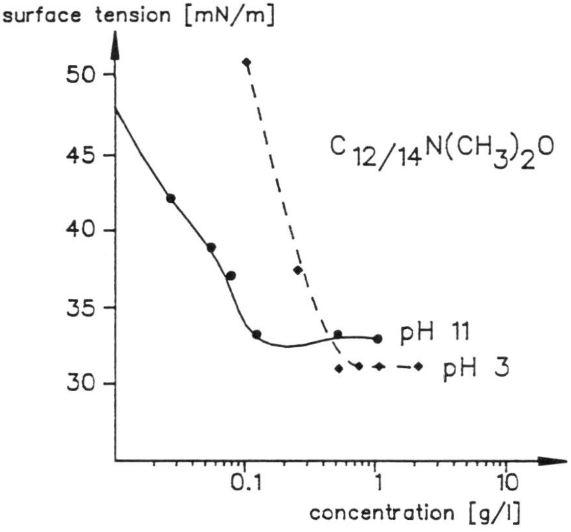

Surface tension *vs.* concentration of coco-amine oxide at 20 °C and different pH-values

These values fit well within those typical for nonionic resp. betaine surfactants. Table 1 shows this for different lauryl-derivatives.

Table 1: Critical Micelle Concentration of Lauric Acid derived Surfactants at 25 °C

Surfactants		CMC [g/l]
Anionic:	$C_{12}H_{25}OSO_3Na$	2.33
Nonionic:	$C_{12}H_{25}O(CH_2CH_2O)_5H$	0.02
Betaine:	$C_{12}H_{25}N^+(CH_3)_2CH_2CO^-_2$	0.22
Cationic:	$C_{12}H_{25}N^+(CH_3)_3Cl^-$	4.16
Amine oxide:	$C_{12}H_{25}N(CH_3)_2O$	0.05
	$C_{12}H_{25}N^+(CH_3)_2OH\ Cl^-$	0.14

Obviously the charge of amine oxide molecules at low pH inhibits micelle formation by electrostatic repulsion. On the other hand the surface activity of these molecules is enhanced as demonstrated by the increased surface tension reduction at pH 3: 30 mN/m vs. 33 mN/m at pH 11.

Naturally the pH-dependent character of amine oxides influences also their compatibility with anionic surfactants as illustrated in figure 4. At medium to high pH-values $\geq$ 8 amine oxides as typical nonionics are completely compatible with anionic surfactants. At lower pH an area of non-compatibility exists resulting in precipitation. Size and form of this area around a 1 : 1 mixing ratio depend on the type of anionic surfactant as shown for lauryl sulphate and lauryl ether sulphate in figure 4. In the past amine oxides were regarded sometimes as incompatible with anionic surfactants [5] at pH below 6. Obviously this is not true. Also under acidic conditions amine oxides can be combined with anionic surfactants.

FIGURE 4: COMPATIBILITY WITH ANIONIC SURFACTANTS

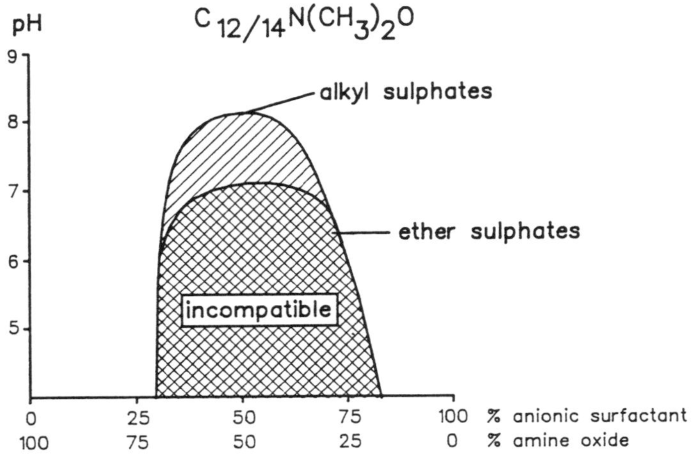

Compatibility of coco–amine oxide with anionic surfactants (10% total surfactant active solution)

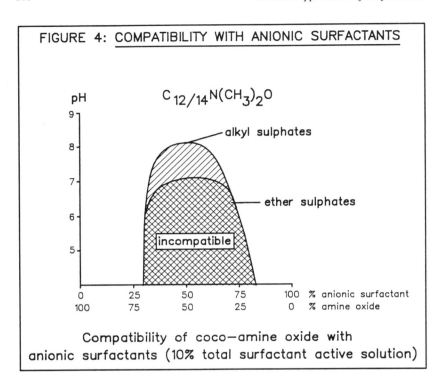

Applicational Properties and Formulations

Typical functions of amine oxides in various formulations are foaming, wetting, emulsifying, cleaning and thickening. These capabilities make amine oxides especially valuable in the field of detergents and cleansers. Therefore they find their main application in this area as illustrated in table 2. This list of application areas is by no means complete. It gives, however, a general picture how amine oxides are employed.

Table 2: Key Functions of Amine Oxides

Application \ Function	Foaming	Wetting	Emulsifying	Cleaning	Thickening
Hair Care	+				+
Foam Bath	+				
Dish Washing	+	+		+	
Creams/Lotions			+		+
Fibre Lubricant			+		
HD Detergent		+		+	
Cleanser		+	+	+	+
Degreaser		+			

Within this chapter we will demonstrate the correlation of chemical structure and applicational properties of amine oxides as well as typical formulations using these properties. According to our list of basic functions: foaming, wetting, emulsifying, cleaning and thickening we will begin with amine oxides as foaming agents.

Foaming

Foam generation is the main function of amine oxides in a lot of formulations. Thus the dependence of foaming on alkyl chain length has been studied for monoalkyldimethyl amine oxides as illustrated in figure 5. The foam heights measured exhibit a clear maximum for chain length of 12 to 14 carbon atoms.

FIGURE 5: FOAMING PROPERTIES vs ALKYL CHAIN LENGTH

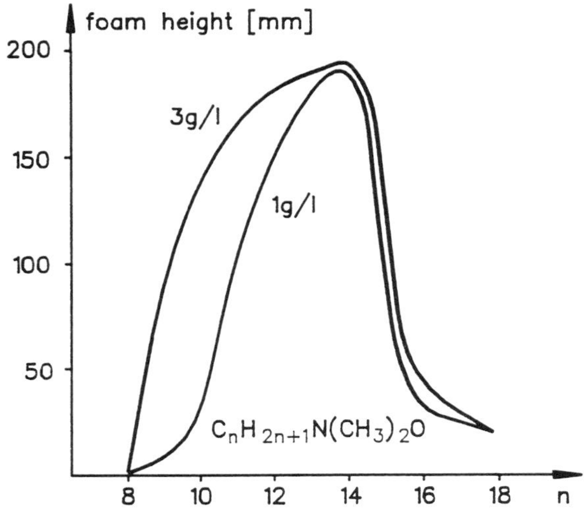

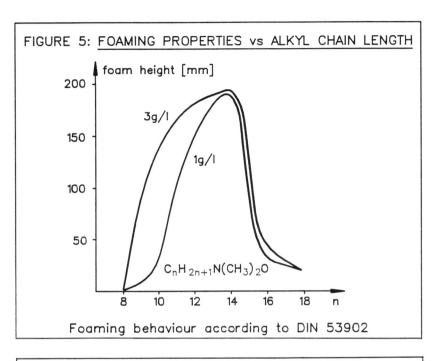

Foaming behaviour according to DIN 53902

FIGURE 6: HAND-MILD DISHWASHING LIQUID

$C_{12/14}N(CH_3)_2O$	4.0 %
$C_{12/14}OSO_3Na$	15.0 %
$C_{12/14}O(CH_2CH_2O)_7H$	4.0 %
water, colour, perfume ...	up to 100 %
pH	6.5

Advantages:
* high foam in presence of fat
* mild to hands
* insensitive to water hardness

Thus a major application area for coco-amine oxide $C_{12/14}\ N(CH_3)_2O$ are hand-mild dishwashing liquids. A frame formulation example is given in figure 6. The combination of alkyl sulphate and amine oxide yields a voluminous creamy foam highly stable in the presence of fat. This is, however, only one of the many synergistic effects exhibited by the combination anionic surfactant plus amine oxide [5, 6]. Also important are:

- mildness: At their normal use level of 1 - 5 % amine oxides are not irritating to skin and eyes. Combined with anionic surfactants they act synergistically mitigating the irritancy potential of the latter [6, 7].

- insensitivity to water hardness: Contrary to the alkyl sulphate alone the combined surfactants are insensitive towards water hardness. Moreover as demonstrated in figure 7 the surfactant combination reacts positively on increasing water hardness with increasing foam heights, especially at pH 7.

Other synergistic effects of amine oxides in this application like
- detergency boosting and
- viscosifying functions

will be discussed later on at different examples. All these synergisms can be ascribed to a strong interaction between amine oxide and nonionic surfactant molecules leading to the formation of mixed micelles [8].

FIGURE 7: DEPENDENCE OF FOAM ON WATER HARDNESS AND pH

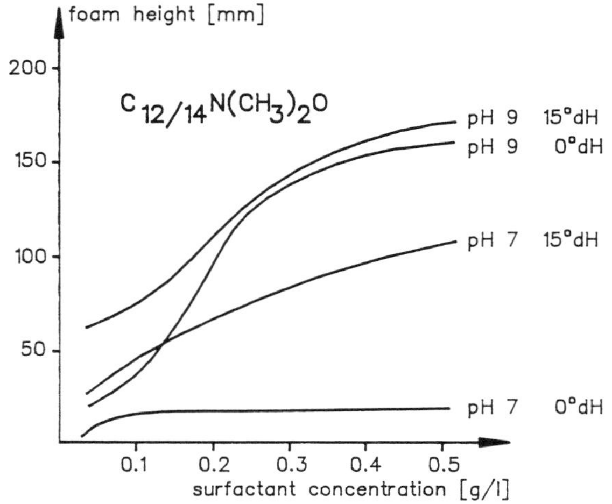

foam height according to Ross Miles of $C_{12/14}N(CH_3)_2O : C_{12/14}OSO_3Na = 3:7$

0°dH = deionised water

15°dH = 267 ppm $CaCO_3$

Wetting

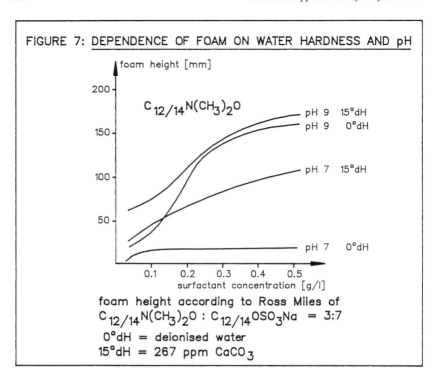

The same optimum as observed for the foaming behaviour, now actually a minimum, is found for the wetting properties (figure 8). Amine oxides with chain lengths around 12 yield the fastest wetting times measured as the sinking time of a standard piece of cotton in a surfactant solution. These results derived from uniform alkyl derivatives hold also for compounds with mixed chain lengths as indicated for a coco-amine oxide (70 % C 12, 27 % C 14, 3 % C 16 = 12.7 medium chain length) and a hardened tallow-amine oxide (1 % C 12, 4 % C 14, 31 % C 16, 64 % C 18 = 17.2 medium length) by the circles MC resp. HT in figure 8.

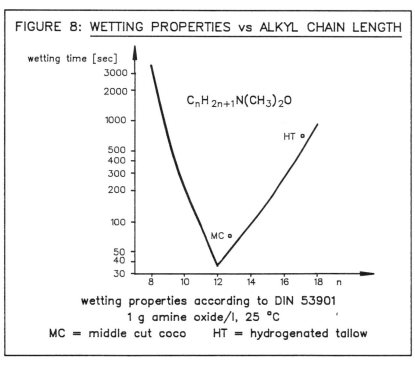

FIGURE 8: WETTING PROPERTIES vs ALKYL CHAIN LENGTH

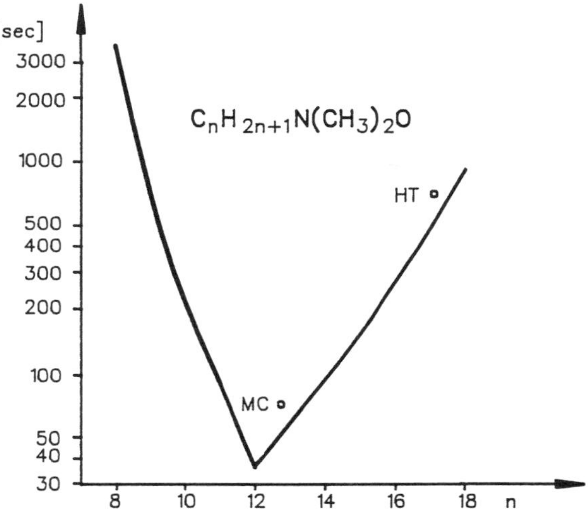

wetting properties according to DIN 53901
1 g amine oxide/l, 25 °C
MC = middle cut coco HT = hydrogenated tallow

This activity as highly effective wetting agents is employed in the formulation of alkaline degreases used in the metal-treating industry. Amine oxides exhibit an outstanding stability in this highly alkaline media containing a.o. sodium hydroxide, metasilicate, phosphate and gluconates.

Emulsifying

The function of emulsification can be divided into two sub groups:

- emulsifying in formulation = hydrotrope properties
- emulsifying in cleaning = solubilizing properties

which will be discussed separately.

As indication for hydrotrope properties the influence of amine oxide addition on the cloudpoint of nonionic surfactant solutions was determined. The results for a nonylphenol ethoxylate (7 mol EO) are given in figure 9. A drastic increase of the cloudpoint is observed especially with C 10 to C 12-amine oxides.

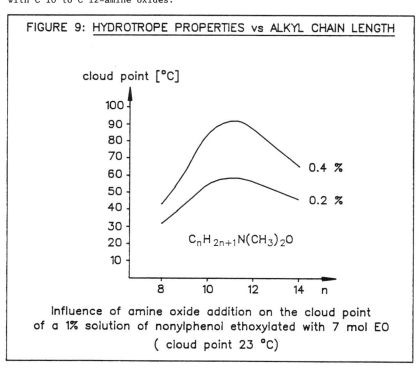

FIGURE 9: HYDROTROPE PROPERTIES vs ALKYL CHAIN LENGTH

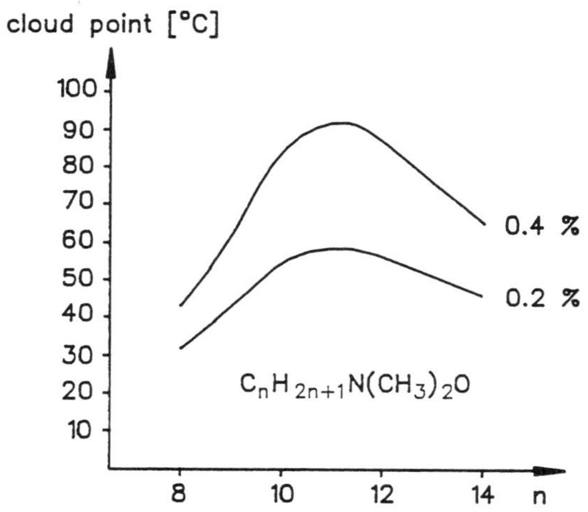

Influence of amine oxide addition on the cloud point of a 1% solution of nonylphenol ethoxylated with 7 mol EO (cloud point 23 °C)

Regarding the closely related solubilising properties the picture is not as clear as above. For the solubilisation of 2-ethylhexanole the same performance maximum is observed as in our former examples (figure 10). The azo-dye Orange PT as hydrophobic compound, however, reveals a completely different diagram. In this case a nearly linear performance increase with increasing chain length is observed. These differences are most probably due to pure geometric restrictions. The dye-molecule is too large for the small micelles of short chain amine oxides. In this test-series the interesting C 16 and C 18-amine oxides could not be studied as they were not sufficiently soluble.

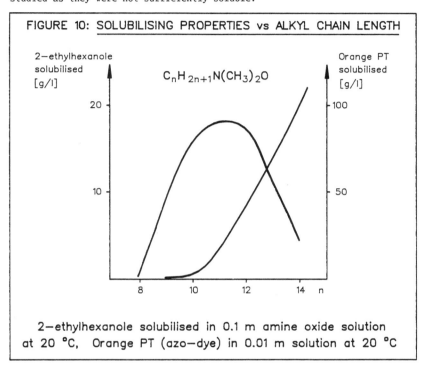

FIGURE 10: SOLUBILISING PROPERTIES vs ALKYL CHAIN LENGTH

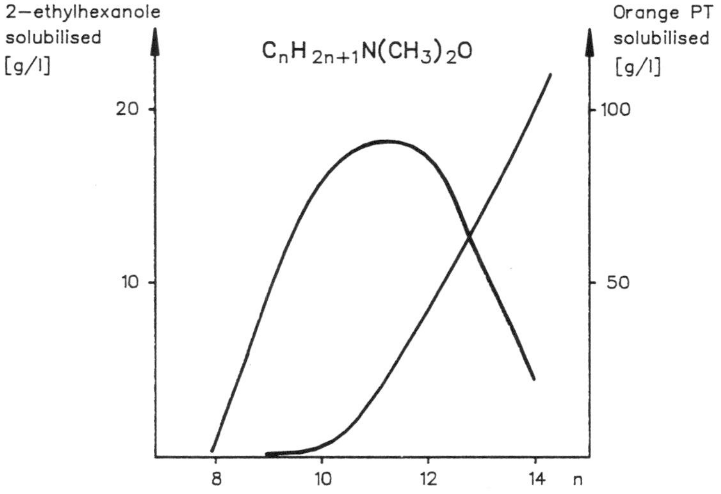

2–ethylhexanole solubilised in 0.1 m amine oxide solution at 20 °C, Orange PT (azo–dye) in 0.01 m solution at 20 °C

Due to these emulsifying properties amine oxides are used in creams and lotions as O/W-emulsifiers. Especially tallow amine oxides exhibit additionally a favourable effect on dry skin [9].

Another application of amine oxides employing the emulsifying capabilities is the formulation of fibre lubricants. Typically these compositions contain more than 50 % oil, e.g. paraffin oil, plus a surfactant-combination. Here the use of amine oxides offers combined antistatic/emulsifying function, thermal stability and low volatility. Furthermore amine oxides are not corrosive and not yellowing.

Cleaning

Synergistic effects of amine oxides with anionic detergents are particularly noticeable with ether sulphates and alkyl sulphates. They are much smaller with linear alkylbenzene sulphonates [10]. It seems, however, unjustified to say there is none [6]. Also in typical household detergents based on LAS amine oxides exhibit a detergency boosting effect [11], once again ascribed to mixed micelle formation [12].

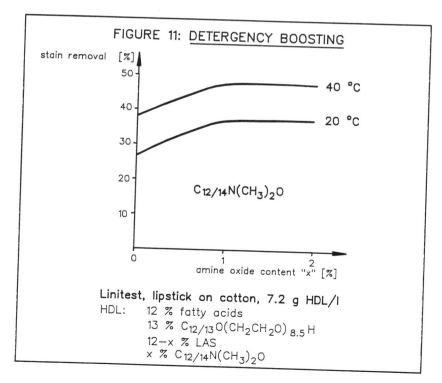

FIGURE 11: DETERGENCY BOOSTING

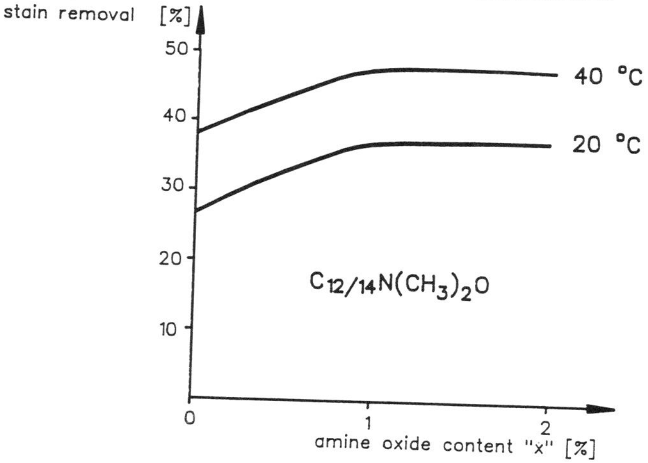

Linitest, lipstick on cotton, 7.2 g HDL/l

HDL: 12 % fatty acids
13 % $C_{12/13}O(CH_2CH_2O)_{8.5}H$
12−x % LAS
x % $C_{12/14}N(CH_3)_2O$

As amine oxides are outstanding fat solubilizers we tested this boosting effect on a greasy soil: lip-stick stains on cotton. Increasing amounts of amine oxide in a simple heavy duty liquid detergent (12 % fatty acids, lauric : oleic = 2 : 1, 13 % nonionic, 12 - x % LAS, x % coco-amine oxide) yielded a significant detergency effect at low wash temperature as shown in figure 11.

Thickening

Another example for synergistic effects exhibited by the combination of anionic surfactants and amine oxides is thickening.

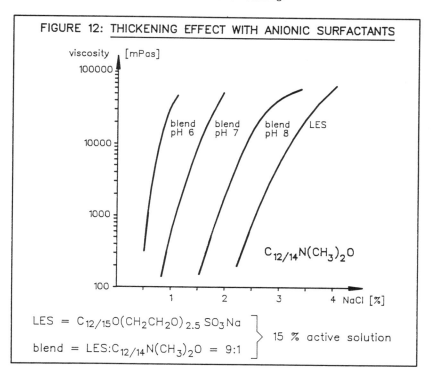

FIGURE 12: THICKENING EFFECT WITH ANIONIC SURFACTANTS

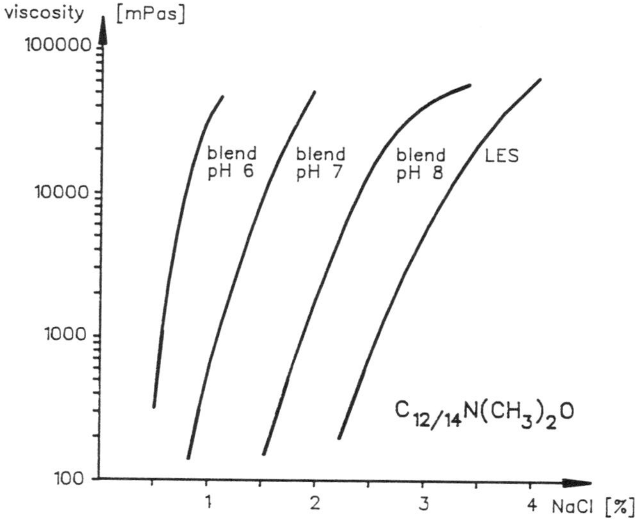

LES = $C_{12/15}O(CH_2CH_2O)_{2.5}SO_3Na$
blend = LES:$C_{12/14}N(CH_3)_2O$ = 9:1
} 15 % active solution

Amine oxides are powerful thickeners for alkyl sulphate and ether sulphate solutions. This thickening performance is particularily strong in acidic media as illustrated in figure 12 for a 9 : 1 blend of ether sulphate and amine oxide. High viscosities can be obtained at very low levels of salt-addition. In conjunction with the excellent hydrotropic properties of amine oxides the dispersions of hydrophobic materials like perfume oils are stabilized.

These viscosifying properties can be employed as already mentioned in dishwashing liquids or for the formulation of highly viscous shampoos. An example is given in figure 13. The combination of ether sulphates and amine oxides yields an outstanding fine creamy foam [10], is mild to hand and eyes [7] and offers additionally a conditioning action resulting in good combing properties.

FIGURE 13: HIGHLY VISCOUS CLEAR SHAMPOO FORMULATION

$C_{12/14}N(CH_3)_2O$	4.5 %
$C_{12/14}O(CH_2CH_2O)_3SO_3Na$	8.5 %
NaCl	4.0 %
water, colour, perfume ...	up to 100 %
pH	8
viscosity at 20 °C	7000 mPas
cloud point	< −5 °C

Advantages:
* fine creamy foam
* mildness
* conditioning action

But not only surfactant solutions can be thickened by amine oxides. A significant part of the amine oxide production goes to the hypochlorite bleach market. In these oxidizing and highly alkaline formulations amine oxides exhibit an outstanding stability. They fulfil two functions: perfume solubilization and if desired thickening. As indicated in figure 14 the viscosifying performance is optimized for pure uniform myristamine oxide $C_{14}N(CH_3)_2O$. The cocoamine oxide $C_{12/14}N(CH_3)_2O$ exhibits no thickening effect but acts as superior fragrance solubilising agent [13] as shown in table 3.

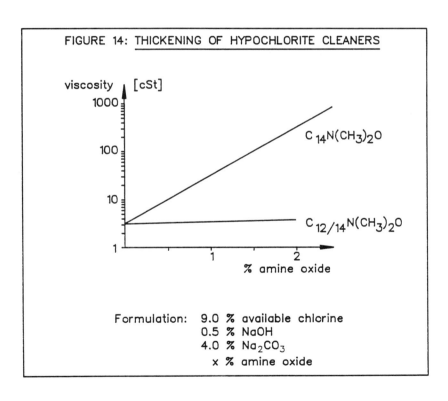

FIGURE 14: THICKENING OF HYPOCHLORITE CLEANERS

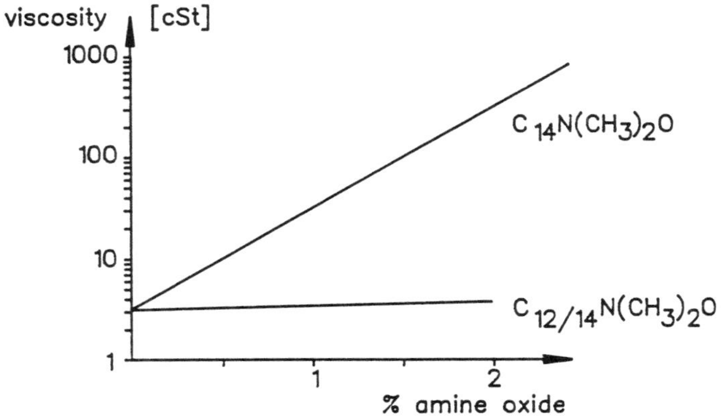

Formulation: 9.0 % available chlorine
0.5 % NaOH
4.0 % Na_2CO_3
x % amine oxide

Table 3: Perfume Solubilisation in Hypochlorite Bleach
Minimum Amount of Amine Oxide for Clear Solution
(Formulation: 7.5 % active chlorine, 1 % NaOH,
0.5 % perfume)

Fragrance	Amine Oxide	
	$C_{12/14}N(CH_3)_2O$	$C_{14}N(CH_3)_2O$
A	0.5 %	1.1 %
B	1.5 %	2.5 %

Summarizing this chapter we can say that amine oxides fulfill following functions:

- foaming
- wetting
- emulsifying
- cleaning
- thickening

in various formulations, mainly in the area of detergents and cleansers.

Ecotoxicological Properties

Besides performance any surfactant has nowadays to fulfil stringent environmental demands like biodegradability, low toxicity and safety of handling. Amine oxides meet all these requirements perfectly.

The biodegradation of amine oxides has been less extensively studied than of other industrial surfactants. Nevertheless all publications report ready biodegradability with biodegradation grades ≥ 95 % for different monoalkyl dimethyl amine oxides [14]. This was established by test in our laboratories. In a Closed Bottle Test coco-amine oxide yielded 88 % biodegradation after 2 weeks, resp. 93 % after 4 weeks, an excellent result within this stringent test design.

Also from a toxicological point of view amine oxides can be used without deliberation. For several different amine oxides tested we found the oral toxicity towards rats (LD_{50}) to be ≥ 2000 mg/kg body weight. Thus these compounds have to be regarded as non-toxic. An important factor in environmental considerations is the toxicity towards fish. For coco-amine oxide a test with rainbow trouts (OECD guideline 203) exhibited LC_{50} value of 42.0 mg/l. No mortality within the test period of 96 h (LC_o) was observed for 33.5 mg/l. These data confirm a comparatively low fish toxicity as typical values for industrial surfactants are about 5.0 mg/l.

Nevertheless amine oxides possess some bacteriostatic activity. Therefore their formulations are self-preserving. Some typical Minimum Inhibition Concentration (MIC) values are given in table 4.

Table 4: Bacteriostatic Activity MIC in [ppm]

Germ	$C_{12/14}N(CH_3)_2O$	$C_{14}N(CH_3)_2O$	$C_{12/14}N(CH_3)_3Cl$
1. Gram-positive bacteria			
Staphylococcus aureus	50	25	< 1
Streptococcus faecalis	50	50	15
2. Gram-negative bacteria			
Pseudomonas aeruginosa	3200	3200	150
Escherichia coli	50	200	30
3. Fungi/yeast			
Aspergillus niger	50	50	5
Candida albicans	50	25	15

Obviously amine oxides cannot match the bacteriostatic performance of quaternary ammonium compounds. For self-preservation, however, even in diluted solutions this activity is sufficient. Some specific amine oxides exhibit even much higher bactericidal effectivity [15].

Summarizing this chapter we can conclude that amine oxides offer a very high level of environmental safety as well as safety of handling.

<u>Different Amine Oxides</u>

To prevent the impression that only coco-amine oxide is a valuable industrial surfactant this chapter will mention some applicational advantages of different amine oxides. In figure 15 alternative modifications of $C_{12/14}N(CH_3)_2O$ are indicated.

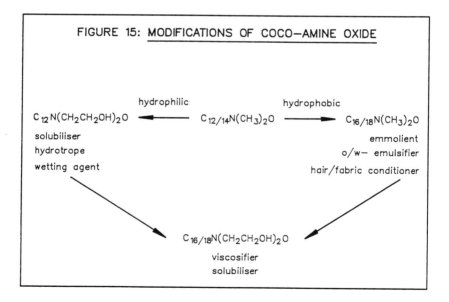

FIGURE 15: MODIFICATIONS OF COCO-AMINE OXIDE

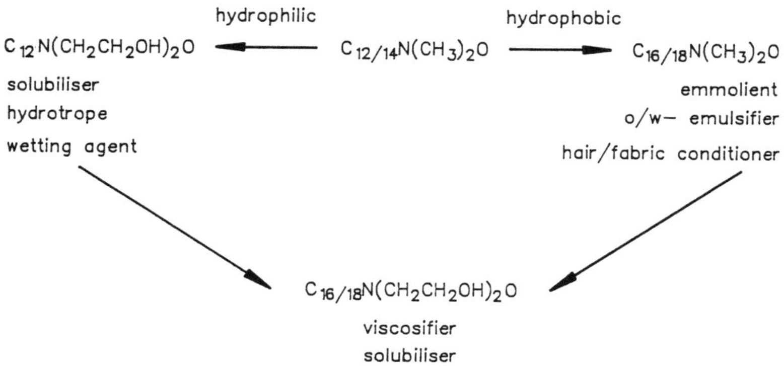

Going to a more hydrophobic amine oxides, i.e. increasing the alkyl chain length, will yield excellent emollients and O/W-emulsifiers. These amine oxides are suited as hair or fabric conditioner and O/W-emulsifier in creams and lotions as already mentioned.

On the other hand a more hydrophilic amine oxide can be achieved by substituting the methyl groups of dimethyl coco-amine oxide by hydroxyethyl groups. The resulting dihydroxyethyl coco-amine oxide is an excellent wetting agent, solubiliser and hydrotrop.

Combining both modifications yields dihydroxyethyl tallow-amine oxide, especially useful as thickening agent with excellent solubilizing properties. For example a combination of 1.5 % dihydroxyethyl tallow amine oxide with 1.5 % tallow trimethylammonium chloride can thicken 10 % HCl-solution to roughly 1000 mPa.s

Further different amine oxides as indicated in figure 1 are possible and valuable for specific applications. Dialkyl methyl amine oxides (structure 1d) are excellent softening agents especially useful in combination with anionic surfactants in a so-called softergent. Molecules with several NO-groups (structure 1e) offer outstanding cleaning, emulsification and solubilising properties. Within this presentation, however, is not space enough to discuss all these molecules in detail. Thus our considerations on amine oxides will end here with a short summary of their applicational properties (figure 16).

Summary

Amine oxides and especially monoalkyldimethyl amine oxides are

- excellent foaming agents
- powerful surfactants and cleaning agents
- effective emulsifiers
- valuable solubilisers and hydrotropes

They offer additional benefits like

- chemical stability against oxidizing agents, acids and alkali
- thermal stability
- mildness to skin and eyes
- synergistic effects with anionic surfactants

Furthermore they are completely safe raw materials regarding

- environmental safety
- safety of handling

FIGURE 16: PROPERTIES OF AMINE OXIDES

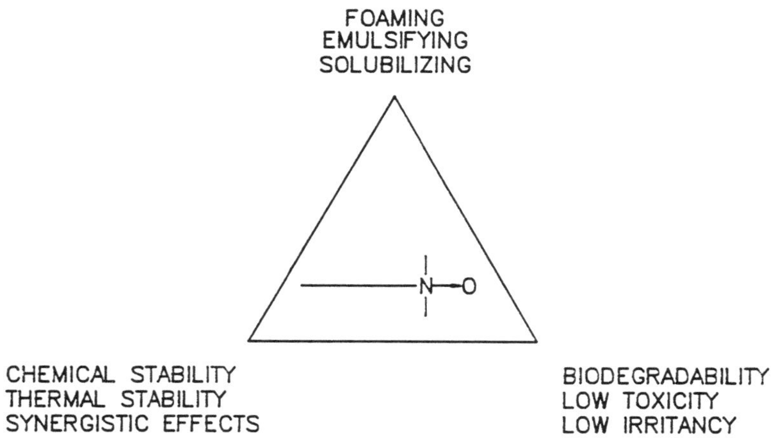

Literature

1) W.R. Dunsting and E. Goulding
J. Chem. Soc. 75, 794 (1899)

2) DRP 664425 (26.01.34)

3) Schweiz. P. 175351 (16.02.34)

4) a) K. Lindner
Tenside 1, 112 (1964)

b) R.J. Nadolsky in
Kirk-Othmer: Encyclopedia of Chemical Technology
Vol. II, 3. Edition, 259-271 (1978)

5) a) G.J. Smith
SÖFW 105, 319 (1979)

b) SÖFW 105, 345 (1979)

6) a) O. Okumura et al.
AOCS 73rd annual meeting 1982

b) -
SÖFW 102, 33 (1976)

7a) US 4033895

7b) U. Zeidler and G. Reese
Ärztliche Kosmetologie 13, 39 (1983)

8) M.J. Rosen in
J.F. Scamehorn (Ed.): Phenomena in Mixed Surfactant Systems
ACS Symposium Series 311, 144-162, Washington (1986)

9) K. Schrader
Parfümerie und Kosmetik 58, 185 (1977)

10) a) H. Watanabe and W.L. Groves

JAOCS 45, 738 (1968)

b) T.P. Matson

JAOCS 40, 640 (1963)

c) L.R. Smith and D.C. Zajac

Household & Pers. Prod. Ind. III, 76, 34 (1976)

11) US 3085982

12) D.G. Kolp et al.

J. Phys. Chem. 67, 51 (1963)

13) a) H. Hoffmann et al.

Progr. Colloid & Polymer Sci. 73, 95 (1987)

b) H. Rörig and R. Stephan

2nd World Surfactant Congress, Paris 1988

14) a) E.A. Setzkorn and R.C. Huddlestone

JAOCS 42, 1081 (1965)

b) SDA, Biodegradation Subcommittee

JAOCS 46, 432 (1969)

c) J. Ruiz Cruz and M.C. Dobarganes Garcia

Grasas Aceites 29, 1 (1978)

15) J. Subite et al.

Antimicrobial Agents and Chemotherapy 12, 139 (1977)

Optimisation of the Performance of Quaternary Ammonium Compounds

R. A. Stephenson
BEROL-NOBEL LIMITED, HENLEY-ON-THAMES, OXFORDSHIRE, RG9 1RY, UK

1. INTRODUCTION

Although the market for cationic surfactants is quite large, the number of cationic products classified as germicides is relatively small. It is a particularly difficult market because it needs a high level of expertise to create new opportunities. The rather complicated chemicals tend to become commodities at a low sales volume, although registration with National Licensing Authorities can be expensive.

To justify Research Projects management now expects very detailed analyses of sales potential and market share, etc. One would be hard put to offer a strong enough argument to obtain approval to research into new molecules in the germicidal field, especially with the enormous costs required for registering a new chemical. Such discoveries will now only arise as spin-offs from projects in larger fields.

Perhaps more mileage can be obtained from gaining a larger share of existing markets by optimising the performance of existing Quaternary Ammonium Compounds used as bactericides.

2. QUATERNARY AMMONIUM COMPOUNDS

Since the germicidal Q.A.C's were introduced around 1935, these chemical agents have received a wide and popular recognition as effective germicides for the destruction of many forms of germ life. The outstanding features of these compounds are that they are odourless, very stable and can be manufactured pure with little or no colour. They are also relatively non-toxic when used at the recommended concentrations.

When they were first studied by early workers, they were considered to be effective against a wide range of bacteria. In the last three decades, exhaustive studies have been made and a range of commercial products have evolved. Because it was discovered that the performance of Q.A.C's in hard water solutions is much better against the gram positive bacteria than the gram negative bacteria, this has been represented as a weakness. In fact, the performance against the gram negative bacteria compares favourably with other germicides, and the performance against gram positive bacteria could be considered exceptional.

Probably the most important work was carried out by DOMAGK in 1935, who considered that ZEPHIROL, a high molecular alkyl dimethyl benzyl ammonium chloride was of particular merit as a hand disinfectant in surgical practice because of its effectiveness against the gram positive cocci; the use of the Q.A.C. reduced the numbers of bacteria flushed from the skin by heavy perspiration.

Now more than fifty years later Benzalkonium Chloride in one form or another still features as one of the most popular Q.A.C's.

A Quaternary Ammonium Salt can be represented by the following general formula:

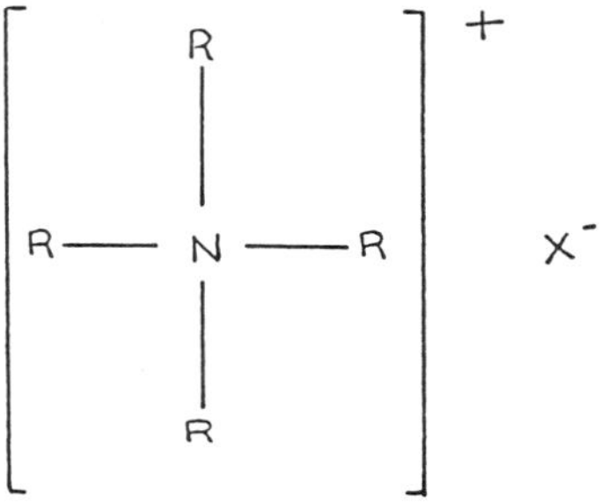

Where R is an organic radical and X is an anion. The number of compounds that can be prepared, therefore, is immense but only comparatively few have any real value as disinfectants.

They can be narrowed down to three alternatives.

1)

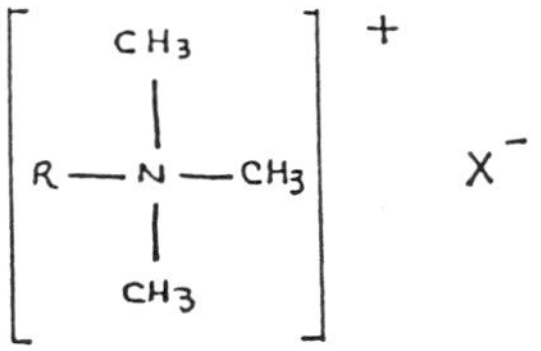

2)

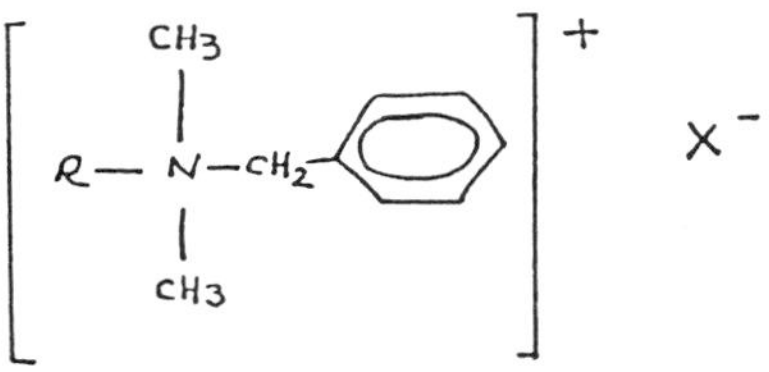

3)

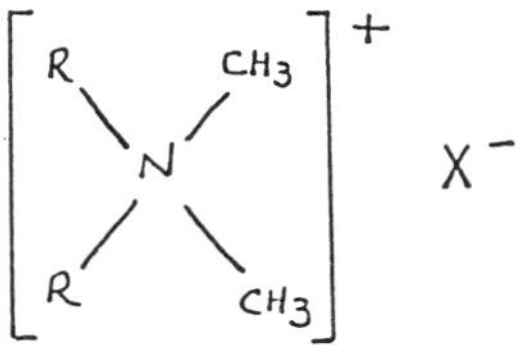

Each of these structures can be found in commerical products and examples would be Cetrimide, Benzalkonium Chloride and Didecyl Dimethyl Ammonium Chloride. From our studies we have found that with the B.K.C. structure the best germicides have a long alkyl chain containing between 12 and 16 carbon atoms. With Dialkyl quats an optimum is reached with the twin ten compound.

It is interesting to note that the adsorption of the B.K.C.'s corresponds to physical adsorption of surfactants at a solid / solution interface, the adsorption increasing with molecular weight reaching a maximum at C16-C18 chain length, the limit of the solubility in water.

With twin chain Q.A.C.'s, by the time one has reached C_{16-18} the compound is virtually insoluble in water and the adsorption of a softener to fabric probably occurs by assuming two processes. One process which occurs is molecular adsorption which will be slow due to the low concentration of the dissolved free quaternary ion in solution, the second process is the deposition of the dispersed colloidal particles on the fabric, which is then followed by spreading of the Q.A.C. over the fibre surface. Hence schematically we have the following:-

2 C_8	2 C_{10}	2 C_{12}	2 C_{14}	2 C_{16}	2 C_{18}
Mild Germicide	Strong Germicide	Very Weak Germicide	Anti Stat	Anti Stat and Softener	

These properties are the clue to various markets that have evolved from this class of compounds.

Although the antistatic and fabric softening Q.A.C's are very important from a commercial point of view, I will confine this discussion to Q.A.C's showing anti-bacterial activity.

Examples of this range of Q.A.C's which is commercially available are given below.

Structure 1

Cetrimide B.P.

An Alkyl Trimethyl Ammonium Bromide have a long alkyl chain of molecular wt. Approximating to Myristyl.

Cetyl Trimethyl Ammonium Chloride

Usually marketed as a 29% solution in water used in cosmetic preparations as an antistat.

Structure 2

Benzalkonium Chloride B.P.C.

Alkyl dimethyl ammonium chloride with alkyl chain approximating to coconut fatty acids.

Benzalkonium Chloride V.S.P.

Alkyl dimethyl ammonium chloride with alkyl chain made from C12 40% C14 50% C16 10%.

Structure 3

Didecyl dimethyl ammonium chloride.

Cetrimide is a well known quaternary which was found wanting very early in its life as a hospital antiseptic, when Pseudomonads were isolated from samples of in use dilutions. In fact Cetrimide agar is now used as a method to purify cultures of Pseudomonas aeruginosa.

Cetrimide is, however, the only Germicidal Q.A.C. which is a detergent in its own right. Other Quats such as the B.K.C's can only claim to be surface active. Cetrimide only survived as a hospital disinfectant by the discovery of Chlorhexidine which gave it another lease of life and in the U.K. prevented the growth of superior B.K.C's. A token amount of Chlorhexidine digluconate 10% with reference to the Cetrimide content, was found sufficient to clean up the bactericidal spectrum sufficient to retain the hospital antiseptic business.

While Cetrimide was being established in the U.K. this market was satisfied by Benzalkonium Chloride in the U.S.A, but particular attention was taken in the quality of the Alkyl dimethyl benzyl ammonium chloride. It was found that the hard water tolerance could be improved by a careful selection of the isomers which make up the alkyl chain. Hence, the U.S.P. specifies a higher molecular weight to the B.P.C. If one makes a B.K.C. from Coconut Fatty Acids it meets the B.P. Specification, the alkyl chain would then be approximately:

C8	2%
C10	5%
C12	50%
C14	20%
C16	13%
C18	10%

To meet the U.S.P. and obtain an optimum performance with the Chambers Hard Water Tolerance Test the akyl chain distribution has to be adjusted to:-

C12	40%
C14	50%
C16	10%

To illustrate the importance placed on hard water tolerance, in 1961 more than twenty brands of B.K.C were submitted for a tender in one of the States in the U.S.A. Only two were approved and both were high Myristyl B.K.C's.

To further illustrate how important hard water tolerance is considered in Food and Catering sectors of the U.S. Industry:-

Dodecyl benzyl trimethyl ammonium has been purported as a B.K.C. equivalent in the U.K. In the U.S.A. because of its poor hard water tolerance, it is only considered as an algaecide.

The main reason why the B.P. product is preferred to the U.S.P. in the U.K. is purely price. Coconut Fatty Acids from time to time become long and cheap. The C14 isomer is much more expensive.

From our bacteriological studies Didecyl dimethyl ammonium chloride always out performs the best B.K.C's.

3. DISINFECTION

Disinfection (Bactericide or Bacteriostat)

A bactericide is an agent which kills bacteria and a bacteriostat is one that prevents the growth of bacteria.

All bactericidal processes first involve the reaction of a chemical agent with the surface of a bacteria. We, therefore, define a bacteriostat as an agent which prevents the reproduction and multiplication of bacteria. It is very important to appreciate the difference between bacteriostatic and bactericidal action, particularly when studying the potency of Quaternary Ammonium Compounds. In soft water solutions most Q.A.C's are bacteriostatic at low concentrations, as little as 2 parts per million of Benzalkonium Chloride B.P. will prevent the growth of Staphylococcus aureus.

Professor Hugo in his book "Sterilisation and Disinfection" gives an excellent picture illustrating the difference between bactericidal and bacteriostatic action.

The following diagram shows the fate of a bacterial population when innoculated into:

A) Nutrient Medium
Normal Growth Curve

B) Bacteriostatic Environment (No change in viable population) after a prolonged time interval the viable population will probably begin to fall.

C) Bactericidal Environment

A Sigmoid Death Curve is shown:

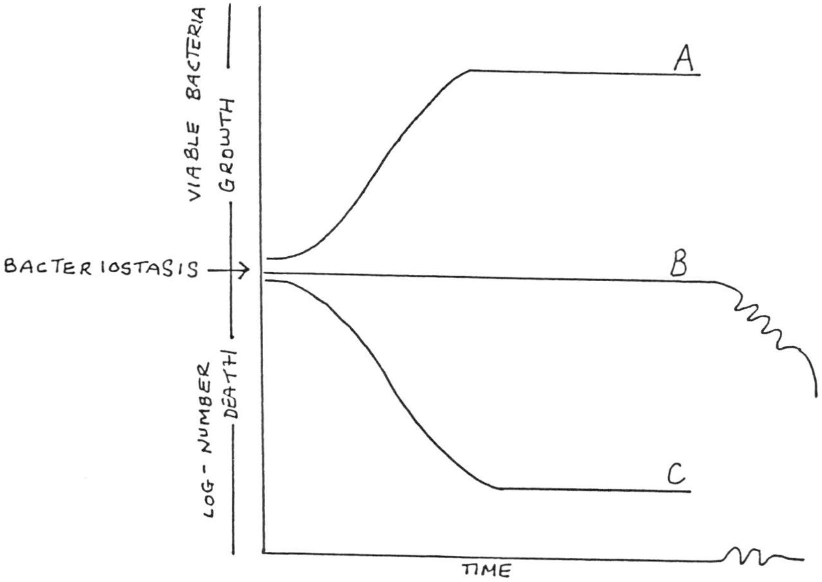

This is a very simple picture which does not tell the whole story and would only apply to those bactericides, which show clearly defined bactericidal and bacteriostatic phases.

In many cases the conditions cannot be divided into clearly defined stages; there is an inevitable overlap, resulting in all three being manifest under some circumstances. This position arises because each cell in a bacterial population is an individual entity, as such its reaction characteristics can apparently be quite different to those of the population as a whole. Thus at selected concentrations of a germicide the more sensitive the cells are easily killed, others being more resistant may be held in suspended animation or inhibited, whilst a few of high resistance may almost be unaffected.

This leads in some bacteriostatic tests to the phenomenon of long term survivors, which must not be confused with the cultivation of resistant strains by continuous application of sub lethal doses of a germicide.

Mode of Action of Q.A.C.'s

As previously stated bacteria can be divided into two types according to the way they stain in the Gram-Reaction. The two types have very different wall structures. Those of the Gram-negative organisms are more complex. [3]Ward and Berkley gave an excellent review of the subject in a symposia on the Microbial Adhesion to surfaces at Reading University on September 3 - 5th 1980.

The following diagram shows the two types of structures.

A. <u>E. Coli</u>

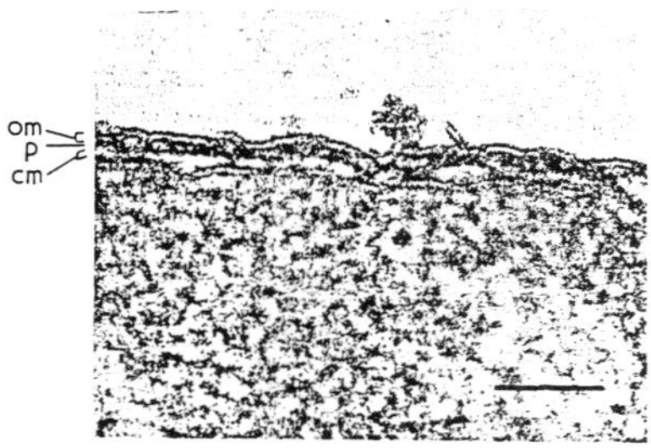

B. <u>B. Subtilis</u>

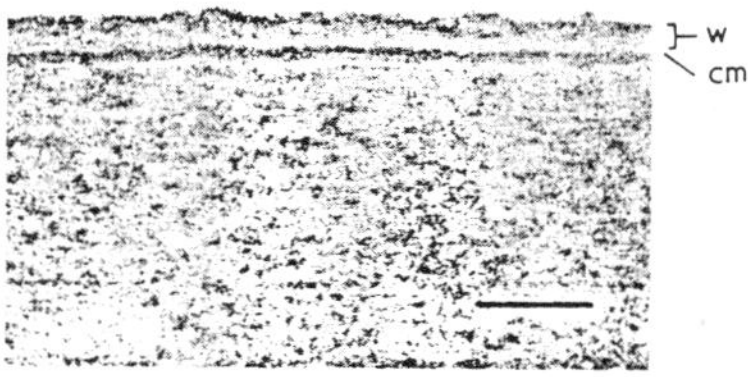

Electron micrographs of sections through the walls of Gram-negative and Gram-positive bacteria.

a) <u>ESCHERICHIA COLI</u> showing the outer membrane (om) with the peptidoglycan (p) closely associated with its inner surface and the cytoplasmic membrane (cm).

b) <u>BACILLUS SUBTILIS</u> showing a much thicker wall (w) containing peptidoglycan and teichoic acids but without any layering of the component polymers and the cytoplasmic membrane (cm). The bars represent 0-1 μm. Figure kindly supplied by Dr. Ian Burdett.

It can be seen that with both classes of bacteria that the cell wall contains peptidoglycan. This with other associated materials confers a negative charge to the bacterial surface. It is well known that once a surface has become charged it will attract oppositely charged ions from the surrounding aqueous phase.

The following diagram illustrates the formulation of an electrical double layer at a planar surface and spherical particle. If the spherical particle represents the bacteria and the charged particle the Quaternary ammonium ion, this diagram illustrates the case of one bacteria in a solution of a Q.A.C.

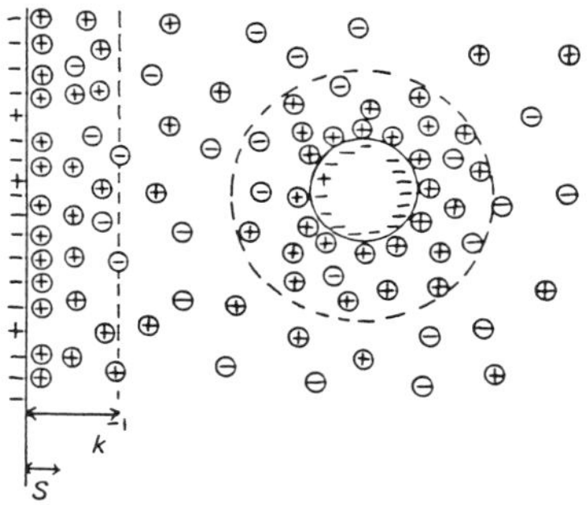

Ref.4 - The electrical double layer associated with a planar surface and a spherical particle. S is the thickness of a layer of adsorbed counter ions referred to as the Stern layer; K^{-1} is the thickness of the diffuse double layer.

Whittet, Hugo and Wilkinson in the book entitled "Sterilisation and Disinfection" give the mode of action for Quaternary Ammonium Compounds as:-

"The positively charged Quaternary Ammonium Compounds are, at physiological pH values, strongly adsorbed onto negatively charged bacterial surface; thereafter damage to the cytoplasmic membrane occurs and an apparently fatal leakage of cytoplasmic material follows".

This is explained in more detail by W.A. HAMILTON in his paper on the "Mode of Action of Membrane Active Antibacterials".

4. EFFECT OF HARD WATER AND PROTEIN ON THE PERFORMANCE OF A Q.A.C.

Every worker in the disinfectant field is aware that Quaternary Ammonium Compounds perform equally in hard or soft water against the Gram-positive bacteria, but their performance against the Gram-negative bacteria can be seriously effected by hard water salts. It is a very complex subject,but it is believed that the divalent cations occupy sites which could have otherwise been available for quaternary ions. It is also only logical that the cation is the dominant particle in such a reaction, hence it can be seen that if some of the negative sites on the cell wall are occupied by divalent cations the concentration of the Quaternary Ammonium Ion on the surface would be less.

The quality of the water can therefore be critical. Although added protein lowers the performance of Q.A.C's I believe that the inhibitive effect of hard waters is more a significant factor than protein.

In the catering and allied industries it is possible in most cases to use the two stage process of cleaning and then disinfecting. Even in hospitals where the two stage process is not possible the amount of water involved does not preclude the use of deionised water. However, in the end at some stage in industrial processes one is obliged to use the local water supply. Well waters can be very hard and even mains supplies can reach 360 ppm hardness measured as Calcium Carbonate.

In England and Wales there are ten water authorities. Map I shows the area they cover. Map II gives a picture of the quality of water supplied in 1968. Currently the Thames Valley Authorities supplies 710 million gallons of water a day to their 11½ million customers. The typical analysis of the water measured as Calcium Carbonate is 290 ppm. 220 ppm temporary and 70 ppm permanent hardness. In this area there are some stations which supply much harder water. For example, water supplied to the districts of Wargrave and Woodley has a typical hardness of 350 all of which is temporary hardness.

Although it is rare to be able to select the source of water the choice of a one stage or two stage disinfection process is usually a matter of economics. Bacteria are invariably associated with protein and even in clean conditions the presence of particulate or soluble organic soil will lower the performance of the Q.A.C. Such organic matter has a surface which is very similar to bacteria, and this surface competes with the bacteria for the Q.A.C.

Water authorities in England and Wales

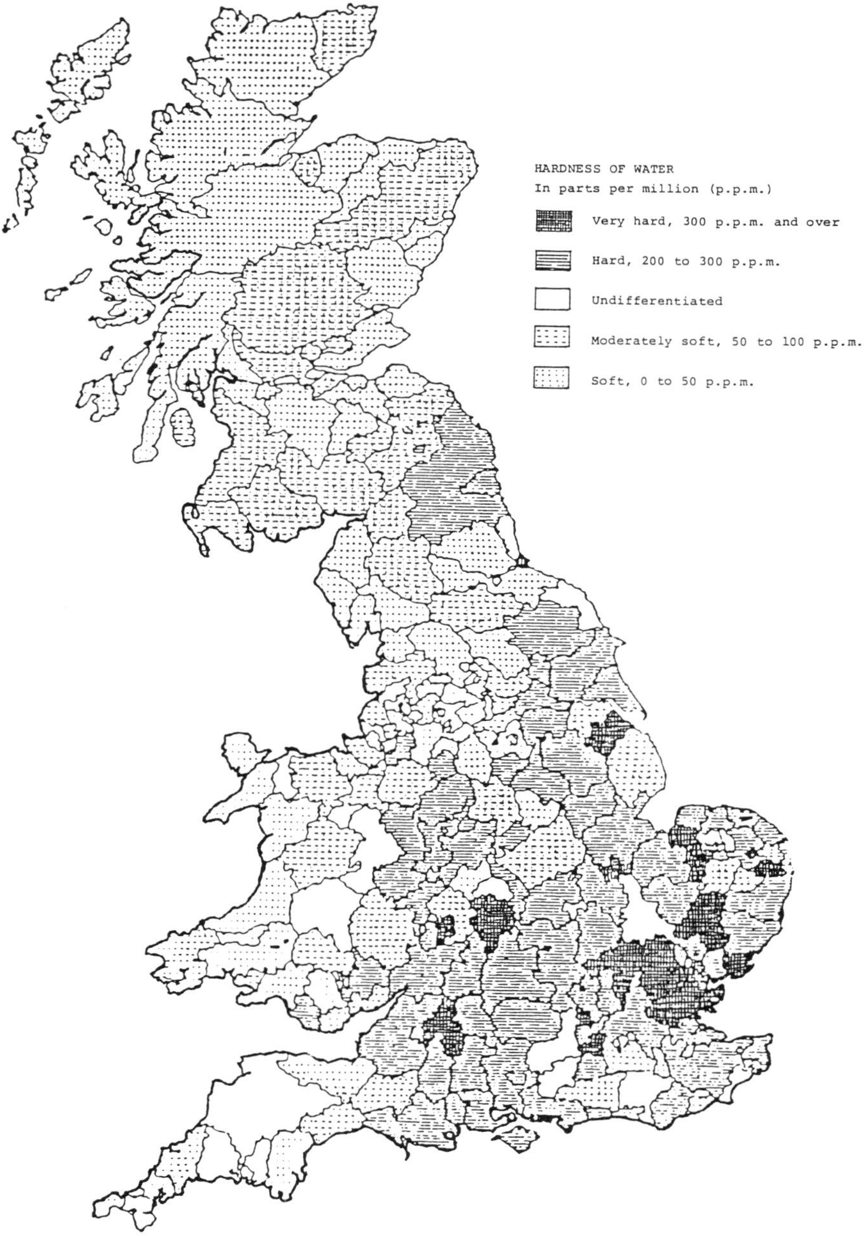

Fig. Degree of hardness of water supplied by direct supply water undertakings in Britain in 1968 (based on data in Water Engineer's Handbook 1968)

It has been previously stated that with Alkyl Trimethyl Quats and Alkyl Dimethyl Benzyl Quats adsorption to protein increases with the increase in the length of the alkyl chain and this continues until solubility becomes the limiting factor. The bactericidal properties appear with the C_{12} isomer and continue up to the C_{16}-C_{18}, when the higher molecular weights are virtually insoluble in water. As the cations cannot differentiate between bacteria and associated protein, the stronger adsorbing higher isomers, such as C_{16} and C_{18}, are preferentially adsorbed to protein and less is available for the bacteria. Hence in distilled water, the performance of the C_{16} isomer is often better than the C_{12} isomer, but in the presence of gross contamination the performance falls off rapidly. For this reason the B.K.C. based on Coconut Fatty Acid is specified by the B.P.

5. BACTERIOLOGICAL TESTS TO ILLUSTRATE THE DIFFERENCE IN PERFORMANCE OF Q.A.C's IN DISTILLED WATER AND HARD WATER.

Tests were carried out to illustrate the difference in the bactericidal performance in hard water of BKC BPC, BKC USP and Didecyl Dimethyl Ammonium Chloride.

In this series of tests the solutions were evaluated by the technique described in the British Standard Specification 3286 : 1960. This specification describes a test procedure for the laboratory evaluation of the disinfectant activity of a Quaternary Ammonium Compound. A suspension of viable micro-organisms is added to solution containing one or more concentrations of Quaternary Ammonium Compounds and the number of organisms surviving after a period or periods of time is recorded.

COMPARISON OF BACTERICIDAL EFFICIENCY OF QUERTON 210Cl, QUERTON KKBCl AND QUERTON 246

Test Compounds

Querton 210Cl Didecyl dimethyl ammonium chloride.

Querton KKBCl A Benzalkonium chloride meeting the requirements of the British Pharmacopoeia.

Querton 246 A Benzalkonium chloride meeting the requirements of the U.S. Pharmacopoeia.

Tests

Comparison of Q.A.C's in the presence of:-

(a) Distilled Water.
(b) Hard Water. (300 ppm Ca^{++} A.O.A.C. formula)

Test Organism

Escherichia coli NCTC 8196 grown on agar slopes and harvested in distilled water.

Contact Time

1, 2, 5 and 10 minutes.

Inactivator

2% Egg Lecithin in 3% aqueous Lubrol W.

Procedure

BS 3286 using double strength dilutions to give a dilution with equal volume of challenge culture, specified concentrations of Q.A.C. in distilled and hard water.

Results

(a) Coli in Distilled Water.

Average of three tests expressed in as % survivors in contact times given above.

Conc^n	QACS	% Survivors in contact times			
		1 Min	2 Mins	5 Mins	10 Mins
10 ppm	210Cl	0.452	0.035	0.004	0.001
	246	14.955	3.257	0.716	0.088
	KKBCl	18.439	4.066	0.650	0.083
25 ppm	210Cl	0.005	0.001	0.000	0.000
	246	0.036	0.003	0.000	0.000
	KKBCl	0.076	0.009	0.000	0.000
50 ppm	210Cl	0.001	0.000	0.000	0.000
	246	0.001	0.000	0.000	0.000
	KKBCl	0.003	0.000	0.000	0.000

Results

(b) Coli and Standard Hard Water

Average of two tests expressed as % solution in contact times given above.

Conc^n	QACS	% Survivors in contact times			
		1 Min	2 Mins	5 Mins	10 Mins
50 ppm	210Cl	0.267	0.025	0.0007	0
	246	3.524	0.200	0.020	0.002
	KKBCl	13.636	2.876	0.016	0
70 ppm	210Cl	0.007	0.001	0	0
	246	0.409	0.014	0.002	0
	KKBCl	0.547	0.004	0.001	0
100 ppm	210Cl	0	0	0	0
	246	0.025	0.003	0	0
	KKBCl	0.021	0	0	0

GRAPH SHOWING RESULTS IN THE 1 MINUTE TEST IN SAME FORM AS ILLUSTRATED IN B.S.3286

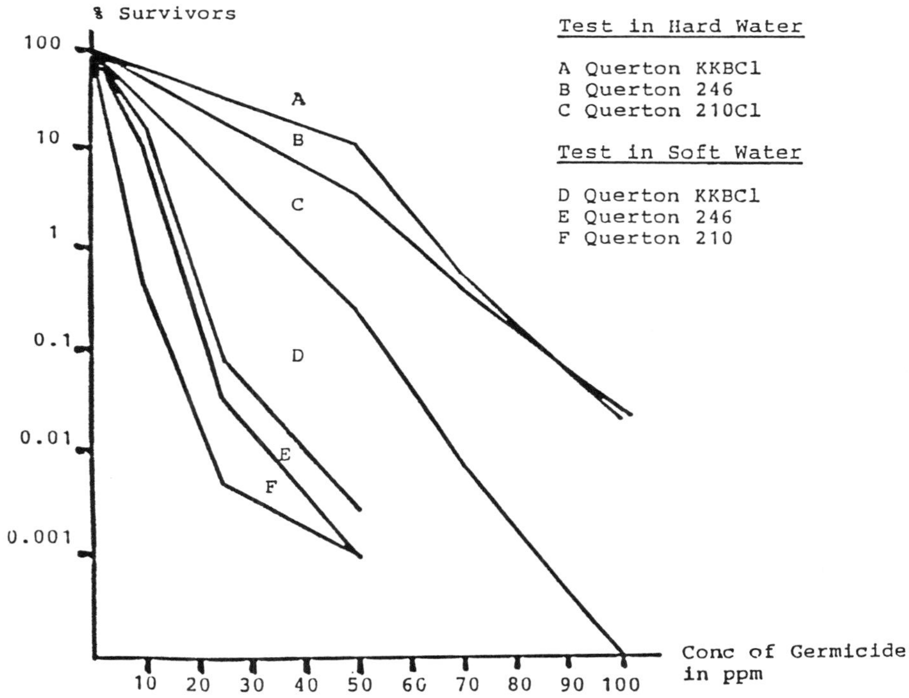

More recently a more comprehensive study was made using the same test to compare the hard water performance of Querton 210Cl and Querton KKBCl against E. Coli in S.H.W. of 300 ppm S.H.W. (AOAC Formula) in contact time of 5 minutes.

The following results were obtained from ten tests:

% SURVIVORS

TEST	Concentration (ppm)					
	50		70		100	
	KKBCl	210Cl	KKBCl	210Cl	KKBCl	210Cl
1	0.436	0.035	0.006	0.002	0.0002	0.0001
2	0.884	0.0035	0.007	0.008	0.0001	0.0001
3	0.038	0.0007	0.012	0.0001	0.0007	NIL
4	0.110	0.0008	0.009	NIL	0.0002	NIL
5	0.035	0.0009	0.012	0.0001	0.0003	NIL
6	0.145	0.0009	0.019	0.0001	0.0010	0.0001
7	0.056	0.0007	0.009	0.0002	0.0010	NIL
8	0.153	0.0011	0.017	0.0002	0.0007	NIL
9	0.070	0.0007	0.002	0.0004	0.0003	NIL
10	0.038	0.0007	0.003	0.0005	0.0003	NIL
Average	0.197	0.005	0.010	0.0012	0.0005	NIL

Although these tests are only with one organism, similar results would be obtained with other gram-negative organisms such as Pseudomonas aeruginosa. More severe conditions would accentuate the differences, e.g; If Querton KKBCl is compared with Querton 210Cl by the 5-5-5 test the following results are obtained:-

STANDARD SUSPENSION TEST FOR DETERMINING THE DISINFECTANT ACTIVITY BY MEANS OF THE "5-5-5" TEST, EXCLUDING BACILLUS CEREUS.

$ME^{20}{}_{5}$ VALUES

SAMPLE QUERTON 210CL				
Conc.	ORGANISM			
	Pseudomonas aeruginosa ATCC 15442	Escherichia coli ATCC 11229	Staphylococcus aureus ATCC 6538	Saccharomyces cerevisiae ATCC 9763
500ppm	4.34	4.58	5.46	2.60
750ppm	4.45	5.12	>7.51	3.43
1000ppm	(1) 4.89	5.75	6.24	4.44
	(2) 4.82	>7.30	>7.51	4.01
1500ppm	5.56	>7.35	>7.54	5.34
SAMPLE QUERTON KKBCL50				
500ppm	2.94	2.97	3.86	2.63
750ppm	3.39	4.44	4.00	3.31
1000ppm	(1) 3.79	>7.35	>7.54	4.34
	(2) 3.63	>7.30	5.66	3.91
1500ppm	5.18	>7.35	>7.54	>6.64

6. USE OF SEQUESTERING AGENTS TO OVERCOME THE DEFICIENCY OF QUATS IN HARD WATER SOLUTIONS.

In 1971 an exhaustive research programme was carried out to improve the performance of Q.A.C's by the addition of sequestering agents such as E.T.D.A. and N.T.A.

In order to examine a very large number of samples a simple and familiar bacteriological test was used which gives the killing dilution in 5 minutes.

It was possible to make a big improvement in the performance of a Quat as measured by the Kelsey-Sykes Test. The solutions were found to be pH dependant and it was only possible to prepare concentrated solutions if the sodium and chloride ions had been removed.

It will be shown later that with equimolar quantities of sequestering agents and Q.A.C's, higher than expected efficencies were obtained.

Bacteriological Technique

The technique used to obtain rapid results from a large number of samples was a modification of the *Staphylococcus* Test specified in the B.S. Spec. 2462. The purpose of the study was to compare performance of samples against *Pseudomonas aeruginosa* in the presence of Hard Water, all culture/disinfectant mixtures contained 300 ppm of hard water made up according to the formula specified by the W.H.O.

Antiseptic Test

(i) Materials - Nutrient Broth

Nutrient Broth No. 2 is used and has the following formula:-

Lab-lemco	5 grams
Peptone	10 grams
Sodium Chloride	5 grams
Distilled water to	1000 mls

This was autoclaved for 15 minutes at 15 lbs per square inch (121°C). The final pH is approximately 7.4.

Inactivator

Nutrient Broth with the addition of 0.5% of lecithin (ex ove) and 2% Labrol W. The pH is adjusted if necessary to 7.2.

Standard Hard Water (W.H.O. Formulation)

10% $CaCl_2 6H_2O$ (wt/vol)	17.5 mls
10% Mg $SO_4 7H_2O$ (wt/vol)	5.0 mls
Distilled water to	3,300 mls

This was autoclaved at 15lbs for 15 minutes.

Bacteria

Pseudomonas aeruginosa (pyocyanea) N.C.T.C. No. 6749 maintained on nutrient agar.

(ii) Preparation of Culture

A Culture from agar of the organism is prepared in nutrient broth and incubated at 37°C for 24 hours. Daily subcultures are continued and cultures between the third and seventh generation are used for the test. The broth culture of *Pseudomonas pyocyanea* is filtered aseptically before use or is shaken, allowed to settle for 15 minutes and the supernatant liquor used for the test.

Culture dilution

One part of the broth culture was diluted with 24 parts of Standard Hard Water, *i.e.* 4 mls broth culture and 96 mls of Standard Hard Water. The average count of such a dilution would be 5×10^7 organisms per ml.

(iii) Method of Test

Serial double strength dilutions of the antiseptica were made in Standard Hard Water and 2.5 mls filled into sterile test tubes. Starting at zero time 2.5 mls of the diluted bacterial suspension is added to each Q.A.C. dilution in the series (N.B. The Culture/Disinfectant mix is so adjusted to contain 300 ppm hardness) after contact times of 5 and 10 minutes a loop full of the culture/disinfectant mix is transferred to 10 mls of the inactivator broth medium. The tubes are incubated at 37°C for 48 hours and examined for bacterial growth. The lowest concentration showing no growth in 5 minutes is recorded. The negative tubes are subcultured into nutrient broth to establish that no bacteriostasis has taken place. A standard Q.A.C. whose killing dilution is known is incorporated with every test. If a phenol coefficient is required a set of phenol dilutions in standard hard water is fitted into the test.

Hence the results can be expressed as a phenol coefficient or as the concentration giving a 100% kill in five minutes at 22°C.

E.g. A typical test on Benzalkonium Chloride B.P.C. is as follows:-

Q.A.C. Dilution	Transfer in minutes 5	10
1-666 or 1,500 ppm	-	-
1-1000 or 1,000 ppm	+	-
1-1000 or 750 ppm	+	-
1-2000 or 500 ppm	+	+
Phenol		
1-60	-	-
1-70	+	-
1-80	+	+

$$\text{Phenol Coefficient} = \frac{666}{60} = 11.1$$

Concentration giving 100% kill = 1,500 ppm.

The following results were obtained using this test in comparing the performance of B.K.C. in distilled and hard water against a range of Gram Negative Bacteria

Benzalkonium Chloride diuluted in	ORGANISM			
	Pseudomonas aeruginosa 6749	*Escherichia coli* 8196	*Bacterium proteus* 4635	*Salmonella typli* 3390
Distilled Water	100 ppm	50 ppm	50 ppm	50 ppm
Hard Water	1,500 ppm	250 ppm	250 ppm	250 ppm

It was hoped by using sequestering agents that killing dilutions approaching that obtained in distilled water could be achieved with hard water.

To sequester all the calcium in hard water containing 300 ppm calcium calculated as $CaCo_3$ one requires approximately 1.12 gms of EDTA Na_2 per litre. For a 100 ppm Solution of B.K.C. the ratio of sequestering agent to Quat is 11.2 to 1. Hence it would not be possible to make a concentrated solution of each mixture as the solubility of EDTA Na_2 is only 12.1% w/w 250°C apart from consideration, only a small amount could be included in a formulation containing a concentrated quantity of Q.A.C.

The literature was full of examples of the enhancement of the bactericidal efficiency of B.K.C. by the addition of E.D.T.A. Solutions. However, in all cases it referred to dilute solutions where a solution of the sodium salt had been added to the diluted

Q.A.C's with no reference to pH. In some cases it specified using the tetra sodium salt.

The solubilities of the three sales of E.D.T.A. are as follows:-

	SOLUBILITY	PH 2% SOLN
EDTA Na_2	12.1g/100 at 25°C	4.7
EDTA Na_3	73g/100 at 250°C	9
EDTA Na_4	Very soluble at 25°C	12

a preference for the more alkaline solutions was the more obvious choice.

Many solutions of mixtures containing Q.A.C's and sequestering agents were prepared balancing the ions to give solutions of Quaternary Ammonium Sequestrates. If we take X to represent the free E.D.T.A. ion, Na as the sodium ion and Bk the cationic radical of B.K.C., then solutions were prepared to contain $BkX Na_3$, Bk_2X, Bk_2XNA, Bk_2XNa_2, Bk_3X, BK_3XNa.

At the first, solutions were made using pure Myristyl B.K.C. powder. Tests were then repeated using commercial grade of BKC B.P. or Querton KKBC150.

The following results were obtained using the Antiseptic Test against Pseudomonas aeruginosa. The results are expressed as the killing dilution of Quaternary Cation.

MBKC \ Na	0	1	2	3
1	-	-	-	$MBkXNa_3$ pH 10.4 500
2	MBk_2X 200 pH 5.1	MBk_2XNa 400 pH 7.9	MBk_2Na_2 750 pH 11	-
3	MBk_3X 300 pH 8.2	MBk_3XNa 1000 pH 10.7	-	-
BKC Reference Sample	MBKC Cl 1000 pH 6.2	-	-	-

Repeat with Querton KKBC150

Bk \ Na	0	1	2	3
2	Bk_2X 250	-	Bk_2XNa_2 1500	-
BkCl	BkCl 1500	-	-	-

Similar results were obtained using N.T.A. but the sequestering effect is not so great as with E.D.T.A.

It was concluded that a solution containing DiBenzalkonium mono EDTAte would be a superior bactericide to B.K.C. It was found, however, that although mixtures containing sodium and chloride ions are equivalent bactericidally to the Quaternary Ammonium Sequestrates, such solutions would contain relatively high quantities of sodium chloride and this limits the concentration of solutions, the solutions separate into two layers on cooling.

The results of this work has not been exploited, Hospitals have preferred to use chlorhexidine, and the food and catering industries seem too ready to buy at price. Without an official standard which recognised that disinfectants are invariably dispensed in tap water solutions it is difficult to make a case for superior products.

7. DETERGENT SANITIZERS

It is standard practice in the Food & Catering Industry to clean and disinfect the plant at regular intervals. It has previously been stated that this can be carried out either by the two stage process of cleaning first then using a sanitizing rinse or by the one stage process using a detergent sanitizer. The Detergent Sanitizer is designed to clean and disinfect at the same time, thus making a considerable saving in labour. Detergent Sanitizers are formulated products which contain a germicide and a compatible detergent. A very popular formulation is a combination of a Quaternary Ammonium Compound and a nonionic detergent.

The most efficient nonionic detergents have a low critical micelle concentration which enables them to solubilise soil at very low concentrations. Unfortunately the bactericidal performance of Q.A.C's falls dramatically as the Critical Micelle Concentration is reached. When Q.A.C's are combined with nonionics mixed micelles are formed. When high ratios of nonionics are used the cations of the Q.A.C's are taken into the nonionic micelle and are not available for disinfection. A rule of thumb is that inhibition occurs when the ratio of nonionic to Q.A.C. exceeds 2:1. There is no need to exceed this ratio as at the recommended dilution for bactericidal activity the disinfectant bath of such a Detergent Sanitizer would be a powerful detergent. For a Q.A.C. with a good hard water tolerance a sanitizing claim could be made with 200 ppm

Anhydrous Q.A.C. and a disinfecting claim at 400ppm.

The figure of the ratio of 2:1 is based in my case of formulating and testing formulations over at least 30 years. The tests were made very much on the evidence of widely spaced dilutions. Recently I have carried out tests comparing Querton 210Cl and Querton KKBCl and have found that even at a ratio of 1:1 the performance of Querton KKBCl was affected by a nonyl phenol ethoxylate, although Querton 210Cl was only slightly affected.

The following results were obtained from a bacteriological test carried out by the technique described in the BS 3286.

Test compounds

1. Querton KKBCl (B.K.C. B.P.C)
2. 1 Part Querton KKBCl50 + 1 part Nonyl Phenol60E
3. 1 Part Querton KKBCl50 + 2 parts Nonyl Phenol60E
4. Querton 210Cl
5. 1 Part Querton 210Cl + 1 Part Nonyl Phenol60E
6. 1 Part Querton 210Cl + 2 Parts Nonyl Phenol60E

Test Conditions

5 Minutes contact with Suspension of *Escherichia coli* NCTC 8196 in 300 ppm SHW (AOAC) at 22°C

Results

Three tests are tabulated giving the % survivors for the concentration of Q.A.C. in Formulae 1 to 6.

SAMPLE	CONCENTRATION (PPM)		
	50	70	100
1	0.014 0.070 0.038	0.005 0.002 0.003	0.0002 0.0003 0.0003
2	>1.042 1.681 0.842	0.035 0.133 0.034	0.032 0.006 0.005
3	>1.042 >2.174 >1.944	>1.042 1.391 0.415	0.444 0.083 0.039
4	0.0007 0.0007 0.0007	0.0004 0.0004 0.0005	NIL NIL NIL
5	0.0008 0.0006 0.0020	0.0005 0.0003 0.0008	NIL NIL NIL
6	0.0030 0.0040 0.0080	0.0020 0.0030 0.0030	0.0001 0.0001 0.0001

The results show quite clearly that inhibition appears at the ratio of 1 part KKBCl and 1 part of nonionic, whereas it is only slight even when a ratio of 2 parts 210Cl and 2 parts of nonionic is reached.

These tests add further force to the argument in favour of the two step process over the use of a detergent sanitizer. Perhaps it makes more sense to use efforts in formulating superior sanitizing rinses rather than refining Detergent Sanitizer.

Recent experience in Food Factories indicates that it may be necessary soon to recommend more effective bactericides. The performance of the Quaternary Ammonium Sequestrates would overcome the problem of the inhibitive effect of hard water salts.

8. EXAMINATION OF 3 ISOLATES FROM FOOD FACTORIES CLAIMED TO BE RESISTANT TO STANDARD QUATERNARY AMMONIUM SALTS

Three cultures were collected from different food factories which appeared to be resistant to Quat formulations. Hitherto these formulae had given a satisfactory performance. The cultures were examined in order to get a positive identification. They were then tested against both standard Q.A.C's and sequestrates to ascertain their resistance.

Identification of Organism

The three organisms labelled A, B and C were subjected to an exhaustive number of tests in order to make a positive identification. The conclusions from these tests were as follows:-

1) Organism A - Doubtful profile, Pseudomonas cepacia
2) Organism B - Good Identification, Serratia marcescues
3) Organism C - Good Identification, Serratia marcesceus

Bactericidal Evaluations

As these organisms had proved to be resistant to Quats, it was decided to test them against some Q.A.C's Sequestrates to confirm the resistance in laboratory conditions. It was hoped that this would be possible, as to often once organisms are taken away from their normal environment they are not so virulent. Tests were also carried out against Pseudomonas aeruginosa 6749.

Tests were made using the BS 3286 in hard water solutions containing 300 ppm Calcium measured as CaC03.

Test Solutions

Querton 210C150
Querton KKBC150
Querton 210NT ate
Querton 210 EDT ate

Disinfectant Performance Tests

Procedure

Using the technique described in BS 3286. All test dilutions were prepared in 300 ppm hard water at double strength concentrations. All organisms were harvested in 300 ppm hard water from 24 hour nutrient agar slopes.

5 ml. of each culture solution was added to 5ml. of each test solution and allowed a contact time of 5 minutes. 1 ml. aliquots were inactivated in Lubrol/Lecithin and serial dilutions prepared in ringers solution. Controls were carried out in the same way. All dilutions were plated on nutrient agar and incubated at 37°C for 48 hours. Kill rates were calculated with reference to the controls.

RESULTS

KKBCl, Querton 210Cl, Querton 210NTA and Querton 210EDTA were tested at 100, 200, 500 and 1000 ppm. The KKBCl and 210Cl as active and the NTA and EDTA Salts as cation.

ORGANISM - Pseudomonas Aeruginosa NCTC 6749 (Kelsey-Sykes strain)

Conc^n	Q210	KKBCL	Q210NTA	Q210EDTA
100 ppm	97.153	96.662	99.828	99.973
200 ppm	99.264	99.823	99.999	99.999
500 ppm	99.998	99.990	100	100
1000 ppm	99.999	99.998	100	100

Results expressed as % kill control 1.6×10^8 Bact/Ml

ORGANISM A

Conc^n	Q210	KKBCL	Q210NTA	Q210EDTA
100ppm	99.999	99.981	99.999	100
200ppm	100	99.999	100	100
500ppm	100	99.999	100	100
1000ppm	100	100	100	100

Control 9.5 x 10^7 Bact/Ml

ORGANISM B

Conc^n	Q210	KKBCL	Q210NTA	Q210EDTA
100 ppm	93.250	Less than 73.0	98.325	98.287
200 ppm	99.244	Less than 73.0	99.970	99.998
500 ppm	99.975	71.892	99.996	99.999
1000ppm	99.989	85.080	99.999	99.999

Control 3.7 x 10^7 Bact/Ml

ORGANISM C

Conc^n	Q210	KKBCL	Q210NTA	Q210EDTA
100ppm	Less than 98	Less than 98	99.693	99.166
200ppm	99.966	Less than 98	99.976	99.964
500ppm	99.999	Less than 98	99.987	99.979
1000ppm	99.999	Less than 98	99.996	99.995

Control 6 x 10^8 Bact/Ml

These results confirm the suspicions expressed in the food factories. Previously only hospitals considered pseudomonads to be a concern, and most testing in food factories was confirmed to Enterobacteriaceae and Escherichia coli in particular. Hospitals had chosen Pseudomonas aeruginosa as a yardstick to measure the efficiency of a germicide, although even in hospitals it may well have outlived its usefulness. Now food factories must study the subject more closely. In these tests the sequestrates of Querton 210 do give a good performance and show a big superiority over B.K.C.

SUMMARY

Quaternary Ammonium Compounds are relatively non-toxic stable salts. Of those members of this large family of chemicals, those which are germicidal are exceptionally active against the Gram-positive bacteria. The performance against the Gram-negative bacteria is lower in the presence of hard water salts. Of the products commercially available Didecyl Dimethyl Ammonium Chloride has the best hard water tolerance. As natural waters can contain up to 360 ppm of hardness the use of Quats as germicides will always be subject to a lowering of performance by hard water salts.

The prospect of new quaternary ammonium salts being researched and registered is not a prospect because of the high cost of registering a new molecule. Therefore, the most profitable approach to overcome the problem is to optimise the performance in the presence of hard water.

Quaternary Ammonium Sequestrates and in particular E.D.T.A. Salt of Didecyl Dimethyl Ammonium Cation has been found to be as active in standard hard water as conventional B.K.C's are in distilled water.

The performance of conventional Q.A.C's is being questioned as more resistant bacteria are being found in food factories. Perhaps the time has arrived when the superior Q.A.C's should be recommended, which are known to be more effective against Gram-negative bacteria in hard water solutions.

REFERENCES

1. G. Domagk, Devt. Med Wochschr., 1935,61,829-32

2. T. D. Whittet, W. B. Hugo and G. R. Wilkinson, "Sterilisation & Disinfection," Heinemann 1965, p.190.

3. J. B. Ward and R. C. W. Berkeley, "Microbial Adhesion to Surfaces," S.C.I. 1980.

4. M. Fletcher, M. J. Latham, J. M. Lynch and P. R. Rutter, "The Characteristics of Interfaces and their Role in Microbial Attachment", S.C.I. 1980.

5. W. A. Hamilton, "Mode of Action of Membrane - Active Antibacterials", Fed. Eur. Biochem Soc. Symp 20 71-9(1970).

6. "Method for Laboratory Evaluation of Disinfectant Activity of Quaternary Ammonium Compounds", B.S.3286:1960.

7. Standard Suspension Test for Determining Disinfectant Activity by Menas of the 5-5-5 Test (Phyto Pharmacy Commission 1974).

Surfactants for Agrochemical Formulation

P. J. Mulqueen
DOW CHEMICAL COMPANY LIMITED, LETCOMBE LABORATORY, LETCOMBE REGIS, WANTAGE, OXFORDSHIRE OX12 9JT, UK

SURFACTANTS FOR AGROCHEMICAL FORMULATION

The global end-user Agrochemicals market in the years 1987 and 1988 was $20,000 million each year, representing a significant opportunity sector for surfactants.

D Karsa (in his overview of Industrial Application of Surfactants Salford 1986) indicated that the Agrochemical markets of USA, Western Europe and Japan consumed about 100,000 tonnes of surfactant, (based on 1982 data), representing only 2-3% of the total surfactant usage for these geographical areas.

If we look at the leading Agrochemical Country Markets (Table 1), USA, Western Europe and Japan account for only 78% of the total cash sales of Agrochemicals. This immediately increases the global probable consumption of surfactants by the Agrochemical market to 130,000 tonnes representing an attractive market for speciality surfactants.

Table 1

The Leading Agrochemical Country Markets
(from County Nat West WoodMac)

Country	1987 Sales ($ million)
USA	4,450
Japan	3,400
France	2,050
Soviet Union	1,075
Brazil	875
Italy	860
West Germany	855
United Kingdom	760
Canada	545
India	510
Spain	485
PRC	465
Australia	390
Hungary	320
South Korea	280
Denmark	240
Netherlands	210
Argentina	205
Indonesia	170
Mexico	155
World	20,000

What governs the choice of surfactant in a pesticide product? A considerable factor is what is generally available from the larger industries. Since the Agrochemical offtake accounts for only 2-3% of the total surfactant consumption, it has to be, to some extent dependent upon what the other major users require. We find therefore, that a large proportion of the components of surfactants used in pesticide formulation tends to be based on the same chemistry as that employed in the Detergent, Textiles, Paint and Mining sectors. Other factors such as whether based on its toxicology, a surfactant will gain approval for use by a registration authority also enter the equation. I shall discuss this point later.

During the rest of this presentation I shall review or discuss the following points :-

1 The major types of pesticide formulation currently manufactured and the types of surfactant employed therein.

2 Where the pesticide industry is going in terms of formulation technology and its impact on surfactant useage.

3 Registration problems associated with new surfactants.

1. Major Pesticide Formulation Types

Pesticides are biologically active in extremely small quantities with as little as 5g/hectare being required to produce the desired biological effect. The chemical has to be prepared in a form that is convenient to use and to spread over large areas.

Unless a pesticide reaches its target, it cannot be active. Once the pesticide reaches its target it can interact with many things as well as the target organism. Pesticide formulation science is therefore a very broad discipline because it must deal with formulation development, production and storage as well as the interaction of the pesticide with plants, insects, mammals and the environment. With the cost of introducing a new pesticide now in the tens of millions of dollars, much greater emphasis is being placed on formulation technology - how to put all the chemical in the right place at the right time, at the right price, with the minimum impact on the environment.

Clearly, if pesticides are to be used more efficiently, the actual target needs to be defined in terms of both time and space. The proportion of emitted pesticide that reaches the target, and is in a form available to a pest must be greatly increased. Definition of the target requires a knowledge of the pest, in order to determine at which stage it is most vulnerable to pesticides. These can be conveniently grouped together as biological target factors. Other factors that influence the formulation chemists choice of formulation are :-

pesticide physical properties
- melting point
- solubility characteristics
- volatility

Pesticide chemical properties
- hydrolytic stability
- thermal stability

Soil application vs. foliar application

Pesticide biological properties
- crop selectivity
- transport

Economics

Pesticide formulations are therefore chosen on a variety of parameters.

There is still one more parameter which is important - the collection of pesticide by the target and specifically the collection of sprayed pesticide droplets on targets.

Pesticide droplets are collected on insect or plant surfaces by sedimentation and impaction, the latter being particularly important for aerosol droplets (<50um). The problem of impaction efficiency can be overcome by spraying the pesticide in a high volume of diluent, in order to completely cover the crop. The variation between aerosol sprays and large volume sprays can be illustrated by the following table:

Table 2

Volume Rates of Different Crops (Litres/Ha)

	Field Crops	Trees and Bushes
High volume (HV)	>600	>1000
Medium Volume (MV)	200 - 600	500 - 1000
Low Volume (LV)	50 - 200	200 - 500
Very Low Volume (VLV)	5 - 50	50 - 200
Ultra Low Volume (ULV)	<5	<50

This potential to apply different volumes of spray liquid also gives the pesticide applicator the ability to alter the spray drop size spectrum in order to optimise collection of those droplets on different targets. A table of optimum spray droplet sizes for selected targets is shown below:

Table 3

Target	Droplet Size Range (um)
Flying insects	10 - 50
Insects on foliage	30 -50
Foliage	40 - 100
Soil (and avoidance of drift)	250 - 500

This potential of varying application targets and spray droplet sizes leads to three main types of application of pesticide:

a) pesticide applied diluted in water
b) pesticide applied diluted in oil
c) pesticide applied as a "ready-to-use" product.

In all of these uses, pesticides tend to be formulated with surfactants, giving in each case a different effect but ultimately allowing the pesticide product to be applied efficiently and evenly by the pesticide applicator.

So, what then are the major types of pesticide formulation available to a pesticide formulation chemist?

These are shown in the following table :-

Table 4

Major Types of Pesticide Formulation
Emulsifiable concentrate
Suspension concentrate
Wettable powder
Water dispersible granule
Aqueous solution
Granule
Ultra-low-volume solution
Controlled release product
Others (dusts aerosols etc.)

Surfactants are used extensively in the first five mentioned pesticide formulations. If we add to these the specific addition of surfactants, as an adjuvant or a compatibility aid, to the farmers spray tank containing a diluted pesticide then these seven uses of surfactants in current pesticide formulations can be discussed individually.

1. Emulsifiable Concentrates

An emulsifiable concentrate is a solution of pesticide(s) in solvent(s) (or without solvent if the pesticide is liquid) which will spontaneously emulsify (ie. with very primitive stirring equipment) regardless of the water hardness or temperature and form an oil-in-water emulsion which remain stable for at least two hours without any further stirring.

An important component in these formulations is the emulsifier. A pesticide dissolved in a suitable organic solvent such as xylene or cyclohexanone cannot be mixed with water, since the two liquids form separate layers. The addition of an emulsifier enables the formation of a homogeneous and stable dispersion of small globules, usually less than 10 µm in size, of the solvent in water. The small globules of suspended liquid are referred to as the disperse phase, and the liquid in which they are suspended is the continuous phase. The concentration of many emulsifiable concentrate formulations is usually 25% w/v active ingredient. One of the lowest concentrations available commercially is 8% w/v tetradifon, but manufacturers prefer to use the highest concentration possible, depending on the solubility of the pesticide in a particular solvent. Some pesticides such as carbaryl, cannot be formulated economically as an emulsifiable concentrate because the solvents in which the active ingredient is soluble are too expensive for field use.

As even minor changes of the solvent properties or of the concentration of the active ingredient will change the polarity of the organic phase and thus will call for an adjustment of the emulsifier blend, this means that each new formulation needs a tailor-made emulsifier system. In order to keep down the number of emulsifier blends the producers will supply a limited number of standard products which then will be mixed by the formulators to achieve the final adjustments.

A typical blend in most cases consists of the calcium salt of dodecyl benzene sulphonate and one or more nonionic surfactants, very often alkyl phenol alkoxylates. However, depending on the polarity of the oil-phase, a wide range of nonionic surfactants can be employed including the readily-available fatty acid ethoxylates, ethylene-oxide-propylene oxide copolymers, alcohol alkoxylates and sorbitan alkoxylates.

The stability of an emulsion is improved by a mixture of surfactants as the anionics increase in solubility at higher temperatures, whereas the reverse is true of nonionic surfactants. An unstable emulsion 'breaks' if the disperse phase separates and oil globules coalesce to form a separate layer. Creaming is due to differences in specific gravity between the two phases, and can cause uneven application. The stability of emulsion is affected by the hardness and pH of water used when mixing for spraying.

Agitation of the spray mix normally prevents creaming. Breaking of an emulsion after the spray droplets reach a target is partly due to evaporation of the continuous phase, usually water, and leaves the pesticide in a film which may readily penetrate the surface of the target.

Choice of solvent may also be influenced by its flash point so as to reduce possible risks of fire during transportation and use, especially with aerial application. For example, naphthenes are too inflammable for use as insecticide solvents.

Such liquid formulations are well liked by farmers since they empty efficiently from containers, emulsify well in water and generally can be applied without problems.

2. Suspension Concentrates

Dr Tadros reviewed the use of surfactants and polymers in pesticide suspension concentrates at Salford in 1986.

These are particulate dispersions of solid pesticide in either water or oil. Commonly called flowables or suspension concentrates these products are produced by reducing the particle size of the solid pesticide to a small size (approximately 1 µm) by a variety of dry or wet milling techniques and then dispersing the milled solid in the continuous medium of water or oil. These products are intended for further dilution, usually into water, although oil flowables may be further diluted with oil for ULV application or emulsified into water in a similar fashion to an emulsifiable concentrate.

Such a formulation will contain the pesticide and the continuous medium eq. water. It will contain a wetting agent (surfactant) to enable the water to wet the surface of the pesticide and therefore reduce the viscosity of the system. It will also contain a dispersing agent (surfactant) to keep the pesticide particles separate from each other, preventing them agglomerating (or flocculating) and subsequently settling to the bottom of the container. Such formulations will also contain (if water based) an antifreeze product like propylene glycol to keep the product liquid below 0°C. These products, because they are dispersions of solid in water (or oil) and the solid often has a density often of 1.3 g/cm^3 and above, all settle on storage. The use is therefore made of thickeners and rheological additives to prevent this tendency to settle (or clay out) on storage.

Such a product would have a typical composition,

		g/L
pesticide		600
wetting agent		10
dispersing agent		30
propylene glycol		100
thickener		2
antifoam		0,1
water	approx	450g

The main advantages of aqueous flowables are the ease of handling compared to the wettable powders from which these formulations were derived. Disadvantages include the tendency to clay (or separate) if not properly formulated, problems of stability if such a product freezes at low temperatures and the lack of utility of such a product with pesticides prone to hydrolysis.

The range of surfactants suitable for use as wetter and dispersants in the preparation of suspension concentrates is large, and closely parallels those used in paint technology including :-

EO/PO copolymers
Lignosulphonates
Phosphate esters
Fatty acid sulphonates
Naphthalene sulphonic acid-formaldehyde condensates
nonyl phenol ethoxylates
fatty alcohol ethoxylates

Also in development are polymeric surfactants which show considerable potential with pesticidal suspension concentrates.

3. Wettable Powders

These formulations sometimes called dispersible or sprayable powders, consist of finely divided pesticide particles, together with surface-active agents that enable the powder to be mixed with water to form a stable homogeneous suspension for spraying. Wettable powders frequently contain 50% active ingredient, but some contain even higher concentrations. The upper limit is usually determined by the amount of inert material such as synthetic silica, required to prevent particles of the active ingredient fusing together during processing in a hammer or fluid energy mill ('microniser'). This is influenced by the melting point of the active ingredient, but an inert filler is also needed to prevent the formulated product from caking or aggregating during storage. The amount of synthetic silica needs to be kept to a minimum as this material is very abrasive. Apart from wear on the formulating plant, the nozzle orifice on sprayers is liable to erosion, thus increasing the volume of applied pesticide. Such a product would have a typical composition of :-

	%wt	
Pesticide	76.5	
Inert Filler (hydrated silicon dioxide)	21.0	- 16.5
Wetting Agent	1.5	- 3.0
Dispersing agent	1.0	- 4.0
	100.0	100.0

Wettable powders have a high proportion of particles less than 5 μm and all the particles should pass through a 44 μm screen. Ideally, the amount of surface-active agents should be sufficient to allow the spray droplets to wet and spread over the target surface, but the particles should not be easily washed off by rain.

Wettable powders should flow easily to facilitate measuring into the mixing container. Like dusts, they have some extremely small-sized particles, so care must be taken to avoid the powder concentrate 'puffing up' into the spray operator's face. Dust can be avoided by compacting the powder into dispersible granules which disintegrate rapidly when mixed with water. Most wettable powders are white, so to avoid the risk of confusing powder from partly opened containers with foods like sugar or flour, small packs containing sufficient formulation for one knapsack sprayer load have been introduced in some developing countries.

Pesticide manufacturers are also adding dyes to distinguish a pesticide product from a food product. The wettable powder should disperse and wet easily when mixed with water and not form lumps. To ensure good mixing most powder should be pre-mixed with about 5% of the final volume of water and creamed to a thin paste. When added to the remaining water the pre-mix should disperse easily with stirring and remain suspended for a reasonable period. The surface-active or dispersing agent should prevent the particles from aggregating and settling out in the application tank. The rate of sediment in the spray tank is directly proportional to the size and density of the particles. Suspensibility is particularly important when wettable powders are used in equipment without proper agitation. Many knapsack sprayers have no agitator. Suspensibility of a wettable powder suspension is checked by keeping a sample of the suspension in an undisturbed graduated cylinder at a controlled temperature. After 30 minutes a sample is withdrawn from the bottom tenth of the graduated cylinder and analysed. The sample should contain no more than 10% of the active pesticide.

Wettable powders should retain their fluidity, dispersibility and suspensibility characteristics, even after prolonged storage. Containers should be designed so that even if wettable powder is stored in stacks, the particles are not affected by pressure and excessive heat, which may cause agglomeration. The World Health Organisation requires tests for dispersibility and suspensibility after the wettable powder concentrate has been exposed to tropical storage conditions. Poor quality wettable powders are difficult to mix and readily clog filters in the spray equipment.

Normally, wettable powder formulations are not compatible with other types of formulation, although some have been specially formulated to mix with emulsions. Mixing wettable powders with an emulsion frequently causes flocculation or sedimentation,owing to a reaction with the surface-active agents in the emulsifiable concentrate formulation. Surfactants include wetting agents such as:

alkyl benzene sulphonates,
alkyl sulphonates,
fatty alcohol ethoxylates and
nonyl phenyl ethoxylates

whilst dispersants include
lignosulphonates,
naphthalene sulphonic acid formaldehyde condensates
and alkyl naphathalene sulphonates.

Wettable powders are increasingly unacceptable owing to the difficulties of producing long term stable, efficient products and also their dustiness which creates hazards on handling. They have in part been superseded by aqueous suspension concentrates and also by the newer water dispersible granules, although they will continue to be widely used where product costs are critical.

4. Water Dispersible Granules

Water dispersible granules or dry flowables have been developed by the pesticide industry as technologically improved wettable powders, but imitating liquids in their handling characteristics. They are complex formulations since, as well as behaving as a wettable powder on dispersion into water, they also must disintegrate rapidly from their granule size (approximately 2mm) to their ultimate particle size (approximately 2 µm) in order to be as biologically active as a wettable powder and also not to block the application equipment filters and nozzles. Such a product will contain an active ingredient and wetting and dispersing surfactants (as with a wettable powder). It will also contain a binder (to bind all the particles of powder together), a disintegrant (to break granules up on addition to water), a diluent and/or sorptive carrier as appropriate and if necessary a dye.

These products have the following advantages over wettable powders and suspension concentrates :-

1. ease of handling
2. increased handling safety
3. flexibility of packaging choice
4. ease of package disposal
5. lack of pack residue
6. improved storage stability

The surfactants required would appear to parallel those commonly employed in the production of wettable powders, although undoubtedly more exotic surfactants may be found effective.

The production process for such a product may involve pan granulation, spray drying, fluidised bed agglomeration or extrusion. Since all of these can be expensive options, this manufacturing capability has been a limiting factor in their wider usage.

5. Aqueous Solution

The pre-requisite for such a formulation is that the pesticide is soluble in water. This is achieved by having a water soluble product or by derivatisaton (eg. neutralisation of a phenoxy acid with alkali).

Whilst not historically a large user of surfactants, aqueous solutions with incorporated adjuvants are increasingly being evaluated as possible finished products.

This gives the pesticide formulator more scope to increase his margins as well as better control of the biological result. Adjuvants incorporated into such systems include mainly the nonionics such as alkyl phenol ethoxylates, and alkylamine ethoxylates.

The advantage of such a formulation is that it is cheap and easily packed. Disadvantages can be low active content and poor frost stability.

6. Adjuvants

It has long been recognised that surfactants play an important role in affecting the biological performance of pesticides. However, pioneering work in the UK by the Weed Research Organisation demonstrated the dramatic improvements in efficacy to be obtained by adding considerable quantities of surfactant to the spray tank with application of foliar absorbed systemic herbicides. This led to the widespread marketing of tallow amine ethoxylate based tank additives to glyphosate formulations and a dramatic reduction in the amount of glyphosate required to produce its biological effect. At the same time, crop oil concentrates have become common place in the USA as a means of obtaining more consistent performance from pesticides.

Both pesticide producers, who do not like to see their profit potential halved at a stroke by a surfactant without their control of the project, and surfactant producers have been quick to develop their technology of adjuvants, so much so that in 1989, there will be the second International Conference on Adjuvants for Agrochemicals, held at Blacksburg, Virginia USA following the first conference in Brandon, Canada in 1986.

Whilst the majority of work with adjuvants has been carried out with herbicides, there are significant moves into the fungicide area where even distribution of pesticide over a crop is seen as an important factor.

The surfactants employed as adjuvants tend to be nonionics, such as nonylphenyl ethoxylates, fatty alcohol alkoxylates, polyamine alkoxylates and fatty amine alkoxylates either alone or formulated with mineral or vegetable oil.

The fact that surfactant choice is related to toxicant, the crop and the target pest to be controlled offers the formulation chemist, plant biologist and surfactant producer a considerable challenge.

This would appear to be potentially a very fruitful sector for interaction between pesticide and surfactant producers since a pesticide producer will charge a price for a biological effect on a hectare basis. If a surfactant mix can halve his application rate, he can do so at the outset by incorporating the surfactant into the finished product or making the tank mix recommendation. Either way, considerable surfactant usage ensues and the pesticide manufacturer does not observe any further dramatic rate reduction by farmers who add yet more 'Adjuvant' to their tanks.

7. Compatibility Agents

It should be remembered that the formulation presented for sale by a pesticide company is usually the result of detailed studies on that pesticide to optimise the biological activity with cost of production. The rates on the label should therefore be strictly adhered to as well as any directions on mixing of the product with other products.

The compatibility of pesticides is assessed in three ways :

a. physical tank mix compatibility
b. biological tank mix compatibility
c. chemical tank mix compatibility

A pesticide company will recommend a tank mix when there is no loss of biological activity of the tank mix components. This, however, can hide physical and chemical incompatibilities which could have undesirable side effects such as phytotoxic responses, different residue profiles. It is therefore essential to check out proposed tank mixes for these three aspects. The adverse consequences of tank mixing untested formulations in a field situation may be to undo all the good work of the formulation chemists in attempting to optimise the individual performance of each pesticide.

Surfactants again offer potential in this area as compatibility agents, usually exhibiting hydrotropic properties coupled with improved emulsion and dispersion characteristics, especially where the tank mix applications include liquid fertilisers.

Whilst phosphate esters have been extensively used in this area, other mixtures, sometimes based on amine oxides, amphoterics and even highly ethoxylated nonionics have also demonstrated utility.

These then are the current major uses of surfactant in pesticide formulations. What now are the new developments or trends in pesticide formulation technology which may require significant surfactant usage?

2. New Trends in Pesticide Formulation Technology

The formulation type chosen for a given pesticide will, as indicated earlier, be influenced by the biological effect required of the pesticide, farming practice and the physico-chemical properties of the pesticide. As well as these factors, three other considerations are becoming important and are beginning to affect the choice of a formulation type. These are :-

- i Operator exposure
- ii Environmentally friendly formulations
- iii Ease of pack disposal

Naturally, a pesticide will have an environmental impact. However, the studies carried out by a pesticide producer are sufficent to demonstrate a suitable risk/benefit ratio for a pesticide and allow a registration authority to give its approval for the use of that pesticide. At the same time, it is recognised that some components of formulation are less desirable than others and that some formulation types exhibit more operator exposure problems than others. Examples of these are aromatic solvents in emulsifiable concentrates which are coming under scrutiny due to their :-

a. flammability
b. toxicity both alone and in the formulation
c. package disposal problems

Wettable powders also cause problems with the toxic dust created during the loading operation to a spray tank.

The ideal product would seem to be one which is solvent free, gives no operator exposure to the operator and produces no pack disposal problem - i.e. a non-contaminated pack which can be disposed of by local burning or with household refuse.

A dry product, powder or dispersible granule, in a water-soluble sachet which can be added directly to a spray tank goes a long way to meeting these requirements and considerable work is indeed being carried out on this option. However, other options which offer part solutions are also being extensively evaluated.

These include

1 suspension emulsions
2 microemulsions
3 emulsions

One important aspect in which the European Agricultural Market differs from the rest of the world is the relatively high price achievable for pesticides. For this reason it is easier to justify special combination products where a company's formulation expertise can combine actives in an effective and novel manner. This allows a manufacturer better control of the market into which a product is sold, and, of course, the pricing policy. This has led to considerable formulation investment in mixture technology, of which suspension-emulsion technology is an example.

1. Suspension-emulsions

These consist of a mixture of three phases; an organic phase dispersed in an aqueous continuous phase, ie. an oil in water emulsion and a solid phase dispersed throughout the continuous phase. These offer a convenient means of combining the tank mix components of an emulsifiable concentrate with an aqueous suspension concentrate in a single pack product.

These systems can offer a reduction in the use of organic solvents, as long as the oil-phase is still non-crystallising, and being aqueous based tend to have no flash point problems. Surfactant choice is critical for such formulations, the newer ether sulphate surfactants showing potential as well as the polymeric surfactants.

2. Microemulsions

As microemulsions can reduce or eliminate the need for solvents, such formulations are currently receiving considerable attention. Microemulsions are generally differentiated from ordinary emulsions by their spontaneous formation, transparency, high stability, small particle size, and the limited range of concentrations over which they exist.

Microemulsions have tended to contain less active ingredients than for example an emulsifiable concentrate and more emulsifier, making them less desirable economically. They do however have important advantages, namely :-

(i) They require less organic solvents than an emulsifiable concentrate.
(ii) They can be largely water based
(iii) Improved flashpoint criteria can be met

Moreover, recent work has shown that by careful choice of surfactants, such microemulsions can also :-

a. display storage stability over a wide range of temperatures (-10°C to +55°C)

b. excellent long term dilution properties in waters of varying degrees of hardness

c. be produced utilising similar levels of surfactant to existing emulsifiable concentrate formulations, reducing the cost of a microemulsion to a more acceptable figure.

With some microemulsion systems, simple nonylphenol ethoxylate/dodecylbenzene sulphonate blends may be employed. However, improved results are often obtained using anionic/nonionic blends where the surfactant components are based on sterically large hydrophobes.

3. Emulsion or Concentrated Emulsions

Emulsions are again receiving considerable attention because of the need to produce stable emulsions for use in suspension emulsion formulations and also as a technique of producing a stable concentrate system which can be largely water based thus reducing the organic solvent contents of the formulation.

No emulsion can be expected to be completely immune to creaming and breaking and a product that has undergone these processes may give poor results. Work has been carried out producing concentrated emulsions of the mayonnaise type but the high viscosities of these products give rise to problems in handling and dilution. There is then, considerable interest in producing low-viscosity, high oil phase oil-in-water emulsions which show no change in emulsion characteristics on storage. Much of this is covered by the proprietary nature of the work and probable patent filings. However, it is probable that nonionic surfactants will play a major part in this formulation type.

Since these are the newer formulation types, what impact will they have on the mix of formulations currently sold by Agrochemical companies.

Within Europe, it is unlikely that the current product mix of emulsifiable concentrates, suspension concentrates, wettable powders and water dispersible granules will change much over the next five years, although with water soluble packaging improvements there may be a shift away from suspension concentrate to their dry product counterpart. In the USA during the same time period, I feel there will be considerable pressure to move to dry products mainly by the pesticide container disposal problems where dry products would appear to have an advantage, especially water dispersible granules.

After this time, environmental pressures may force the industry to move to dry or water based systems. This will pose many problems for optimisation of pesticide biological activity, especially for oils or low melting solids which will otherwise be formulated as emulsifiable concentrates. At this point I would envisage much more use of adjuvants, either tank mixed or built into the formulation in order to optimise biological activity. This will obviously allow and require considerable interaction between Agrochemical companies and surfactant producers, beginning right back at the invention stage of the molecule. Since it can take seven or more years to bring a new pesticide to market, I would venture to suggest this detailed interaction should be starting now.

4. Registration of Pesticides-Surfactant Considerations

Finally, I consider it is pertinent to comment on the registration procedures which impact upon surfactants. The ingredients used in agricultural formulations are divided into two classes by government regulations. The first class contains those materials which provide the biological activity of the formulation. The second includes the other, so-called 'inert' ingredients which make up the remainder of the formulation and convert the active ingredient into a usable form. In the United States, with a few exceptions, the active ingredients are subject to the requirement that a tolerance level be set and met for the allowable residue on any crop at harvest following application of pesticide. This means that the total amount of the compound and its derivatives which remain as a residue on an agricultural product is regulated. In consequence, an active ingredient must generally be separately tested and approved for each possible application.

In theory, this procedure could also be followed for an inactive ingredient, but in practice the cost would in most cases be prohibitive. In consequence, the materials used to formulate active ingredients are limited to the ones that have been exempted from this requirement by the Environmental Protection Agency. These materials are listed in the Code of Federal Regulations (CFR) Title 40, Section 180.1001 [2], commonly called the exempt list. This section consists of several subsections. Subsection (a) concerns the requirements for placing materials on the list. Subsection (b) lists a few pesticides, mostly copper salts and natural products, which are, under some circumstances, exempt from the requirement of a tolerance. It should be noted that petroleum oils are listed in this section, and they can be useful as solvents as well as insecticides. Subsection (c) contains materials which are exempt both on growing crops and on raw agricultural commodities after harvest, while those in subsection (d) are exempt only on growing crops. Subsection (e) lists the materials that can be used in formulations which are applied to animals.

In Europe, we do not currently have such a limitation. However, all new chemicals now need to be EINECS (European Inventory of Existing Commercial Chemical Substances) listed, an action dependent upon the producer. The only exemption would appear to be polymers which contain less than 2% of the initial monomer. Exemptions or approvals are obtained by discussion with the appropriate competent authority - the equivalent to the Health and Safety Executive in different countries. The toxicological test requirements are staged dependent upon the tonnage.

In admixture with pesticides as a formulation, there are currently little or no registration requirements. Some local authorities such as the UK and German pesticide registration regulatory authorities are making lists of regularly used surfactants but it is not well coordinated.

A positive aspect, however, is that all pesticide, ie. formulations, which are registered for use have been evaluated both for their toxicology and for the residue of the pesticide following field usage. The contribution by the surfactant in the product to the overall effect of the product is thus covered.

With 1992 just around the corner and discussions abounding about a unified Pesticide Registration Scheme it is probably a fair guess that pesticide formulation inerts such as surfactants will have to be regulated in a manner similar to the way it is done by the E.P.A. in the United States.

Dow in Europe is currently adopting compliance with EINECS as a condition of purchase of chemicals.

Within a local region such as Europe, then there would not appear to be a gross problem with the usage of any surfactant so long as it is EINECS listed. However, the larger Agrochemical companies are taking an increasingly global view of the situation especially with regard to new pesticides. If a new pesticide has a market in the USA and Europe, it is only sensible to formulate it in manner whereby it can be registered in all areas simultaneously. It also makes sense in terms of comparison of biological data. The EPA exempt list then becomes significant. Unfortunately many of the new exciting surfactants such as the polymeric surfactants and the nonionics based on a large hydrophobe such as polysubstituted phenols are not EPA approved. This seriously impacts on the use of such surfactant systems especially where global products are concerned. Whilst it is expensive to obtain E.P.A approval, I think it is up to the surfactant industry as a whole to decide how it wants to handle such an issue and start discussions with a representative group of major agrochemical producers on a strategy for the future in this area.

In summary, the global Agrochemical market is large with considerable usage of surfactants. The range of application is diverse, requiring a wide range of surfactants.

Continuing development of technology and new pesticides will result in a continued market for surfactants but environmental challenges may require a rationalisation of the formulation types employed and a consequent shift in the types and mixtures of surfactant currently employed.

Factors Affecting the Activation of Foliar Uptake of Agrochemicals by Surfactants

Peter J. Holloway and David Stock
DEPARTMENT OF AGRICULTURAL SCIENCES, UNIVERSITY OF BRISTOL, AFRC INSTITUTE OF ARABLE CROPS RESEARCH, LONG ASHTON RESEARCH STATION, BRISTOL, BS18 9AF, UK

Introduction

Surface active agents or surfactants are widely used as spray adjuvants in formulations of agrochemicals to improve their effectiveness following application to foliage. The inclusion of surfactants, therefore, offers considerable economic and environmental benefits to both manufacturers and farmers alike, because in practical terms the dose of a potentially toxic or expensive active ingredient can often be substantially reduced without sacrificing the biological effect required.

Surfactant adjuvants can be conveniently divided into two broad categories according to their use and primary mode of action;[1,2] more than one type of adjuvant may be included in the same formulation. Firstly, there are spray modifiers that can be added to improve the wetting/spreading properties of formulations, their activity as adjuvants being attributable mainly to their ability to lower the surface tension of the spray liquid. This type of adjuvant is probably more important for non-systemic active ingredients which invariably require complete and uniform target coverage for maximum effectiveness. Activators are the second major category of surfactant adjuvant and these are added specifically to aid the foliar absorption of systemic active ingredients and to increase their ultimate biological activity. Such effects, however, cannot be related to the intrinsic surface activity of the compounds since they usually occur only at concentrations far in excess of those required to produce maximum lowering of surface tension, the critical micelle concentration. Their activity must therefore be the net result of specific physicochemical interactions between the surfactant, the active ingredient and the target plant. These may occur in the spray droplet before evaporation, the residue eventually deposited on the leaf surface as well as in the cuticle and internal tissues of treated plants during the course of foliar uptake from the formulation.

Unlike spray modifers, the performance of which can usually be predicted from simple physical measurements before formulation, the selection of the best surfactants to use as activators is still largely a matter of trial and error in commercial practice. The choice is usually made from the biological performance of different formulations of the particular active ingredient screened in the laboratory and in the field; parallel assessments of uptake are not often carried out. At the present time it is not known precisely what physicochemical properties in a surfactant are needed in order to induce the uptake of an agrochemical compound when both are applied to a leaf surface. Reference to technical literature, for example Harvey,[3] is usually uninformative since understandably the precise composition of most commercially available activators is never disclosed. Most of them would appear to be recommended for use only in specific crop/pesticide situations. However, according to McWhorter[2] and Hellsten,[4] alcohol (A), alkylphenol (AP), sorbitan (S) and alkylamine (AM) polyoxyethylene surfactants with Hydrophile-Lipophile Balance (HLB) values ranging between 10 and 13 are most commonly employed as activator adjuvants. Such products are added to give concentrations of between 0.1 and 0.5% in the final spray mixtures; they may have already been added to the formulation concentrate by the manufacturer or they may be supplied as a separate item for addition at the time of spray application.

The main purpose of our contribution, therefore, is to review the current state of knowledge about the part that surfactants play in the process of foliar absorption of agrochemicals when the two are applied together. This will be discussed mainly in relation to physicochemical parameters and the possible modes and sites of action of surfactant activators added to aqueous formulations such as simple solutions, suspensions or emulsions. Not all of these aspects have been considered in earlier reviews of surfactant action.[1,2,5-18]

The work described herein forms part of a long-term research programme at Long Ashton Research Station (LARS) aimed at providing scientific guidelines for formulation design to aid the selection of potential adjuvants for optimising uptake of foliage-applied compounds.

Foliar Uptake Behaviour of Surfactants

A fundamental prerequisite for the elucidation of any role that surfactants might play in foliar absorption is knowledge about their behaviour and fate when they are themselves applied to foliage. Definitive information has only been provided in recent years from the use of radiolabelled compounds although the range of plants and surfactant types examined in detail so far is not comprehensive. Some supportive information has also been obtained from HPLC[19] and colorimetric[20-22] analysis of unlabelled materials but such methods generally lack sensitivity and are really suitable only for determining residues on the leaf surface.

Entry through the cuticle. Several A and AP surfactants are capable of penetrating into leaves through the cuticle. Work from the LARS laboratory[23-27] in particular has provided uptake data for a range of plant species and several ^{14}C-labelled materials including the homogeneous model compounds C12E8, C13E6 and octylphenol (OP)E7 as well as oligomeric mixtures such as nonylphenol (NP)E$\overline{5.5}$, NPE$\overline{9.5}$, OPE$\overline{9.5}$ and C18E$\overline{8.5}$, more representative of commercial surfactant products. It was found that the penetration rates and total amounts being taken up varied according to surfactant composition and plant species; there was no relationship with surfactant HLB which ranged from 10.5 to 13.5. Some results for four species and three of the surfactants are shown in Table 1. Foliar

Table 1. Uptake from 0.1% w/v aqueous solutions of three nonionic surfactants following application of 0.2 μl droplets to the adaxial leaf surface of four plant species.

Surfactant/Plant	Time after application (h)					
	2	8	16	24	48	144
C13E6						
Bean	43	80	90	93	95	94
Beet	75	92	96	92	97	96
Chickweed	34	55	55	61	69	84
Barley	65	96	93	90	89	85
OPE7						
Bean	5	12	10	22	26	53
Beet	37	69	76	73	85	88
Chickweed	40	49	43	69	61	77
Barley	18	35	47	47	48	58
C18E$\overline{8.5}$						
Bean	3	13	11	13	23	25
Beet	7	12	13	15	19	24
Chickweed	18	23	27	31	32	35
Barley	11	22	11	17	25	43

Mean uptake values (to nearest whole number) expressed as a percentage of the applied radioactivity recovered from the treatment area of leaves;[22] standard errors of these values were never greater than ± 6.

absorption was most rapid with C12E8 and C13E6, particularly so on waxy leaves like pea and rape, where > 90% of the radiolabel penetrated within 1 h after application. In comparison, the uptake of the AP surfactants tested was much slower being intermediate between that of the C12 and C13 compounds and the oligomeric C18 surfactant. The increase in hydrophobe chain length in the latter reduced uptake considerably, the value 5 - 6 d after application rarely exceeding 50% of the radioactivity applied to any of the test plants. Additional experiments[28] with NPE$\overline{5.5}$ and C18E$\overline{8.5}$ using epidermal stripping techniques confirmed that these surfactants were penetrating into the internal tissues of treated leaves. A significant proportion of the radioactivity from the latter surfactant, however, was retained in the epidermis throughout the course of uptake.

Substantial foliar entry of dilute solutions of ^{14}C-labelled OP surfactants has also been reported for barley,[29] balsam fir[30] and maize;[31] in the latter case uptake of radiolabel after 1 d was inversely related to E content in the range $\overline{40}$ to $\overline{9.5}$. Rapid absorption by field bindweed leaves was also found following droplet application of a 1% solution of AME$\overline{15}$, a weakly cationic surfactant adjuvant used for the herbicide glyphosate, ca 50% of the radioactivity applied being found in the plant tissue 6 h after treatment.[32]

However, not all classes of nonionic surfactants are absorbed well by leaves. ^{14}C-Labelled S polyethoxylates, in particular, would appear to be poorly taken up, e.g. the monolaurate E$\overline{20}$ by barley, bean and cotton[33] and the monooleate E$\overline{20}$ by bean and tobacco,[34] and johnson grass and soybean.[16] Similar behaviour was noted for a labelled high MW EO/PO polymer on bean leaves.[35] It is likely, therefore, that any activator activity exhibited by such products will have a mode of action different from that of surfactants which penetrate into leaves.

Entry through stomata. Another potential route for the entry of surfactants into leaves is by infiltration as a solution directly through the pores of stomata if they are open. Although there is conflicting evidence it is likely to occur only with surfactants that possess exceptionally high surface activity. Work by Bukovac and coworkers[36,37] demonstrated that spontaneous penetration through open stomatal pores occurs only with liquids having an equilibrium surface tension less than 30 mN/m. Few nonionic surfactants can reduce the surface tension of water to values as low as this apart from silicone polyalkyleneoxide copolymers.[38] Surface tensions approaching 20 mN/m can often be achieved by adding such surfactants to water; this usually causes spontaneous spreading of the solution on leaf surfaces accompanied by some degree of stomatal infiltration.[39-42]

Fate following foliar application. Radiotracer experiments[23-25,27] have demonstrated that nonionic surfactants usually remain in the treated area of tissue if they penetrate into leaves. Significant translocation of radiolabel (ca 10% applied dose after 5 d) from applied A and AP surfactants could only be found in barley leaves[23,27] where it occurred mainly in an acropetal direction.

Following uptake, A and AP surfactants are metabolised in the leaf tissues immediately underlying the site of application, the rate at which metabolites are formed and their composition varying considerably according to plant species.[23-25,27] Both nonpolar (esters with short chain acids) and polar (probably glycosides) metabolites may be formed but substantial degradation of the parent compound leading to loss of radiolabel does not appear to take place even 14 d after penetration. Polar metabolites of ^{14}C12E8 were the only translocated form of radioactivity detected after foliar penetration into barley.[25] More extensive conjugation and metabolism of OP surfactants, including oxidation and de-ethoxylation, was found after radioisotope feeding experiments using excised barley leaves.[29] The adjuvant properties of surfactant metabolites are unknown; presumably they are inactive. Radio-TLC profiles of the metabolism of ^{14}C13E6 by three plants are shown in Figs 1 - 3.

There was evidence, however, of degradation of the molecules of S ester polyethoxylates on the leaf surface of several species a few days after applications of the corresponding fatty acid or E labelled compounds.[16,33]

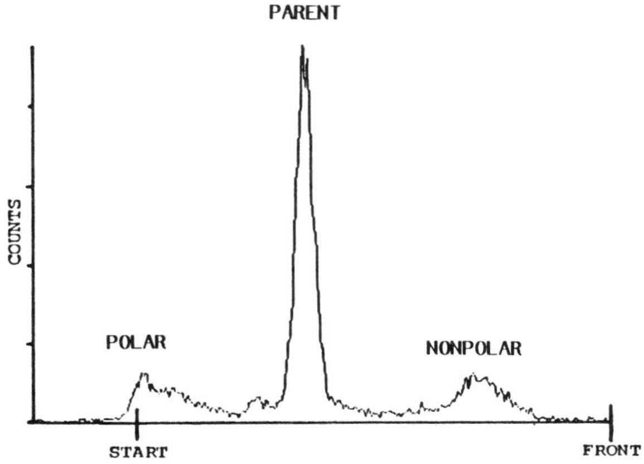

Fig. 1. Radio-TLC profile of the metabolism of ^{14}C13E6 6 d after droplet application of a 0.1% w/v aqueous solution to field bean leaves. The surfactant and its metabolites were extracted from the treated area of leaf tissue with hot methanol and analysed without further purification. Silica gel: one development in water saturated methyl ethyl ketone (MEK).

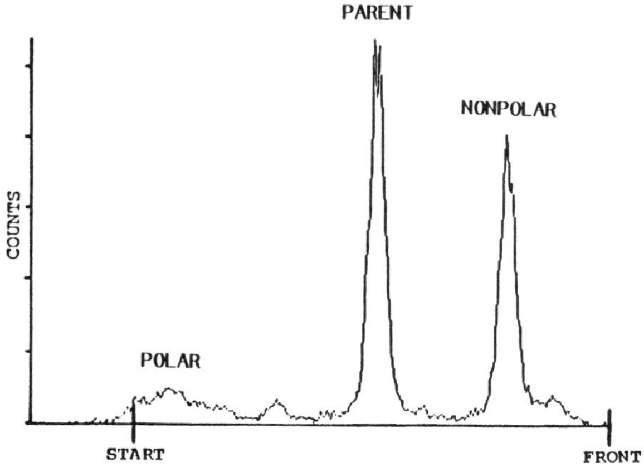

Fig. 2. Radio-TLC profile of the metabolism of ^{14}C13E6 2 d after droplet application of a 0.1% w/v aqueous solution to sugar beet leaves. Preparation and chromatography as in Fig. 1; two developments in MEK.

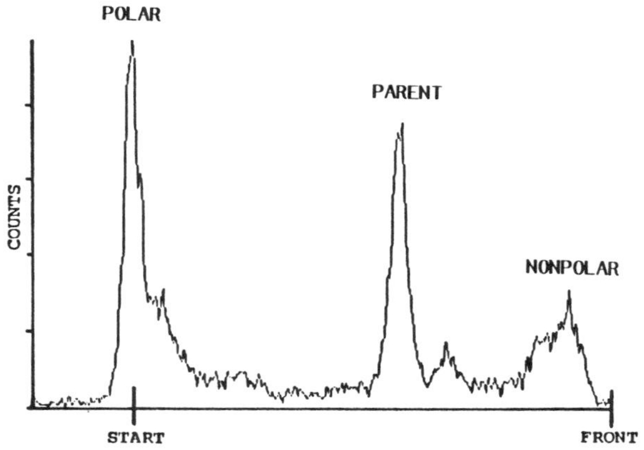

Fig. 3. Radio–TLC profile of the metabolism of ^{14}C13E6 5 d after droplet application of a 0.1% w/v aqueous solution to barley leaves. Preparation and chromatography as in Fig. 2.

Foliar Uptake Behaviour of Agrochemicals and Other Compounds in the Absence of Surfactants

The ability of a compound to penetrate into a leaf unaided will be an important factor in determining any effect that an added surfactant can have on the uptake process. Such information, therefore, is essential for quantifying such effects. Obviously if a compound is taken up well there may be no need to add an activator to the formulation; indeed addition of a surfactant may even impede foliar entry.

Organic compounds with a wide range of physicochemical properties are used as agrochemicals. In experiments with selected ^{14}C-labelled compounds Baker and coworkers[43-46] have examined the influence of such intrinsic properties on foliar absorption by a number of test species. For this study the materials were not formulated but sprayed onto leaves as solutions in aqueous acetone or methanol at 1 g/l. The work demonstrated wide variations in the abilities of plants to take up agrochemicals (Table 2), uptake into waxy leaves such as rape and strawberry usually being greater than that into less waxy ones such as sugar beet; the same penetration phenomenon has also been noted for surfactants.[25,26] Foliar absorption of the active ingredients studied, however, could not be correlated readily with fundamental properties such as octanol-water partition coefficients (P) or water solubilities.

Although polar herbicides, e.g. glyphosate and maleic hydrazide, were poorly absorbed, other less polar compounds of disparate properties, e.g. metalaxyl and diclofop-methyl, could be taken up equally well, or conversely equally poorly, e.g. simazine and bitertanol. Only after multiple regression analyses were some inverse relationships demonstrated between uptake, and melting point of the active ingredient for three species, and water solubility for two species.[45]

Table 2. Uptake from 0.1% w/v solutions of 22 agrochemicals formulated without surfactant following spray application of ca 300 μm diameter droplets to the adaxial leaf surface of four plant species.

Compound	log P	Total uptake after 24 h			
		Maize	Rape	Strawberry	Beet
CHLORMEQUAT CHLORIDE	-3.8	10	17	12	14
GLYPHOSATE ISOPROPYLAMINE	-3.7	7	16	2	6
MALEIC HYDRAZIDE	-0.9	3	5	4	4
PHENYL UREA	0.8	4	8	21	8
METALAXYL	1.6	53	82	93	14
CHLOROTOLURON	2.0	5	10	32	11
SIMAZINE	2.0	3	5	6	7
ATRAZINE	2.1	3	15	16	5
CARBARYL	2.2	8	19	19	13
ETHIRIMOL	2.2	2	5	8	3
ISOPROTURON	2.3	1	3	29	4
BUPIRIMATE	2.7	6	84	60	8
TRIADIMEFON	2.9	21	81	84	19
VINCLOZOLIN	3.0	25	55	62	39
PACLOBUTRAZOL	3.2	2	16	50	7
FENARIMOL	3.7	1	5	11	3
DICLOBUTRAZOL	3.8	9	23	15	21
FLUROXYPYR	3.8	16	53	42	29
CHLOROXURON	3.8	2	15	5	6
PROCHLORAZ	4.1	6	18	60	6
BITERTANOL	4.3	1	2	3	2
DICLOFOP-METHYL	4.6	24	87	49	18

Mean uptake values (to nearest whole number) expressed as a percentage of the applied radioactivity.[45]

On the other hand, similar work by Chamberlain et al.[47] indicated that compounds of intermediate lipophilicity (log P 1-3) were generally taken up by barley leaves better than more polar or more lipophilic ones, the uptake of neutral compounds being greater than that of weakly acidic compounds. These findings were in accordance with those previously observed for movement in the transpiration stream of barley following root uptake of the same compounds.[48]

From the information available it is clear that it is difficult to predict the uptake behaviour of an agrochemical after it is applied to a leaf surface from purely physicochemical considerations. The problem is confounded by the specific nature of the interactions between compounds and plants, and therefore generalisations cannot be made. It can be anticipated that the situation will become even more complex when surfactants are included in the equation. However, the physicochemical properties of active ingredients should not be ignored completely since they need to be examined in relation to those of surfactants as a possible factor influencing the activation potential of the latter. The dose applied to the leaf surface is yet another factor that may modify the uptake of agrochemicals.[41,49]

Foliar Uptake Behaviour of Agrochemicals and Other Compounds in the Presence of Surfactants

Although there is an extensive literature on the biological activation effects induced by mixing surfactants with agrochemicals, especially herbicides, comparatively little information is available about activation of their uptake in similar circumstances. It is often assumed that the two types of activation are coincident. In the past little attempt has been made to examine the influence of surfactant structure or composition in a very systematic manner; many uptake investigations have been conducted with only one or two surfactants in specific situations.

Extensive surveys[50-54] carried out in the USA in the 1960s using maize and soybean as test plants and sprays of water-soluble herbicides (salts of dalapon, 2,4-D, dinoseb, amitrol and paraquat) demonstrated the profound effects that addition of a wide range of different surfactant classes could have on the ultimate phytotoxic action of the compounds. Although enhancement was commonly observed, occasionally there was suppression and sometimes little difference from the activity attributable to herbicide alone. Of more importance, the work identified a number of key interacting factors, each of which could affect the level of herbicidal activation and thus probably also the concomitant absorption of the active ingredient. These were:-

(i) the concentration of surfactant added to the formulation

(ii) the chemical composition of the hydrophilic and hydrophobic portions of the surfactant molecule

(iii) the particular herbicide used

(iv) the target plant

The dominant factor was concentration with enhancement generally being found to increase progressively up to 1% w/v. An overall influence of surfactant structure on activation was not apparent since it could vary according to both herbicide and plant species. Significant differences in herbicide toxicity were most often noted with surfactant hydrophile content, an optimum E range usually being observed for maximal activation. For example, for dinoseb it was found to be $\overline{3}$–$\overline{5}$E[51] but for dalapon, amitrol and paraquat $\overline{10}$–$\overline{15}$E[54], the latter value becoming smaller at higher surfactant concentrations. Variations attributable to hydrophobe composition were observed infrequently as in the case of dinoseb where A surfactants gave good activation of its phytotoxicity whereas AP and S derived ones were ineffective.[51] The surfactant requirements for biological activation of a given herbicide on the two test plants also were not always the same. It is unfortunate that simultaneous radiotracer studies were not carried out in conjunction with any of this American work.

Similar survey work[55] with another water soluble herbicide, glyphosate, and two weeds, common milkweed and hemp dogbane, confirmed the importance of surfactant type, its E content and concentration for maximising herbicidal action of the spray. For this herbicide a general increase in uptake enhancement was found with increasing degree of ethoxylation but ethoxylated AM surfactants were by far the most effective class of activator. Subsequent experiments with ^{14}C-glyphosate using other plants[21,49,56,57] have shown that uptake of its radiolabel can be improved several-fold by adding 0.5 - 1.0% w/v AME$\overline{15}$ to the formulation. The inclusion of C13E$\overline{6}$[58] and NPE$\overline{8}$[21], on the other hand, limited foliar absorption of the labelled herbicide in barley and wheat, respectively, but NPE$\overline{9.5}$ promoted it in soybean and johnson grass.[59] Sorbitan monolaurate E$\overline{20}$ was an effective uptake activator for glyphosate in barley[58] but not in field bindweed.[57] Although the surfactant-induced enhancement of glyphosate phytotoxicity would appear to be closely related to increased uptake of the radiolabelled compound, the herbicide is systemic and translocation of the radiolabel is often markedly reduced in comparison to that observed without surfactant.[57,59]

Several other publications describe improved uptake of radiolabelled compounds in the presence of an added surfactant but few general predictions or meaningful structure-activity correlations can be made from such data. Most of the surfactants are nonionics and the compounds include deoxyglucose,[40,41] picloram,[41,60] chloramben,[61] dicamba,[62,63] difenzoquat,[26,64] aminotriazole,[65] bentazon,[66] MCPA,[67] mefluidide,[68] 2,4-D,[69,70] NAA,[71] paraquat,[62] atrazine[62] and aryloxyphenoxy-propionates.[72-75] Differences between plants and effects of different surfactant concentrations often have not been evaluated; application methods vary considerably.

The strong influence of the properties of both the compound and the plant species on the magnitude of enhancement is well illustrated by the work of Stevens and Baker[43-46] who have examined the uptake of a wide range of compounds on four species of plants (see Table 2 for listing) after spraying in admixture with an equal concentration of NPE$\bar{8}$ (1 g/l in an aqueous organic solvent formulation). However, there was no obvious relationship between NPE$\bar{8}$-induced uptake after 1 – 3 d and the physico-chemical properties of the compounds, enhancement ranging from $<$ 1- to $>$ 30-fold according to the species used. For compounds that did not penetrate readily in the absence of surfactant, greatest uptake activation with NPE$\bar{8}$ tended to occur on the waxy plants, rape and strawberry. Other investigators,[47] however, using an identical formulation system and other model compounds (log P range –1 to 4) at 1 mM, report a general enhancement effect on barley leaves, with the entry of more polar compounds being facilitated more than lipophilic ones.

Recent work with compounds other than ionised water-soluble herbicides has now confirmed that surfactant type, E content and concentration are also important for efficient uptake activation, the same factors that apply equally well to biological activation. Results from the LARS laboratory[76-78] with dilute aqueous suspensions of ^{14}C-ethirimol provide a good example of the first factor (Fig. 4). For this systemic fungicide applied to wheat leaves, NP surfactants (E$\bar{5}$–E$\overline{20}$) generally perform badly as activators when added at 0.1% to the formulation whereas A surfactants of similar E content at the same concentration give much better enhancement of radiolabel uptake. An S surfactant showed an intermediate response which was slow to develop. The same surfactant effects were also observed on acropetal movement of radiolabel (Fig. 5); these are a direct consequence of the increasing amount of ^{14}C-ethirimol being able to penetrate into the treated areas of leaf and are not specific for translocation <u>per se</u>. However, in parallel experiments on wheat leaves with ^{14}C-diclobutrazol,[77,78] another systemic fungicide, significant differences in activation performance between A and NP surfactants were not apparent.

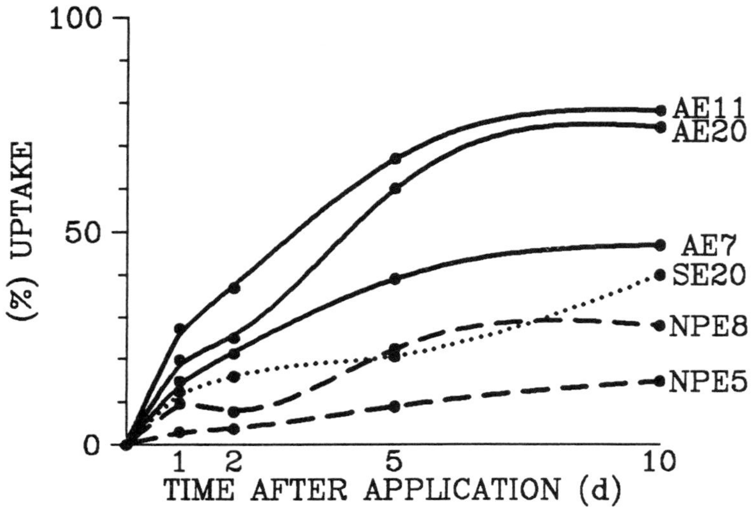

Fig. 4. Total uptake of radiolabel by wheat leaves following droplet application of six 0.05% aqueous suspensions of ^{14}C-ethirimol containing different nonionic surfactants added at 0.1% to the formulation. Values are the means of 5 replicates expressed as % of radioactivity applied; their standard errors do not exceed ±6.

The importance of the ethoxylation factor for uptake activation has been demonstrated on maize leaves with simple formulations containing ^{14}C-labelled 2-deoxyglucose, atrazine and DDT at 1 g/l and OP surfactants (E$\bar{5}$-E$\overline{40}$) at twice this concentration. Uptake after 1 d was optimal at E$\overline{16}$ for the water soluble glucose derivative but at ca E$\bar{5}$ for the two more lipophilic pesticides. Evidence for the latter, however, is less convincing because of the low levels of activation observed after this time. A requirement for low E is also reported by Steurbaut et al.[22] for effective penetration of isolate apple leaf cuticles by the fungicide propiconazole in the presence of NP polyethoxylate surfactants. Although a ca three-fold increase in penetration could be achieved with NPE$\bar{6}$ the effects were only significantly different from those of higher E at concentrations of 1%. On barley leaves,[79] an E content of $\overline{10}$ was found to be optimal for the uptake activation of three compounds, uracil, sucrose and 2,4-D, in the presence of C_{16} and C_{18} A surfactants; the dose of each compound applied was 1 μg and the ratio with surfactant was varied between 1 : 1 and 1 : 200. E content was much less important for activation of these compounds by monounsaturated C_{18} A surfactants and for that of the herbicide flamprop-methyl[19] by both OP and A surfactants in the concentration range 0.1 - 1.0% when applied to wheat leaves.

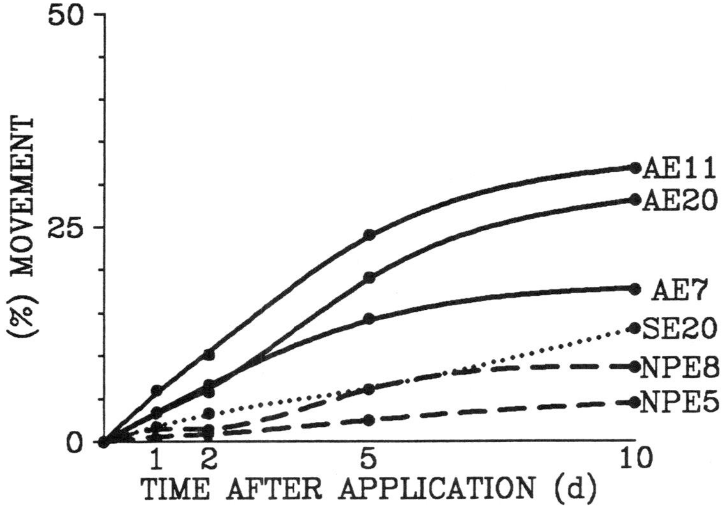

Fig. 5. Acropetal movement of radiolabel in the ^{14}C-ethirimol experiment as described in Fig. 1. Values are the means of 5 replicates expressed as % of radioactivity applied; their standard errors do not exceed ±5.

Good examples of the E effect on uptake activation from our work are shown in Figs 6 - 8, using a C_{13}/C_{14} A surfactant series ($\bar{E}6$–$\overline{E20}$). For ^{14}C-3-O-methylglucose applied at 0.05% to wheat leaves very efficient activation of radiolabel uptake could be obtained 1 d after application and this increased progressively with increasing E content (Fig. 6); such a pattern of enhancement is similar to that observed for deoxyglucose[31] on maize with a different series of surfactants and for ethirimol[76-78] on wheat with a similar series of related A polyethoxylates. The opposite effect of E, however, is found with A surfactant containing solutions of an experimental tetrazole herbicide applied to field bean leaves (Fig. 7). The same low E preference is shown in the uptake activation of diclobutrazol when the fungicide is applied to wheat leaves in suspension formulations containing either A or NP surfactants.[77,78] A different pattern of activation again is seen with the herbicide cyanazine after application to rape leaves (Fig. 8) using formulation conditions identical to those used for methylglucose and the tetrazole. Here, variations in E content appear to have much less influence on the magnitude of activation than surfactant concentration. We have observed comparable, but less marked, E effects with this particular surfactant series for all three test compounds; field bean, rape, wheat and chickweed are used routinely in our screening programme.

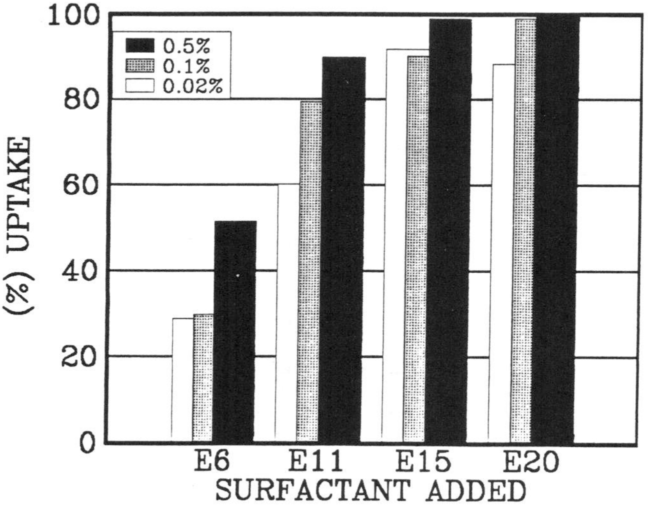

Fig. 6. Influence of E content and concentration of an A series of surfactants on uptake from a 0.05% aqueous solution of ^{14}C-3-O-methylglucose 1 d after droplet application to wheat leaves. At this time there was ca 10% uptake of radiolabel in the absence of surfactant. Values are the means of 4 replicates expressed as % of radioactivity applied; the standard error of the difference of the means for the 12 treatments is 4.5.

The uptake data illustrated in Figs 6 - 8 also reveal some important information about the effects of surfactant concentration and the requirements for effective activation. Surfactant concentration plays an important role in controlling both the rate and total amount of uptake of the compound being absorbed - this has obvious practical implications for pesticidal applications in the field. Whereas very little A surfactant is needed to induce uptake of the radiolabel from ^{14}C-methylglucose into wheat leaves (Fig. 6), much greater amounts (> 0.1%) are required to produce significant activation of the tetrazole in bean (Fig. 7) and cyanazine in rape (Fig. 8). The contrasting activation thresholds and profiles that may occur for different compounds are well illustrated by the two fungicides ethirimol (Fig. 9) and diclobutrazol (Fig. 10) on wheat.[76-78] Diclobutrazol has a high surfactant threshold requirement of *ca* 0.5% when applied as a 0.05% suspension. Although uptake of radiolabel with A surfactants added is essentially complete 1 d after application of the formulation, that with NP surfactants added is much more gradual and it continues for up to 10 d. A much lower threshold concentration of A surfactant is needed for ethirimol activation at the same concentration in a suspension and enhancement would appear to be maximal in rate and amount

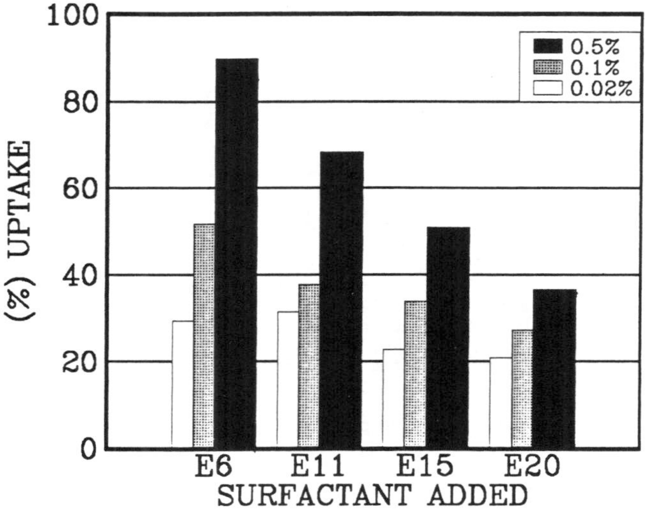

Fig. 7. Influence of E content and concentration of an A series of surfactants on uptake from a 0.05% acetone-water solution of a ^{14}C-tetrazole herbicide 1 d after droplet application to field bean leaves. At this time there was *ca* 15% uptake of radiolabel in the absence of surfactant. Values are the means of 4 replicates expressed as % of radioactivity applied; the standard error of the difference of the means for the 12 treatments is 5.9.

at 0.5%. The activation response is much more gradual than that observed with diclobutrazol using the same surfactant series reaching a maximum usually 5 d after treatment. The inferior activation performance of NP surfactants with ethirimol on wheat (Fig. 4) can be substantially improved by increasing their concentration in the formulation.

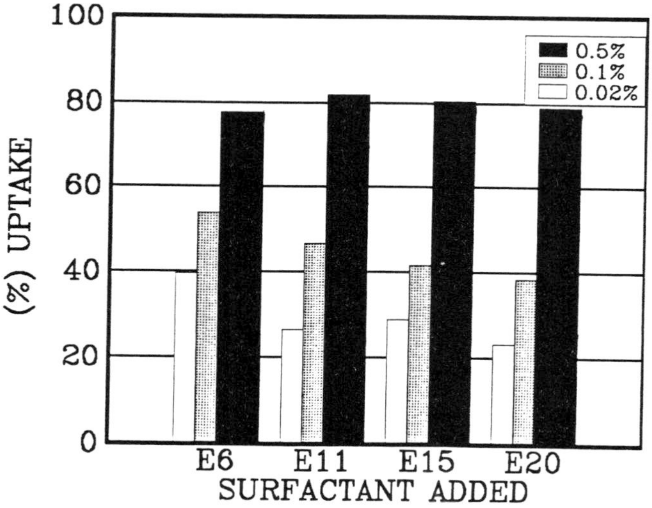

Fig. 8. Influence of E content and concentration of an A series of surfactants on uptake from a 0.05% acetone-water solution of ^{14}C-cyanazine 1 d after droplet application to rape leaves. At this time there was ca 25% uptake of radiolabel in the absence of surfactant. Values are the means of 4 replicates expressed as % of radioactivity applied; the standard error of the difference of the means for the 12 treatments is 4.1.

One more factor has been shown to play an important part in both the biological and uptake activation responses of agrochemicals to surfactants and that is environment (recent review by Wills and McWhorter[80]). Conditions that are unfavourable to the growth of the plant prior to treatment, such as low temperature, low light intensity and moisture stress, have all been shown to affect activation adversely. Temperature and relative humidity at the time of application, and thereafter, also have a considerable influence on the magnitude of the response. For example, the activation of absorption and translocation of many systemic herbicides is encouraged by high relative humidities (95 - 100%). Temperature effects on the degree of activation of these pesticides tend to be more variable but there is usually an optimum for maximal activity which differs from species to species.

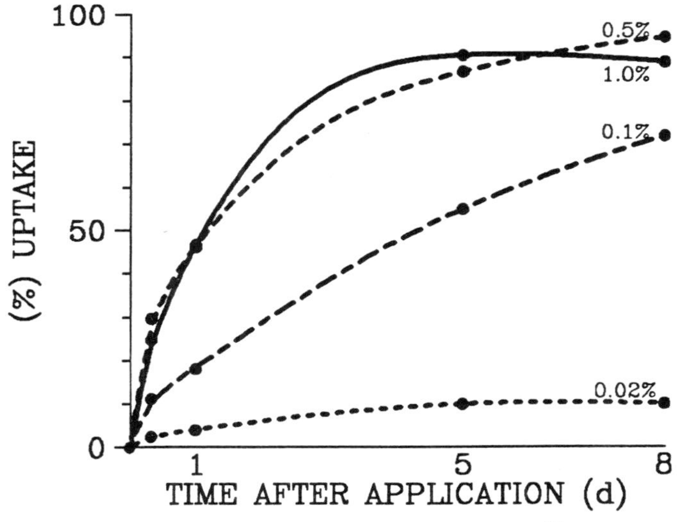

Fig. 9. Effect of increasing concentration of an AE11 surfactant on the uptake of radiolabel from a 0.05% aqueous ^{14}C-ethirimol suspension applied as droplets to wheat leaves. Values are the means of 5 replicates expressed as % of radioactivity applied; their standard errors do not exceed ±5.

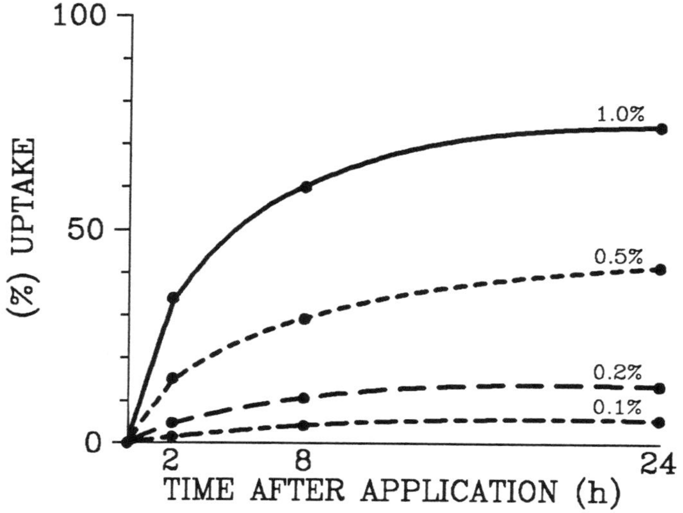

Fig. 10. Effect of increasing concentration of an $A\bar{E}7$ surfactant on the uptake of radiolabel from a 0.05% aqueous ^{14}C-diclobutrazol suspension applied as droplets to wheat leaves. With this surfactant there is little further increase in absorption of ^{14}C-fungicide beyond the 24 h period. Values are the means of 5 replicates expressed as % of radioactivity applied; their standard errors do not exceed ±6.

Sites and Mechanisms for Surfactant Activator Action

Results from radiotracer and other experiments have established what the basic patterns of foliar absorption of surfactants are likely to be (Fig. 11) and thus serve to identify potential sites of action for the promotion of foliar uptake of agrochemicals. These are:-

(i) on the leaf surface
(ii) in the external tissues, the cuticle and epidermis
(iii) within the internal tissues

Such sites will be located mainly in the immediate vicinity of the position of application. Surfactants may act at one or more of the three sites in treated leaves.

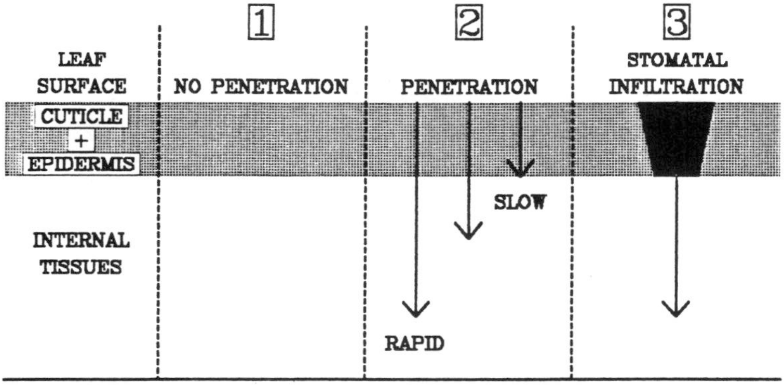

Fig. 11. Summary of the possible behaviour of an activator following its application to a leaf surface. 1. The compound is unable to enter the treated leaf. 2. The compound is able to penetrate into the treated leaf; if slow not all of the compound applied may reach the internal tissues. 3. The compound is able to enter in solution via stomatal pores.

Possible Mechanisms of Action on the Plant Surface. Adding surfactants to formulations of agrochemicals usually improves their wetting/spreading properties often leading to better coverage of the leaf surface by the deposit after evaporation of the droplets. Other beneficial effects on spray retention are also produced; these aspects are discussed elsewhere.[81,82] For A and AP polyoxyethylene surfactants, the efficiency of wetting is inversely related to their E content. Droplet evaporation rates are also inversely related to their spread areas in the presence of added surfactants.[40,83]

However, many reports in the literature indicate that whatever effects that a surfactant added to a formulation might have on coverage, or other surface factors, they can seldom be related to its observed biological or uptake activation performance. Wetting/spreading is readily quantified by measurements of contact angles, deposit areas, spread factors, etc., as well as surface tension. Uptake and spreading data for different plants and different surfactants from some LARS work (Tables 3 and 4) provide an illustration of this lack of correlation. Although a dilute aqueous solution of NPE$\bar{5}$ provides exceptionally good surface coverage on some water repellent waxy leaves (cereals, rape) the amount of droplet spread is much less on other plants of a similar nature (pea, corn marigold) and on less waxy species (field bean, beet, chickweed). Uptake of radiolabel from the ^{14}C-surfactant, however, can be high (pea) or low (chickweed) from deposits of small surface area as well as from those of large surface area (low uptake on wheat, high uptake on barley, rape) (Table 3).[28] On wheat leaves treated with an aqueous ^{14}C-ethirimol suspension, the size of the deposit area is inversely related to the surface tension of the surfactant solution added to the formulation (Table 4). Uptake of radiolabel, on the other hand, cannot easily be related to deposit area although it is generally better from formulations that spread little on the treated leaf surface.[76,78] However, for similar suspension formulations of ^{14}C-diclobutrazol effective activation of radiolabel uptake is achieved by adding low E content A or AP surfactants; both of these cause a substantial amount of droplet spreading following application of the corresponding formulation to wheat leaves.[77,78]

Although there is undoubtedly a strong physical attraction between the hydrocarbon chains present in surfactant molecules and those of the constituents of plant epicuticular waxes, there is little evidence from scanning electron microscope studies that microcrystalline wax deposits on leaf surfaces are dissolved or disrupted very much following droplet application of surfactant solutions.[25,31,32,83-87] A detergent-like action was originally thought to be the primary mode of action of surfactants in increasing the wettability of leaf surfaces. Disruptive effects have been reported only after dipping or immersing waxy leaves in surfactant solutions.[79,88]

Evaporation of droplets of agrochemical formulations containing added surfactants on any leaf surface will initially leave a deposit containing the adjuvant in a concentrated or "neat" micellar form.[18] Since surfactants in this form may vary in consistency from a viscous liquid for compounds of low E, to semi-solids and hard waxes for those of intermediate and high E, respectively, they have a considerable influence on the physical form of residues. Crystal growth and coalescence can be prevented during and after evaporation[89] to leave an amorphous deposit from an otherwise crystalline compound. Changes in the distribution of agrochemical within the deposit may also occur; in many cases the presence of surfactants result in residues being preferentially precipitated near the edge forming an annulus.[18,43,44,90] It is not clear whether such surfactant-induced changes play a key role in activating the uptake of an agrochemical. Although Baker and coworkers[43,44] could find no relationship between uptake and deposit characteristics of a variety of agrochemicals applied with and without $NPE\bar{8}$, Hess <u>et al</u>.[91] using a number of different emulsion preparations of flamprop-ethyl on cereals found that both uptake and herbicidal activity was greater with formulations that dried to form an amorphous rather than a crystalline deposit on treated leaves.

Table 3. Dry-down deposit areas and uptake values observed for 0.2 μl droplets of a 0.1% aqueous solution of NPE5.5 on ten plants.[25,26]

Plant*	Mean deposit area (mm^2)**	Mean uptake after 24 h***
Barley	8.9	59
Field Bean	1.6	38
Beet	1.3	25
Chickweed	1.8	17
Corn marigold	2.6	83
Maize	9.7	53
Pea	1.8	94
Rape	13.6	57
Wheat	10.4	16
Wild Oat	7.0	35

*Droplets applied to adaxial leaf surface.

**Area of leaf surface covered after dry-down of droplets of surfactant solution containing 0.05% w/v of added fluorescent tracer Uvitex 2B.

***Determined after a 10 droplet application of ^{14}C-labelled surfactant, expressed as % of radioactivity applied.

Equilibrium surface tension of solution 29.7 mN/m.

Table 4. Equilibrium surface tension, deposit area and uptake data for a 0.05% ^{14}C-ethirimol suspension formulation containing 0.1% of six different nonionic surfactants.[76,78]

Surfactant	Surface tension (mN/m)*	Deposit area (mm^2)**	Mean uptake of radiolabel 5d after application to wheat leaves***
AE7	28.6	3.5	29
AE11	32.8	1.5	53
AE20	41.3	1.0	53
NPE5	29.4	8.3	13
NPE8	30.0	2.7	16
SE20	34.7	0.8	27

*Determined by Wilhemy plate method.

**Determined using method described in Table 3.

***Expressed as % of radioactivity applied after application of ten 0.2 μl droplets.

If a sufficient quantity of a surfactant persists on the leaf surface after application it may facilitate uptake of the agrochemical by aiding its dissolution in the deposit. In these circumstances dissolution would almost certainly be achieved mainly by solubilisation involving incorporation of the compound into surfactant micelles.[18] The process would be assisted by any humectant activity that the surfactant possessed enabling water to be attracted to and retained in the deposit. It is generally agreed that micelles are too large to penetrate through plant cuticles but, because their life time is short (< 1 ms),[18] they would still function as effective "transport packages" releasing their dissolved contents near the sites of absorption. Micelles could reform, however, in the aqueous phases of the treated plant if a sufficient quantity of the monomer reaches the internal tissues. Solubilisation of the agrochemical could then take place also within the plant. Although considerable increases in the water solubilities of a number of pesticides have been demonstrated in vitro in the presence of concentrated surfactant solutions,[8,13,31,77,78,92] differences between the solubilising capacities of nonionic surfactants of different classes and E contents are generally small. Differences between the relative solubilisation efficiencies of surfactants for agrochemicals per se are therefore unlikely to account for any differences in their activation performance despite the fact that some degree of solubilisation is likely in the deposit.

The moisture absorption capacity of surfactants varies according to chemical class and E content, and with the ambient relative humidity (r.h.);[8,22,83,93] many nonionic compounds retain more than 50% of their own weight of water under humid conditions. Hydrophilic surfactants of high E content are the most hygroscopic. There is some evidence that surfactant adjuvants with good humectant properties are important for the activation of water soluble compounds. On bean leaves maintained under low r.h. conditions, uptake of the herbicide amitrole[65] was enhanced by adding SE$\overline{2}$0 to the formulation; activation was concentration dependent. Confirmation of the principal mode of action was obtained from parallel experiments with a known humectant, glycerol. Surfactant effects on the phytotoxicity of another herbicide maleic hydrazide[93] were also closely related to their moisture absorption capacity and these too could be duplicated by the addition of a number of highly hygroscopic organic compounds such as ethyl lactate. A similar relationship between the efficiency of uptake activation of 2-deoxyglucose on maize and the hygroscopicity of added OP surfactants has also been observed by Stevens and Bukovac.[31] In field conditions, especially at dewfall, hygroscopic surfactants in foliar deposits of agrochemicals can act as condensation nuclei[94,95] just below water vapour saturation. Rewetting then occurs and a water droplet may eventually be reformed over the original deposit. This phenomenon, therefore, could provide a further mechanism for increasing the uptake of an agrochemical.

Possible Mechanisms of Action in the Cuticle and Epidermis. It is generally agreed that foliar penetration of xenobiotics is a partition-diffusion process and, therefore, it can be considered in terms of a compartmental model described using fundamental physicochemical principles. A mathematical treatment is beyond the scope of the present article; full details are contained in the original papers.[14,19,96,97] The system has three main components (Fig. 12), the predominantly aqueous external phase on the cuticle surface, the heterogeneous middle cuticle phase having both hydrophobic (lipid) and hydrophilic components,[98-100] and the internal aqueous compartment corresponding to the cell wall phase, the plant apoplast or free space. The main factors governing the rate of penetration through such a system are:-

(i) the concentration of substance in the outside compartment
(ii) the concentration of substance in the inside compartment
(iii) the diffusion coefficient of the substance in the middle compartment
(iv) the partition coefficient for transfer of the substance from the outside to the middle compartment (deposit ---→ cuticle)
(v) the partition coefficient for transfer of the substance from the middle to the inside compartment (cuticle ---→ cell wall)

Other variables include thickness and tortuosity of the middle compartment, effective area of tissue accessible to solute, molecular radius of solute, viscosity and temperature.[14] The driving force for the whole process is the concentration gradient between the outside and inside compartments; movement between compartments is by partition. Transport mechanisms serving to deplete the innermost compartment of its solute will favour continued uptake by maintaining the concentration gradient.

Although there is no experimental evidence, a simple model of foliar uptake provides important clues for solving mechanisms of surfactant activator action. If a surfactant has penetrant properties, it could be present in some or all of the three compartments and thus would be expected to have some influence on the rate-determining steps for uptake of any agrochemical also present. Some of the possibilities for activation of the outer compartment have already been discussed. The major effects of surfactants on the other two compartments of the model are likely to be on their partition and diffusion characteristics.

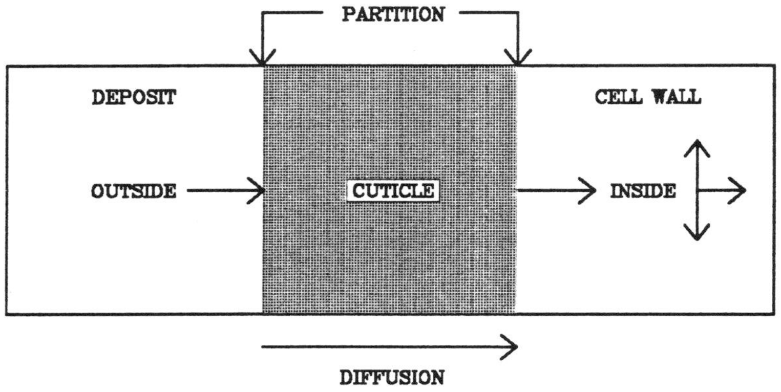

Fig. 12. Simplified compartmental partition-diffusion model for foliar uptake (adapted from refs[14,19]).

Van Valkenburg[1] has proposed that activator surfactants function primarily by altering the solubility relationships of the agrochemical during the course of its foliar penetration and uptake. Such changes arise from cosolvent effects of the adjuvant on the two phases of any partitioning system. Thus, for partitioning of an agrochemical between lipid components, such as those constituting the cuticle, and water, the modified system becomes water + surfactant/lipid + surfactant. Consequently, both the surfactant type and its concentration in each phase will be particularly important in influencing the relative proportions of agrochemical eventually distributed between the two immiscible phases. Uptake and biological activity of systemic agrochemicals is rationalised by Van Valkenburg in terms of an effective partition coefficient, there being an optimum for each compound and each plant system. The relative efficiency of activation by an adjuvant is then ascribed to its ability to produce changes of this hypothetical partition coefficient towards or away from the optimum. It is not clear from this proposal, however, what relationship, if any, these values have to conventional octanol-water coefficients for agrochemicals. For a range of lipophilic chemicals

Kerler and Schönherr[101] have found good agreement between their cuticle-water and octanol-water partition coefficients in predicting their sorption by isolated plant cuticles; the latter is effectively the middle compartment of the model system. Uptake experiments on intact leaves, on the other hand, strongly suggest that the plant and octanol coefficients are not clearly related.

Some authors[19] are of the opinion that the main mechanism for surfactant-induced activation of foliar uptake is directly on the cuticle compartment, lowering its diffusional resistance to passage of the agrochemical. To exert this mode of action the surfactant presumably must be retained within the cuticle and the amount present at the activation sites will be especially important. It is generally agreed that waxes embedded within the cuticle constitute the major barrier to its permeability; this view is supported by work on intact and dewaxed isolated cuticles (review[102]). Geyer and Schönherr[103] have also shown that the water permeability of isolated cuticles can be increased up to two-fold by application of AP surfactants (E $\bar{1}$-$\overline{10}$) and up to eight-fold with A surfactants (E $\bar{1}$-$\overline{20}$). To explain these increases the authors propose two modes of action, swelling of the cutin polymer matrix and solubilisation of cuticular waxes. There is no experimental data to support either mechanism.

Although there is little direct microscopical evidence, the results of foliar uptake studies with agrochemicals and other model compounds indicate that there may be separate routes or pathways through the cuticle for penetration of different compounds. A mainly aqueous pathway is often proposed for the facile foliar entry of certain polar water soluble substances like urea and sucrose, and cationic compounds such as paraquat, 1-methylpyridine and difenzoquat. A predominantly lipid route is proposed for the passage of other less polar compounds. This does not preclude the possibility that an agrochemical might enter by more than one route. Both of these plausible pathways could be activated, that is, made more permeable, by interaction with surfactants but their requirements for activation may not be the same. Activation of an aqueous route probably requires the presence of a more hydrophilic surfactant than that required for activation of the lipid pathway. If this was true it might provide some explanation for the differences in activator performance between surfactants of differing E content for some agrochemicals. The mechanism of activation in both pathways could be the removal of adsorption sites within the cuticle thereby increasing the diffusion rate for an agrochemical. Experiments with radiolabelled OP surfactants have already shown that they are rapidly sorbed by isolated plant cuticles at concentrations above the critical micelle concentration.[104,105] However, the degree of sorption expressed in absolute terms varied little with E content in the range $\overline{3}$–$\overline{40}$.

From the above account it is clear that the rate and amount of surfactant penetration in the presence of an agrochemical will play an important part in governing the site and extent of any interactions that might occur with the agrochemical during the course of its foliar uptake. Uptake experiments at LARS[26,77,78] with radiolabelled A and AP surfactants of medium E content added to agrochemical formulations have demonstrated that substantial quantities of surfactant still enter leaves following foliar application of such formulations. In these circumstances the ratio between the weights of compound and surfactant strongly influence the rate of penetration and the amount of surfactant taken up. Large surfactant doses in a formulation usually produce a correspondingly high uptake of the adjuvant; this may or may not be accompanied by a concomitant rise in agrochemical uptake.

Parallel studies with radiolabelled formulations containing either ^{14}C-surfactants or ^{14}C-agrochemicals[26,77,78] has provided new insights into mechanisms of uptake activation. This information is obtained by inspection of individual uptake profiles obtained for surfactant and active ingredient in such experiments. Results obtained so far have indicated two mechanisms for nonionic surfactants exhibiting penetrant properties:-

(i) copenetration, where activation is associated with closely related uptake rates for the two components
(ii) prepenetration, where activation does not occur until the surfactant has penetrated into the leaf

The efficiency of both mechanisms would appear to be concentration dependent in terms of both surfactant and agrochemical. In addition a surfactant may exhibit a different mode of interaction according to the particular plant species and agrochemical. For example, with C12E8 there was copenetration activation on field bean leaves with difenzoquat solutions[26] using an active ingredient/surfactant ratio of 5 : 1 but prepenetration activation on wheat leaves with ethirimol and diclobutrazol suspensions[77,78] at ratios varying between 1 : 2 and 1 : 5. However, for the latter two fungicides on the same plant in the presence of NP$\overline{E5}$ there was copenetration activation at an active ingredient/surfactant ratio of 1 : 5; formulations with a lower ratio were not activated because only small amounts of surfactant were able to penetrate into treated leaves.

Further work is needed to identify the sites of action in relation to the compartmental model. A prepenetration requirement could indicate that there is a requirement for the activating surfactant to be located mainly in the inner cell wall compartment.

Possible Mechanisms of Action within Internal Tissues. A popular hypothesis often used to explain the mode of action of activator surfactants in plants is that such compounds facilitate the uptake of an agrochemical by increasing the permeability of cell membranes in treated tissues. This proposal presupposes that the activator actually penetrates as far as membranes in the cell following its foliar application.

The effects of surfactants on membrane permeability and integrity are well documented for animal, bacterial and artificial systems and have established the nature of their interaction and structure-activity relationships (reviews[106-108]). For A and AP polyoxyethylene compounds a parabolic relationship between membrane activity and lipophilicity is frequently observed, the optimum for the hydrophobic moiety most often being C_{12}, that for the hydrophile being around $\overline{10}$E units. At low concentrations such surfactants penetrate into cell membranes altering their fluidity and increasing their permeability but at higher concentrations the membrane may be disrupted following solubilisation and extraction of important structural lipid and protein components. The apparent lack of biological activity of low E compounds is thought to be due to their high lipid solubility, that of high E compounds to their molecular size and high water solubility.

Although only a limited amount of permeability data is available for plant cell membranes (review[109]) it would appear to follow the general pattern found in other biological systems with regard to structure and concentration of nonionic surfactants. Such results, therefore, are poorly correlated with activation performance observed following foliar application of agrochemicals with the same surfactant, where maximal activity can occur across a wide range of hydrophobe and hydrophile compositions. *In vitro* membrane activity of nonionic surfactants, however, is more closely related to their phytotoxicity,[109] such damage being the result of disruption or death of cells at the site of application.[25,86] The rate of surfactant uptake is also implicated in this phenomenon influencing the speed of development and severity of the symptoms.[25,86] Such adverse effects clearly can limit the usefulness of some surfactants in agricultural practice especially for systemic pesticides where phytotoxicity may restrict translocation of the active ingredient.

Experiments with radiolabelled formulations of systemic agrochemicals containing A and AP compounds[26,77,78] have confirmed that few surfactant activators exert any influence inside the plant other than in the immediate zone of application on the leaf. Enhanced translocation of radiolabel from the active ingredients was not accompanied by a similar pattern of movement of radioactivity from any of the surfactants tested. Although copenetration with a surfactant activator may occur during uptake of a systemic agrochemical, the two part company inside the treated area of leaf tissue.

An agrochemical will be able to gain rapid access to internal leaf tissues if a formulation of it can infiltrate in liquid form through stomatal pores. Convincing evidence that this can provide an important mechanism for uptake activation has been obtained recently for perennial ryegrass and ^{14}C-glyphosate formulations containing an added organo-silicone surfactant.[42] Ryegrass, however, is somewhat unusual in possessing large numbers of stomata on its upper surface.

Conclusions

One of the main objectives in formulating an agrochemical is to produce a physically and chemically stable product; the basic principles and technology involved with this aspect of formulation strategy are generally well understood. Although it is recognised that the nature of the formulation can also have a profound influence on the uptake and biological activity of an agrochemical we still know very little about the precise mechanisms by which formulation produces such an effect. Surfactants play a key role in both of these aspects of formulation design. In our contribution we have considered just one of these, that of uptake activation by surfactants, and have shown it to be a complex interactive process between active ingredient, surfactant and plant which may or may not involve foliar penetration of the adjuvant. The principal factors influencing the efficiency of activation are:-

(i) the dose and physicochemical properties of the agrochemical
(ii) the dose and physicochemical properties of the surfactant
(iii) the nature of the target plant

On the basis of results from work at LARS and elsewhere on nonionic polyethoxylate surfactants we are proposing a response surface model (Fig. 13) for the uptake activation of neutral compounds. It is thought that a predictive model of this nature would offer considerable potential for optimising the performance of formulations. The model is based on the

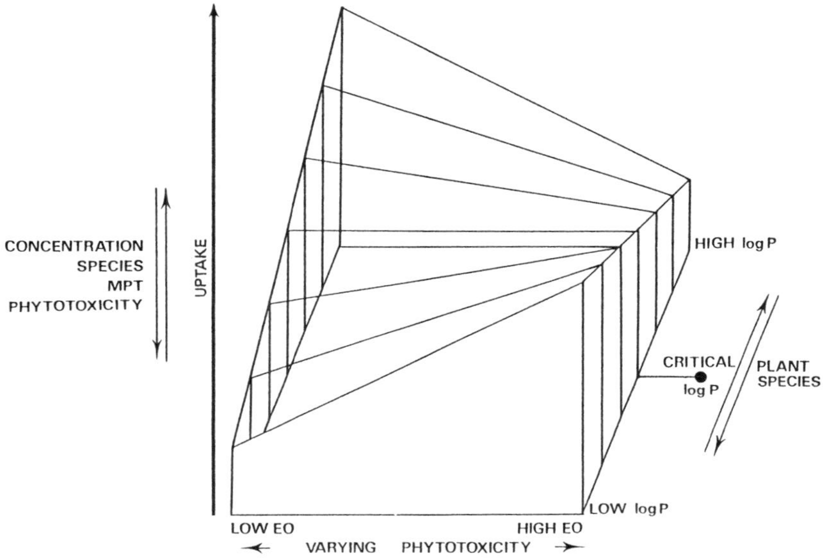

Fig. 13. Model relating the foliar uptake of neutral compounds to their log P and the interaction of this variable with the E content of polyethoxylate surfactants applied in combination with the compounds.

observation that there is a shift in optimum promoting properties from high E content for low log P compounds (Fig. 9) to low E content for high log P compounds (Fig. 10). From the model a critical log P of ca 2 is predicted for which there will be essentially no optimum E content for promotion of foliar uptake. However, it is likely that this value will vary between plant species. In addition, the effects of numerous other factors may be superimposed on the idealised model surface to alter its position in the vertical dimension, that is absolute uptake. These factors include physicochemical properties of the agrochemical such as melting point, surfactant concentration and phytotoxicity, and their relative importance may vary in different zones of the model. The use of the general term "surfactant" must be avoided in the context of activation because products may possess markedly different properties according to the particular situation.

Since the development costs for new agrochemicals are so expensive, research into spray adjuvants including surfactants is a profitable and exciting area for future development. It is possible that the performance and selectivity of many existing compounds could be improved substantially by a more rational approach to formulation design.

Acknowledgements. DS is grateful to Shell Research Limited for the award of a postgraduate studentship tenable at LARS. PJH wishes to thank ICI Agrochemicals for their financial support of the activator work on ethirimol and diclobutrazol. We are indebted to Mrs. J. Mizen for preparation of the typescript.

References

1. J.W. Van Valkenburg, in "Adjuvants for Herbicides", ed. R.H. Hodgson, Weed Science Society of America, Champaign, 1982, p.1.
2. C.G. McWhorter, in "Adjuvants for Herbicides", ed. R.H. Hodgson, Weed Science Society of America, Champaign, 1982, p.10.
3. L.T. Harvey, "A Guide to Agricultural Spray Adjuvants in the United States", Thomson Publications, Fresno, 1988.
4. M. Hellsten, in "Industrial Applications of Surfactants", ed. D.R. Karsa, Special Publication No. 59, The Royal Society of Chemistry, London, 1987, p.179.
5. J.F. Parr and A.G. Norman, Bot. Gaz. 1965, 126, 86.
6. C.L. Foy and L.W. Smith, in "Pesticidal Formulations Research, Physical and Colloidal Aspects", ed. R.F. Gould, Advances in Chemistry Series 86, American Chemical Society Publications, Washington DC, 1969, p.55.
7. K. Holly and D.J. Turner, in "Advances in Pesticide Science", ed. H. Geissbühler, IUPAC/Pergamon Press, Oxford, 1978, p.726.
8. C.E. Price, in "Herbicides and Fungicides - Factors Affecting their Activity", ed. N.R. McFarlane, Special Publication No. 29, The Chemical Society, London, 1977, p.42.
9. G.T. Merrall, in "Microbial Ecology of the Phylloplane", ed. J.P. Blakeman, Academic Press, London, 1981, p.265.
10. H.M. Hull, D.G. Davis and G.E. Stolzenberg, in "Adjuvants for Herbicides", ed. R.H. Hodgson, Weed Science Society of America, Champaign, 1982, p.26.
11. R.F. Norris, in "Adjuvants for Herbicides", ed. R.H. Hodgson, Weed Science Society of America, Champaign, 1982, p.68.

12. D. Seaman, in "Solution Behaviour of Surfactants: Theoretical and Applied Aspects", eds K.L. Mittal and E.J. Fendler, Plenum Press, New York, 1982, p.1365.
13. G.D. Wills and C.G. McWhorter, in "Pesticide Chemistry: Human Welfare and Environment", eds J. Miyamoto and P.C. Kearney, Pergamon Press, Oxford, 1983, p.289.
14. C.E. Price, in "The Plant Cuticle", eds D.F. Cutler, K.L. Alvin and C.E. Price, Linnean Society Symposium Series No. 10, Academic Press, London, 1982, p.237.
15. Y. Sugimura, K. Kawashima and T. Takeno, Shokubutsu no Kagaku Chosetu 1984, 19, 34.
16. C.G. McWhorter, in "Weed Physiology, Volume II: Herbicide Physiology", ed. S.O. Duke, CRC Press, Boca Raton, 1985, p.141.
17. R.C. Kirkwood, in "Pesticides on Plant Surfaces", ed. H.J. Cottrell, Critical Reports on Applied Chemistry, Vol. 18, Society of Chemical Industry/John Wiley, Chichester, 1987, p.1.
18. T.F. Tadros, Aspects Appl. Biol. 1987, 14, 1.
19. A.W. McCann, Ph.D. thesis, Liverpool Polytechnic, 1982.
20. N.H. Anderson and J. Girling, Pestic. Sci. 1983, 14, 399.
21. H. de Ruiter, M.A.M. Verbeek and A.J.M. Uffing, in "Pesticide Formulations, Innovations and Developments", eds B. Cross and H.B. Scher, ACS Symposium Series 371, American Chemical Society, Washington DC, 1988, p.44.
22. W. Steurbaut, G. Melkebeke and W. Dejonckheere, in "Adjuvants and Agrochemicals", eds P.N.P. Chow and C.A. Grant, CRC Press, Boca Raton, 1989, in press.
23. P.J. Holloway and D. Silcox, Proc. 1985 British Crop Protection Conference - Weeds, p.297.
24. D. Silcox and P.J. Holloway, Aspects Appl. Biol. 1986, 11, 1.
25. D. Silcox, Ph.D. thesis, University of Bristol, 1988.
26. D. Silcox and P.J. Holloway, in "Adjuvants and Agrochemicals", eds P.N.P. Chow and C.A. Grant, CRC Press, Boca Raton, 1989, in press.
27. S.M. Parker, B.Sc. project report, Huddersfield Polytechnic, 1986.
28. D. Silcox and P.J. Holloway, Aspects of Appl. Biol. 1986, 11, 19.
29. G.E. Stolzenberg, P.A. Olson, R.G. Zaylskie and E.R. Mansager, J. Agric. Fd Chem. 1982, 30, 637.
30. A.K. Garg and G.E. Stolzenberg, Proc. 12th Annual Workshop on Chemistry and Biochemistry of Herbicides, Saskatoon, 1985, p.5.
31. P.J.G. Stevens and M.J. Bukovac, Pestic. Sci. 1987, 20, 37.
32. S.L. Sherrick, H.A. Bolt and F.D. Hess, Weed Sci. 1986, 34, 811.
33. L.W. Smith and C.L. Foy, J. Agric. Fd Chem. 1966, 14, 117.

34. Y. Sugimura and T. Takeno, J. Pestic. Sci. 1985, 10, 233.
35. L.A. Norris and V.H. Freed, Res. Prog. Rept. West Weed Cont. Conf. 1962, p.92.
36. J. Schönherr and M.J. Bukovac, Plant Physiol. 1972, 49, 813.
37. D.W. Greene and M.J. Bukovac, Amer. J. Bot. 1974, 61, 100.
38. G.L.F. Schmidt, in "Industrial Applications of Surfactants", ed. D.R. Karsa, Special Publication No. 59, The Royal Society of Chemistry, London, 1987, p.24.
39. P.M. Neumann and R. Prinz, J. Sci. Fd Agric. 1974, 25, 222.
40. J.A. Zabkiewicz, D. Coupland and F. Ede, in "Pesticide Formulations, Innovations and Developments", eds B. Cross and H.B. Scher, ACS Symposium Series 371, American Chemical Society, Washington DC, 1988, p.77.
41. P.J.G. Stevens and J.A. Zabkiewicz, Proc. EWRS Symp. Factors Affecting Herbicidal Activity and Selectivity 1988, p.145.
42. R.J. Field and N.G. Bishop, Pestic. Sci. 1988, 24, 55.
43. P.J.G. Stevens and E.A. Baker, Pestic. Sci. 1987, 19, 265.
44. E.A. Baker and G.M. Hunt, in "Pesticide Formulations, Innovations and Developments", eds B. Cross and H.B. Scher, ACS Symposium Series 371, American Chemical Society, Washington DC, 1988, p.8.
45. P.J.G. Stevens, E.A. Baker and N.H. Anderson, Pestic. Sci. 1988, 24, 31.
46. E.A. Baker, unpublished results.
47. K. Chamberlain, G.G. Briggs, R.H. Bromilow, A.A. Evans and C.Q. Fang, Aspects Appl. Biol. 1987, 14, 293.
48. G.G. Briggs, R.H. Bromilow and A.A. Evans, Pestic. Sci. 1982, 13, 495.
49. C.M. Boerboom and D.L. Wyse, Weed Sci. 1988, 36, 291.
50. L.L. Jansen, W.A. Gentner and W.C. Shaw, Weeds 1961, 9, 381.
51. L.L. Jansen, J. Agric. Fd Chem. 1964, 12, 223.
52. L.L. Jansen, Weed Sci. 1965, 13, 123.
53. C.L. Foy and L.W. Smith, Weeds 1965, 13, 15.
54. L.W. Smith, C.L. Foy and D.E. Bayer, Weed Res. 1966, 6, 233.
55. J.B. Wyrill and O.C. Burnside, Weed Sci. 1977, 25, 275.
56. D.J. Turner, in "The Herbicide Glyphosate", eds E. Grossbard and D. Atkinson, Butterworths, London, 1988, p.221.
57. S.L. Sherrick, H.A. Bolt and F.D. Hess, Weed Sci. 1986, 34, 811.
58. J.T. O'Donovan, P.A. O'Sullivan and C.D. Caldwell, Weed Res. 1985, 25, 81.
59. C.G. McWhorter, T.N. Jordan and G.D. Wills, Weed Sci. 1980, 28, 113.
60. M.P. Sharma and W.H. Vanden Born, Weed Sci. 1970, 18, 57.

61. C.A. Ozair, L.J. Moshier and G.M. Werner, Weed Sci. 1987, 35, 757.
62. D.W. Jones and C.L. Foy, Weed Sci. 1972, 20, 81.
63. P.J. Petersen, L.C. Haderlie, R.H. Hoefer and R.S. McAllister, Weed Sci. 1985, 33, 717.
64. I.D. Clipsham, Aspects Appl. Biol. 1985, 9, 141.
65. G.T. Cook, A.G.T. Babiker and H.J. Duncan,Pestic. Sci. 1977, 8, 137.
66. S.M. Irons and O.C. Burnside, Weed Sci. 1982, 30, 255.
67. S.J. Midgley, Aspects Appl. Biol. 1982, 1, 193.
68. C.G. McWhorter and G.D. Wills, Weed Sci. 1978, 26, 382.
69. H.D. Coble, F.W. Slife and H.S. Butler, Weed Sci. 1970, 18, 653.
70. G. Singh and V.M. Bhan, Ind. J. Weed Sci. 1984, 16, 213.
71. N.K. Lownds, J.M. Leon and M.J. Bukovac, J. Amer. Soc. Hort. Sci. 1987, 112, 554.
72. G.D. Wills and C.G. McWhorter, Aspects Appl. Biol. 1983, 4, 283.
73. F.N. Keeney, R.L. Noveroske and T.D. Flaim, in "Pesticide Formulations, Innovations and Developments", eds B. Cross and H.B. Scher, ACS Symposium Series 371, American Chemical Society, Washington DC, 1988, p.102.
74. A. Hamburg and P.J. McCall, in "Pesticide Formulations, Innovations and Developments", eds B. Cross and H.B. Scher, ACS Symposium Series 371, American Chemical Society, Washington DC, 1988, p.56.
75. N.R. Chandrasena and G.R. Sagar, Weed Sci. 1986, 34, 676.
76. W.W.-C. Wong, B.Sc. project report, Liverpool Polytechnic, 1987.
77. H.J. Partridge, B.Sc. project report, Liverpool Polytechnic, 1988.
78. P.J. Holloway, D. Seaman and R. Perry, manuscript in preparation.
79. E.M. Hill, Ph.D. thesis, Liverpool Polytechnic, 1986.
80. G.D. Wills and C.G. McWhorter, in "Pesticide Formulations, Innovations and Developments", eds B. Cross and H.B. Scher, ACS Symposium Series 371, American Chemical Society, Washington DC, 1988, p.90.
81. N.H. Anderson and D.J. Hall, in "Adjuvants and Agrochemicals", eds P.N.P. Chow and C.A. Grant, CRC Press, Boca Raton, 1989, in press.
82. H. de Ruiter and A.J.M. Uffing, Proc. EWRS Symp. Factors Affecting Herbicidal Activity and Selectivity 1988, p.163.
83. P.J.G. Stevens and M.J. Bukovac, Pestic. Sci. 1987, 20, 19.
84. M.J. Bukovac, R.F. Whitmoyer and D.R. Reichard, HortSci. 1983, 18, 618.
85. N.K. Lownds and M.J. Bukovac, HortSci. 1983, 18, 619.
86. N.K. Lownds, Ph.D. thesis, Michigan State University, 1987.
87. N.K. Lownds and M.J. Bukovac, J. Amer. Soc. Hort. Sci. 1988, 113, 205.

88. I.J. Kuzych and W.F. Meggitt, Micron Microscop. Acta 1983, 14, 279.

89. T.F. Tadros, in "Industrial Applications of Surfactants", ed. D.R. Karsa, Special Publication No. 59, The Royal Society of Chemistry, London, 1987, p.102.

90. C.A. Hart and B.W. Young, Aspects Appl. Biol. 1987, 14, 127.

91. F.D. Hess, J.R. Goss, D.L. Buchjoltz and R.H. Falk, in "Pesticide Science and Biotechnology", Proc. Int. Congr. Pestic. Chem. 6th 1986, eds R. Greenhalgh and T.R. Roberts, Blackwell, Oxford, 1987, p.209.

92. R.E. Temple and H.W. Hilton, Weeds 1963, 11, 297.

93. K. Otsuji, J. Pestic. Sci. 1986, 11, 387.

94. B. Thaveau, R. Serpolay and S. Piekarski, Atmos. Res. 1987, 21, 83.

95. C.E. Price, Aspects Appl. Biol. 1983, 4, 157.

96. R.J. Hamilton, A.W. McCann, P.A. Sewell and G.T. Merrall, in "The Plant Cuticle", eds D.F. Cutler, K.L. Alvin and C.E. Price, Linnean Society Symposium Series No.10, Academic Press, London, 1982, p.303.

97. R.C. Bridges and J.A. Farrington, Pestic. Sci. 1974, 5, 365.

98. P.J. Holloway, in "The Plant Cuticle", eds D.F. Cutler, K.L. Alvin and C.E. Price, Linnean Society Symposium Series No. 10, Academic Press, London, 1982, p.1.

99. P.J. Holloway, in "The Plant Cuticle", eds D.F. Cutler, K.L. Alvin and C.E. Price, Linnean Society Symposium Series No. 10, Academic Press, London, 1982, p.45.

100. E.A. Baker, in "The Plant Cuticle", eds D.F. Cutler, K.L. Alvin and C.E. Price, Linnean Society Symposium Series No. 10, Academic Press, London, 1982, p.139.

101. F. Kerler and J. Schönherr, Arch. Environ. Contam. Toxicol. 1988, 17, 1.

102. A. Chamel, Physiol. Vég. 1986, 24, 491.

103. U. Geyer and J. Schönherr, in "Pesticide Formulations, Innovations and Developments", eds B. Cross and H.B. Scher, ACS Symposium Series 371, American Chemical Society, Washington DC, 1988, p.22.

104. W.E. Shafer and M.J. Bukovac, Plant Physiol. 1987, 85, 965.

105. W.E. Shafer, M.J. Bukovac and R.G. Fader, in "Adjuvants and Agrochemicals", eds P.N.P. Chow and C.A. Grant, CRC Press, Boca Raton, 1989, in press.

106. A. Helenius and K. Simons, Biochim. Biophys. Acta 1975, 415, 29.

107. D. Attwood and A.T. Florence, "Surfactant Systems, Their Chemistry, Pharmacy and Biology", Chapman and Hall, London, 1983.

108. A.T. Florence, I.G. Tucker and K.A. Walters, "Structure/Performance Relationships in Surfactants", ed. M.J. Rosen, ACS Symposium Series 253, American Chemical Society, Washington DC, 1984, p.190.

109. D. Silcox and P.J. Holloway, Aspects Appl. Biol. 1986, 11, 149.

Cationic Surfactants in Road Construction and Repair

A. D. James and D. Stewart
AKZO CHEMICALS, RESEARCH CENTRE, HOLLINGWORTH ROAD, LITTLEBOROUGH, LANCASHIRE OL15 0BA, UK

Introduction

The history of road construction in the United Kingdom is well described in the introduction to "Roadwork Theory and Practice".[1] The evolution of roadways from log-raft construction in 2500 B.C. through Roman pavements made from large paving slabs placed on broken rocks, to the development of modern highway design in the 18th and 19th centuries is charted. Roads built by John Macadam in the early 19th century were constructed of layers of small broken stones, but without binder. Later in the 19th century coal tar binders began to be used. In a typical technique the binder was hand-sprayed onto compacted aggregate with an open structure and allowed to soak in. Then the gaps in the structure were filled with smaller stones[2].

The 20th century has seen the development of separate mixing plants in which hot binder and aggregates are mixed before laying on the road. Bitumen, the residue from the vacuum distillation of crude oil, has largely replaced coal tar as a binder. With the growth in car travel has come the increasing mechanisation of road construction and the expansion of the road network. But despite the introduction of Portland cement concrete roads around the end of the 19th century, 70-80% of new highways are of "flexible" bituminous construction,[3] and 2.5 million tons of bitumen are used in road construction and repair in the UK each year.

Road Construction and Repair

A typical flexible road is built of layers of aggregate and bitumen-aggregate mixtures (see Figure 1). Large aggregate is used in the lower layers, only the upper layers normally contain bitumen. The mixture are usually made in separate 'hot-mix' plants where bitumen is heated to 140-180C until its viscosity is low enough for mixing with hot dry aggregate. These mixtures contain 3-8% binder and are transported and laid onto the roadbase while still hot. The function of the roadway is to diffuse the stresses caused by traffic so that the underlying soil is not deformed. The energy of traffic is relieved by friction between the aggregate particles and viscous effects in the binder.

Figure 1

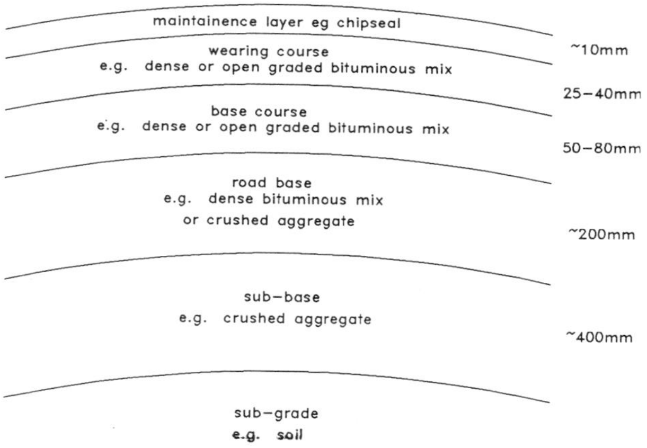

Bituminous materials are also used in the repair of both flexible and rigid (i.e. Portland cement concrete) roads. Repair is often carried out by further layers (5 cm) of hot-mix. Another common repair technique is surface dressing (chip sealing) in which liquid bitumen (hot bitumen or emulsion) is sprayed onto the old surface, then chippings (5-25 mm) are distributed over and rolled in.[4] (Figure 2.) Rolling and the action of traffic creates a mosaic of chippings bound together by bitumen which provides a good wearing surface. The process can be repeated to give thicker layers. Slurry sealing is a repair technique in which fine aggregate, water and bitumen emulsion are mixed in a mobile plant to a slurry consistency and spread over the road surface

(Figure 3), to a thickness of typically 3-15 mm. Both slurry sealing and surface dressing restore the skid resistance of the road surface and seal minor cracks and fissures in the old surface. These two techniques will be used in the paper to illustrate the applications of cationic surfactants.

Figure 2

Surface Dressing

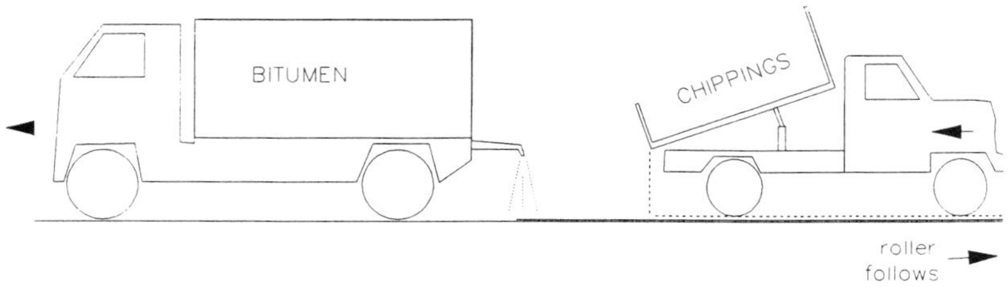

Figure 3

Slurry Sealing

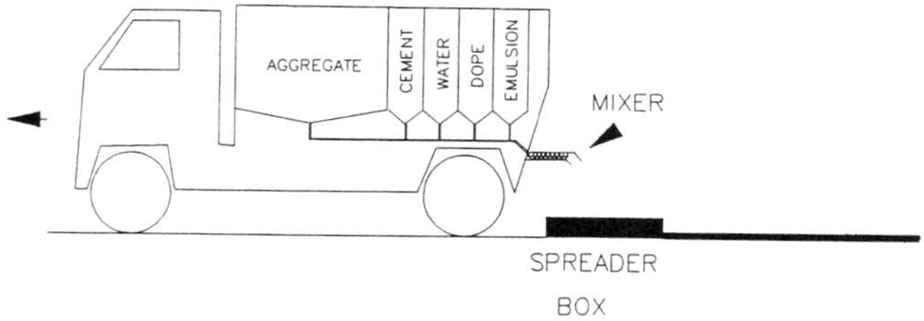

Adhesion Agents

Bitumen is essentially a non-polar hydrocarbon and does not stick well to the hydrophilic aggregates used in road construction, particularly in the presence of water. This leads to two problems : a failure of "active" adhesion - the inability of bitumen to displace water from the surface of wet stones; and a failure of "passive" adhesion - the stripping of bitumen from the surface of a coated aggregate under the influence of water.

Active adhesion is relevant to mixing processes with soft binders at relatively low temperature which may be carried out with damp aggregates, and to surface dressing in wet weather. Failure of active adhesion leads to uncoated aggregate surfaces in mixes and early loss of chips in surface dressing.[5] Failure of passive adhesion leads to loss of chips during the life of a surface dressing and loss in stability of structural layers.[6,7]

Cationic surfactants are used to solve these problems. They are known as "wetting agents", "adhesion agents" or "antistripping agents" and usually are added to the bitumen at 0.2-2%. The surfactants reduce the (advancing) contact angle of bitumen on mineral surface. For example, from 120° to 18° for bitumen on wet glass.[8,9] The bitumen-water interfacial tension is also reduced e.g. from 34 to 17 dyne/cm.

Several studies have confirmed the efficacy of cationic agents under both laboratory and field conditions and their use is specified by many national and local authorities. Passive adhesion is more easily achieved than active adhesion. Once cooled, the viscous bitumen is only displaced by water with difficulty. Also, polar components in the bitumen may slowly diffuse to the interface, strengthening the bond. 0.2-0.5% adhesion agent is usually recommended to ensure passive adhesion. Active adhesion is generally impossible without adhesion agent and recommended use levels are 0.5 - 2.0%.

Table 1: Structures of Typical Emulsifiers and Adhesion Agents

Adhesion Agents	Emulsifiers
RNH_2	
$RNHCH_2CH_2CH_2NH_2$	$RNH_2CH_2CH_2CH_2NH_32Cl$
$R - N(CH_2CH_2OH)_2$	$RNH[(CH_2CH_2O)_xCH_2CH_2OH]_2Cl$
$RCONHCH_2CH_2CH_2N(CH_3)_2$	$RCONHCH_2CH_2\dot{C}H_2NH(CH_3)_2Cl$
$RCONHCH_2CH_2NHCH_2CH_2NH_2$	$RN(CH_3)_3Cl$

$R = C_{12}-C_{20}$ alkyl

The chemical structures of typical products are shown in Table 1. The products are supplied as liquids or free-flowing flakes or pellets for easy dosing to bitumen. The most widely used solid products are based on hydrogenated tallow propylene-diamine; liquid products are often based on oleylpropylene diamine or on amidopolyamines and substituted imidazolines derived from tall acids.

<u>Degradation of Adhesion Agents</u>. A consequence of the chemistry of cationic adhesions leads to one of their main disadvantages - amines degrade slowly in hot bitumen. The rate of degradation depends on temperature, the chemical types of adhesion agent, and the binder.

<u>Figure 4</u>

Degradation of Adhesion Agents in hot Bitumen at 140C

Analysis by GC

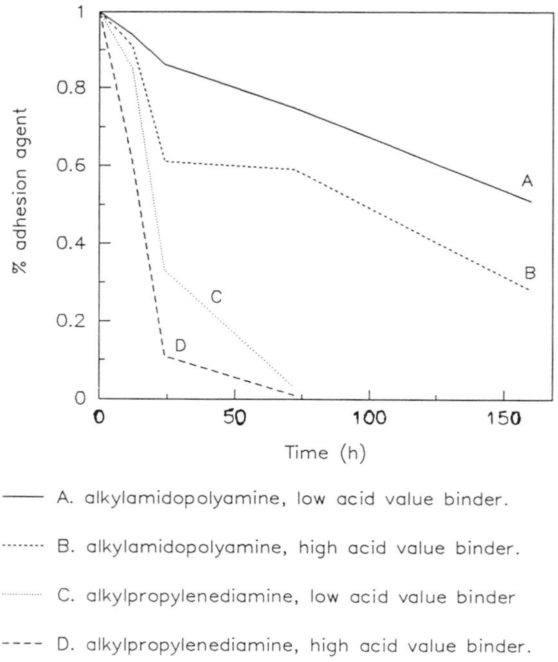

The mechanism of degradation at moderate temperatures (up to 170-180C) is the reaction of amine groups with acid groups in the binder to form amido-compounds which are less effective adhesion agents :

$$RNH(CH_2)_3NH_2 + R'COOH \rightarrow RNH(CH_2)_3NHCOR' + H_2O$$

Bitumen contains small amounts of carboxylic acid and anhydride groups which form part of macrocyclic molecules (high MW analogues of naphthenic acids). The acid value of the binder depends on its source and viscosity, values between 0.2 - 4.0 mg KOH/g are common.

Primary and secondary amine groups are lost more rapidly than tertiary amines and high acid value binders cause a more rapid loss of adhesion agent efficiency (Figure 4). As the amine is degraded, the acid value of the binder is reduced confirming reaction with acids, but not all the components contributing to binder acid value are reactive to amines.[11] At high temperatures, oxidation of adhesion agent is an important degradation route.

Degradation can be minimised by adding adhesion agent to the bitumen just before use, and by choosing a "heat-stable" agent. Another approach is to avoid the storage of adhesion agent in the binder altogether by adding the agent separately to the aggregate in water or oil solution.[12] Water-soluble agents are based on hydrochloride or acetate salts of alkyldiamine types. In surface dressing they may be sprayed onto the hot binder before chip spreading. Another advantage claimed for these agents is that they migrate to the bitumen-aggregate interface more rapidly than products added to the bitumen.

<u>Testing and Selection of Agents</u>. The most suitable adhesion agent can be selected on the basis of laboratory tests. It is important to use the bitumen and aggregate intended in practice, because the relative performance of different amine types depends on the aggregate studied. It is also important to store the treated binder for a period at the temperature expected in the field, to allow for the possibility of degradation.

Laboratory tests are summarised in Table 2. Tests may be specific to a particular application, like surface dressing, or a general measure of antistripping relevant to several applications. Many of the methods form part of national or local authority standards, and test methods have been recently reviewed.[13] Critical factors to control are binder viscosity and test temperature especially in "active" tests.

Table 2: Test Methods for Adhesion Agents

Test Method	Relevant Application	Summary of Method	Reference to Typical Procedure
Static Immersion (Passive)	hot mix, surface dressing	dry aggregate is coated with binder, immersed in hot or cold water, stored, then coverage estimated	DIN 52006 (III) ASTM D1664
Static Immersion (Active)	cold mix, road oil mixes, surface dressing	as above, but using wet aggregate	ASTM D1664, DIN 52006 (II) Method 262 SVL Norway
Dynamic Immersion (Passive)	hot mix, surface dressing	dry aggregate is coated with binder, covered with water then shaken or rolled before estimating covering	Rolling Bottle Test, VTI, Sweden
Dynamic Immersion (Active)	road oil mixes, surface dressing	as above, but using wet aggregate	Method 263, SVL Norway
Plate/Tray Tests (Active)	surface dressing	chippings are pushed through a water layer into binder. After a time, the chips are removed and examined for coverage	Ref. 31.
Plate/Tray Tests (Passive)	surface dressing	chippings are pushed or rolled into binder layer, then immersed in water before removal. In the Vialit test the stones are knocked off.	Ref. 32, Method NRBM/1. M.D.W. New Zealand Method Q212A MRD Queensland Australia
Marshall Stability or Modified Stability	hot and cold mixes	stability is determined before and after moisture conditioning optionally incl. freezing.	ASTM D1075
Immersion Wheel Tracking	hot and cold mixes surface dressing	slabs of compacted macadam are subjected to abrasion under water.	Ref. 33.

Cationic Emulsifiers

Because of the high viscosity of bitumen, mixing with aggregate is only possible at high temperatures - typically 140-180C. This means that the aggregate needs to be dried and heated which is expensive, and if overheated the bitumen can oxidise leading to a reduction in its ductility.[14] The mix also has to be kept hot during transportation to the site, which presents problems in countries with few hot-mix plants. If the mix does cool then it cannot be compacted to its design density and will not develop its full strength. Lower temperatures can be used if the bitumen is diluted with solvents like kerosene (to give "cut-back" binders) and this is the usual binder for surface dressing. But it may take several months for the solvent to leave the binder after road construction. The bitumen remains soft even after cooling, and the road may fail in hot weather.[15]

An alternative is to use bitumen emulsions. Bitumen in water emulsions can be formed with anionic, nonionic or cationic emulsifiers. Droplet sizes in typical commercial emulsions range from 1-20 microns and bitumen content from 40-70%.

Figure 5

Bitumen Deposition from Cationic Emulsions

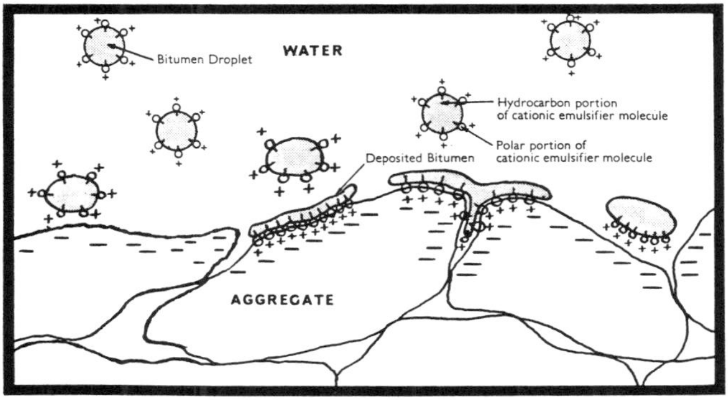

Bitumen emulsions are quite fluid at low temperatures and can be used for similar mixing processes and repair techniques as bitumen itself[16,17] except that heating binder and aggregate is unnecessary. Western Europe production of emulsions was estimated at 2.5 million tons in 1984[18]. The use of emulsions has advantages which are summarised in Table 3. The main uses of emulsions in the United Kingdom are in surface dressing, patching mixes, slurry sealing and tack coats. Total usage was 140,000t in 1986.[20] Tack coats are thin layers of bitumen sprayed between layers of hot-mix to prevent slippage.

Table 3: Advantages and Disadvantages of Emulsions

Advantages	Disadvantages
Low fire hazard compared to cut-back binders	Need (clean) water supply
Low burn hazard	Need Emulsion Quality Binders
Low dust/fume emission from mix-plants	Need Emulsion Plant and Quality Control
Low energy consumption in hot mix-plants	Limited storage stability sensitive to overheating and freezing
Built in adhesion agent not affected by storage (cationics)	Slower strength build-up
Can be used with damp aggregates	
Less binder oxidation than hot processes	
Special Techniques possible like in-place mixing, slurry seal	

Cationic emulsions have advantages over anionic and nonionic emulsions. Most aggregates including limestone, have a negative surface charge.[23] Cationic bitumen droplets react with the mineral surface leading to breaking and coalescence of the bitumen with the aggregate, (Figure 5). By correct emulsion formulation, the rate of this breaking can be controlled. The cationic emulsifier remaining in the cured binder is concentrated at the bitumen-aggregate interface and acts as an adhesion agent. Use of emulsion eliminates the of problems associated with degradation of adhesion-agents in hot bitumen. Nearly all bitumen emulsions used for road construction in W. Europe are cationic.

Emulsions are classified according to their breaking rate. "Rapid-setting" emulsions break rapidly even with aggregates of low surface area, such as the chippings used in surface dressing.

Medium-setting emulsions break sufficiently less readily so that they can be mixed with aggregates of moderate surface area; slow-setting emulsions can be mixed with aggregates of even higher surface area.

Typical emulsifier types for rapid-setting cationic emulsions are based on fatty diamines and alkylamidoamines. Emulsion formulations include acid, usually hydrochloric or acetic acid, to protonate the amines, forming cationic emulsifier :

$$RNHCH_2CH_2CH_2NH_2 \quad + \quad 2HCl \quad \rightarrow RNH_2CH_2CH_2CH_2NH_3{}^{++}2Cl^-$$

Similar emulsifiers, at higher concentration, can be used for medium-setting emulsion; quaternary amines are also effective. Slow-setting emulsions can be formulated with ethoxylated and quaternary amines. Although emulsifiers are chemically similar to adhesion agents they are rarely identical since the requirements are quite different, (Table 4).

Table 4: Requirements of Cationic Emulsifiers and Adhesion Agents

CATIONIC EMULSIFIER	ADHESION AGENT
- MUST WITHSTAND ACID WATER PHASES AT 60°C	- MUST WITHSTAND STORAGE IN HOT BINDERS AT 90-160°C.
- MUST GIVE GOOD QUALITY EMULSIONS.	
- CAN BE FORMULATED IN WATER OR WATER/ALCOHOL SOLVENTS.	- MUST NOT CONTAIN VOLATILE/HARMFUL COMPONENTS UP TO 160°C.
- USED IN FACTORY SITUATION.	- MUST BE CONVENIENT TO USE IN THE FIELD.
- SOLUBLE IN (ACID) WATER.	- BITUMEN SOLUBLE.
- MUST GIVE GOOD ANTISTRIPPING EFFECT.	- MUST GIVE GOOD ANTISTRIPPING EFFECT.

Mechanism of Breaking. Adsorption of cationic emulsifier onto the negative minerals leads to a gradual destabilistion of bitumen droplets as emulsifier is lost from the bitumen-water interface.[21] Higher emulsifier levels give slower breaking and slower deposition of bitumen on the stones.[22] Typical rapid-setting emulsions are stabilised by 0.2% emulsifier, medium-setting by 0.3-0.5%, slow-setting by 0.8-2%. Aggregates with high surface area or high cation exchange capacity cause more rapid breaking. The presence of metal ions such as Al^{3+} compete with emulsifier for adsorption sites and slow the break.

Electrophoresis of cationic bitumen droplets to the negative mineral surface is also thought to be a factor, and the breaking rate of emulsions has been related to the zeta potential of the emulsion droplets, high positive zeta potentials leading to rapid breaking.[23] A similar mechanism has been proposed for the adsorption of cationic latex particles onto negatively charged fibres.[24] Emulsions with small bitumen droplets break faster than coarse emulsions prepared with the same emulsifier concentration.[22]

Rapid pH rises occur on mixing acid emulsions with most mineral types, and this pH-shock destabilises cationic emulsions by a reduction in droplet charge as the emulsifiers are deprotonated and as acid components in the bitumen are ionised. Cement and lime are deliberately added as filler in some applications, such as slurry seal, to accelerate breaking by causing even more dramatic pH changes. The spraying of sodium hydroxide solution together with emulsion has been proposed to give very rapid breaking in surface dressing at low temperatures.[25]

Breaking of emulsion has been considered as a two-step process involving initial coagulation of bitumen droplets followed by a slower coalescence step during which all water is displaced and the final cohesive strength is developed. Agitation and compaction of the mixture accelerates breaking and development of cohesive strength. High temperature accelerates breaking.

<u>Testing emulsions and selection of emulsifiers</u>. Emulsions are tested for their storage stability and physical properties - viscosity, binder content, particle size, settlement <u>etc</u> and these form the basis of national standards (Table 5). Performance tests are used to classify emulsions according to their breaking rate, and their adhesion to aggregates. (Table 6)

The national standards are not alone sufficiently comprehensive to guarantee good performance in the field. Additional performance tests specific to particular applications are used during the development of new emulsifiers.

<u>Table 5:Test Methods for Cationic Emulsions - Physical Properties</u>

Property	Method	Reference to Typical Procedures
Particle Charge	Electrophoresis	BS434, DIN 52044, ASTM D244
Water + Binder Content	i) by Dean-Stark distillation	BS434, ASTM D244, DIN 52048
	ii) by evaporation residue	AS 1160
	iii) by Karl Fisher titration	Ref. 28
Solvent Content	i) by distillation	ASTM D244
	ii) by difference in water content and evaporation residue	AS 1160
Binder Viscosity	i) Ring and Ball on residue	DIN 52011
	ii) Penetration on residue	ASTM D244
Binder Composition	solubility in trichloro-ethylene	ASTM D244, AS 1160
Particle size	i) sieve residue	BS434, ASTM D244, DIN 52040
	ii) Coulter Counter	Ref. 29
Viscosity	Flow-cup methods	BS434, AS1160, DIN 52023, ASTM D244
Storage Stability	i) sieve residue after storage	DIN 52042
	ii) water content after storage	BS434
	iii) centrifuge test	BS434
Low Temperature Stability	i) sieve residue after cooling	BS434, DIN 52043
	ii) appearance after cooling	ASTM D244
Stability to Transport	sieve residue after shaking	Ref. 30
Settlement Rate	i) difference in water content of upper and lower layers	ASTM D244, AS 1160
	ii) Portion of bitumen free liquid on surface	Ref. 30
pH	glass/calomel electrode	Ref. 30

Table 6: Test Methods for Cationic Emulsions - Performance

Property	Method	References to Typical Procedures
Setting Rate	i) The weight of quartz powder needed to break emulsion is determined	DIN 52047, NFT 66-017
	ii) The extent of coagulation caused by added cement is determined	ASTM D244
	iii) The extent of coagulation caused by anionic surfactant is determined	ASTM D244
	iv) Ability to coat cement/sand mix	ASTM D244
	v) mixing time with aggregate	AS1160, Ref.
Coating Tests	i) mixing with coarse aggregate followed by washing	ASTM D244
Adhesion of Cured Binder	i) immersion of cured coating in water and inspection	DIN 52006, Ref. 30
	ii) modified plate test (see Table 1)	Ref. 25
Wetting Power of Tack Coats	The Emulsion must soak into quartzsand in less than 20 minutes	DIN 52046
Cohesion	i) cohesiometer for slurry seals measure effect of torque on small pad of seal	Ref. 35
	ii) Mini-fretting test for chip seals measures chip loss after rubbing	Ref. 22
Final Cohesive Strength of Mix	i) Wet-Track abrasion test for slurry seals measures aggregate loss after rubbing under water	BS 434
	ii) cohesiometer (see above) test on cured seal	ASTM D3910
	iii) Marshall Stability on cured mix	Ref. 36

The actual requirements of the various applications can be quite severe. For example, in slurry sealing the emulsion must be sufficiently stable to allow mixing with fine aggregate to form the slurry and spread over the road surface, but must then break quickly so the road can be opened to traffic. The slurry viscosity must be low enough to allow spreading over the road surface, but high enough to prevent classification of the coarse and fine components. The slurry seal technique must also be able to cope with a range of temperatures (high temperatures accelerate breaking) encountered during the year and variations during the day. One solution is to add small quantities of additional emulsifier or "dope" to the slurry on the mobile slurry sealing machine in order to delay the break in hot conditions (Figure 3).

Emulsions for surface dressing must de-stabilise in contact with aggregate of relatively low adsorption capacity, (surface area $<1m^2/g$) and so are classified rapid-setting. So they must be stabilised with relatively low concentration of emulsifier. To some extent the requirements for rapid-setting emulsions are hard to reconcile because factors favouring rapid-breaking such as low emulsifier level also have an adverse effect on storage stability and adhesion. The result is a compromise. Relatively low emulsifier levels (0.15-0.25%) are used, depending on the storage stability required. In the UK rapid-setting emulsions for surface dressing are used within a day or so of production with only 0.15% emulsifier and emulsions are sprayed hot to accelerate breaking. In Australia, because of the time needed to transport emulsions to the site, 0.25-0.3% emulsifier is not unusual. To meet the requirements of storage-stability and a small bitumen particle size a particularly efficient emulsifier is required. It is also an advantage that the emulsifier is pH-sensitive so that the pH rise which generally occurs on mixing with aggregate will cause breaking. An emulsifier which gives high droplet zeta potential is preferred. N-alkylpropylenediamines are the most common emulsifier types used for rapid-setting emulsions (Table 7).

By a similar argument, slurry sealing needs slow-setting emulsion which can contain a large amount of emulsifier, because of the high adsorptive capacity of the aggregates with which they are mixed (5-10 m^2/g). The emulsifier need not be particularly efficient, provided there is sufficient present in excess over the amount needed to give a storage-stable emulsion. The emulsifier need not be pH-sensitive, and quaternary amines are widely used, at a level of 0.6 - 1.2% (Table 7).

Table 7: Typical Emulsion Formulations

1) Rapid Setting Emulsion for Surface Dressing			2) Slow Setting Emulsion for Slurry Sealing		
bitumen	:	67%	bitumen	:	60%
kerosene	:	3%	quaternary amine	:	0.8%
tallow diamine	:	0.2%	$CaCl_2$	:	0.1%
HCl (33%)	:	0.15%	water	to	100%
$CaCl_2$	:	0.10%			
water	to	100%			

Future Uses of Cationic Surfactants

The increasing sophistication and quality control of road construction techniques and the demands of higher traffic density has focussed attention on the bitumen, and its shortcomings as a binder. For example, the last 10 years has seen a tremendous increase in the use of polymer-modified binders designed to improve the rheological properties of the bitumen.

Now the perceived shortcomings of bitumen in emulsion applications are being studied, and cationic surfactant bitumen additives are being promoted. It is clear in the emulsion manufacturing industry that bitumens vary in the details of their composition, and that different bitumens give emulsions with different properties. These may include differences in viscosity or particle size but can also be more subtle effects concerning the breaking and curing properties of the emulsions.[26]

These differences may arise from the composition of the original crude oil sources from which the bitumen were derived or the details of the refining and processing of the bitumens. Some differences can be detected by chemical analyses of the binders. However these analyses have not yet been exactly correlated to emulsion performance.

Cationic surfactants added to bitumen at 0.2-0.5% have been shown to reduce emulsion particle size, (Figure 7) and increase emulsion viscosity, without any adverse effect on the breaking behaviour of the emulsions.[27] Performance in the final application can also be improved. For example, the ability of tack coat emulsions to wet and penetrate dusty surfaces (as measured by DIN 52046) is improved (Figure 8).

Figure 7

Effect of Cationic Surfactant Additive on Emulsion Particle Size (Tack coat emulsion 40% bitumen)

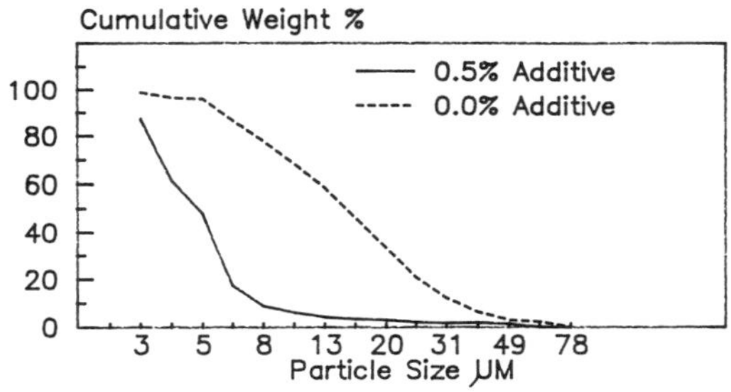

Figure 8

Effect of Cationic Bitumen Additive on the Ability of Tack Coat Emulsion to wet quartz powder. (DIN 52046)

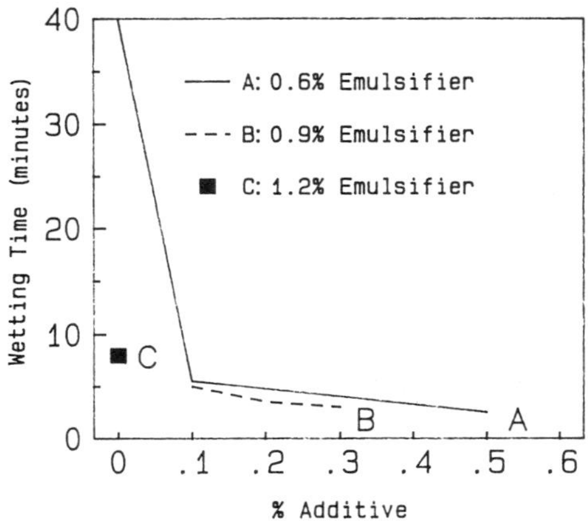

The trend towards cationic emulsion systems on performance, environmental and health and safety grounds, and new applications as bitumen improvement additives will lead to a continued growth in the market for cationic surfactants in road construction and repair.

References

1. A. Wignall, P.S. Kendrick and R. Ancill "Roadwork" 2nd Edition, Heinemann, Oxford 1988 Chapter 1.
2. E.J. Dickinson, "Bituminous Roads in Australia" Australian Road Research Board, Victoria 1984. p.2.
3. M. Heslop, Program and Papers, Bitumen Emulsion Workshops, Melbourne and Sydney, October 1985, 63-64.
4. "Recommendations for Road Surface Dressing", Road Note 39, HMSO 1972.
5. G.L. Dalton and H. Karins, J. Institute Petroleum, 1964, 50, (481), 1-14.
6. Sabry Shihata and Omar Aburizaiza, Proceedings of the 3rd IRF M. East Regional Meeting, Volume 1, 1988, 597-621.
7. Finn Christensen, Stripping. A Problem in Road Construction?, paper presented to S.C.I., Institution of Highway Engineers and Institution of Asphalt Technologists, November 17, 1977, London.
8. Ilan Ishai and Joseph Craus, "Asphalt Paving Technology 1977" Volume 46, Association of Asphalt Paving Technologists pp. 228-258.
9. Stuart D. Cameron, "The Performance of Surface Active Cationic Agents in Improving Adhesion between Road Binders and Aggregates in the presence of water", Thesis, University of Strathclyde, June 1978.
10. D.H. Matthews, J. Inst. Petroleum, 1958, 44, 423-32.
11. A.D. James, R. Senior and D. Stewart, "Proceedings of the 3rd IRF M. East Regional Meeting", Volume 5, 1988, 193-214.
12. J.N. Dybalski, "Cationic Surfactants in Asphalt Adhesion", presented to the Association of Asphalt Paving Technologists, Annual Meeting, Kansas City Missouri 1982.
13. A.M. Ajour "Le Probleme de L'adhesivite liants Hydrocabones-Granulats" RILEM, Cahier 17 BM No.3, Laboratoire des Ponts et Chaussees, Paris 1979.
14. V. Potschka, "3rd Eurobitume Symposium", Volume 1, 1985, 38-40.
15. J.E. Patrick, "Proceedings of the New Zealand Roading Symposium Vol.2", National Roads Board, Wellington, New Zealand 1987, pp223-232.
16. "A Basic Asphalt Emulsion Manual, 2nd Edition", AEMA, Washington, 1986.
17. "Bitumen-Emulsion", Fachverband der Kaltasphaltindustrie, Norderstedt, W. Germany 1983.
18. W.T. Hulshof, Program and Papers, Bitumen Emulsion Workshops, Melbourne and Sydney October 1985, pp 45-62.
19. "Bituminous Emulsions for Highway Pavements", NCHRP Synthesis 30, Transportation Research Board, National Academy of Sciences, Washington 1975.
20. R.W.J. Tosh, Road Research Unit Occasional Paper, Emulsion Sealing and Open-Graded Emulsion Mixes, National Roads Board, Wellington, New Zealand 1988, pp1-8.

21. A.R. Lane and R.H. Ottewill, "Theory and Practice of Emulsion Technology", Proceedings of the Symposium at Brunel University 1974, publ. S.C.I., 129-151.
22. J.A.N. Scott, W.J.M. Stassen, R.J.M. Tausk and W.C. Vonk, Colloids and Surfaces, 1981, 3, 13-36.
23. Redicote Reference Manual, Akzo Chemicals Inc.
24. B. Alince, A. Robertson, M. Inoue, J. Colloid and Interface Science, 1978 65 (1), 98-107.
25. UK Patent Application GB 2167975A.
26. D. Pedro Ferre Franquet, paper presented to the Emulsion Technology Session of the 15th Annual Convention of ISSA and 1st World Congress on Slurry Seal, Madrid, Spain February 1977.
27. A.D. James and D. Stewart, to be presented at 4th Eurobitume Symposium Madrid, 3-6 October, 1989.
28. Highways Newsletter, Akzo Chemicals America, Fall 1981, p.3.
29. R.J.M. Tausk, W.J.M. Stassen and P.N. Wilson, Colloids and Surfaces, 1981, 2, 89-99.
30. "Les Emulsions de Bitume", Syndicat des Fabriquants d'Emulsions Routieres de Bitume, 1978.
31. T.R.R.L. Road Note 14, HMSO, 1964.
32. M. Brossel, Bull. Liaison Laboratoire Ponts et Chaussee, 1973, 149 (4), 1-28.
33. Bituminous Materials in Road Construction, HMSO 1962, Chapter 5.
34. C.R. Benedict, Uses of the Modified Cohesion Test for Emulsion Formulation and Mix Design of Performance Cold Mix Systems, paper presented at the 12th Annual Convention of AEMA, March 5-8, 1985, New Orleans.
35. Jack Dybalski, Proposed Mix Design Methods for Emulsion Cold Mixes, presented at 11th Annual Meeting of AEMA, 1984, Florida.

Surfactants in the Paper and Board Industry

R. A. Morland and N. Morgan*
HENKEL-NOPCO, LEEDS, UK

The paper industry utilises a vast range of chemicals, some of which are identified within the industry as follows;

Defoamers

Drainage aids

Pitch control agents

Coating lubricants

Wetting agents

Cleaners

Flocculants

Sizing agents

Biocides

Dispersants

This list is not comprehensive, but it does represent the typical terminology that a chemical company may come to expect, when dealing with the paper industry. To a paper-maker the chemicals all perform a specific duty, but to a surfactant chemist one thing becomes immediately apparent. They all conform to the following definition;-

"A surfactant is a substance which modifies the properties at a surface or interface".

Conventional surfactants, as defined by the chemical manufacturers, are often formulated together with other chemicals, not normally classified as surfactants, but which nevertheless effect the surface properties. The paper-making process can be simplified into three distinct areas;-

1. Stock preparation
2. The paper-machine
3. Coating

STOCK PREPARATION

Cellulose fibres represent the basic raw-material for all types of paper or board. By mechanical treatments and the judicious choice of additives, this one raw material can produce a multitude of finished products (eg tracing paper to cardboard boxes, and from electrical insulating papers to high quality printing paper).

The UK paper industry has never been self-dependant, in terms of supply, of basic cellulosic fibres. Virgin paper pulp is imported from Scandanavia, Canada, S. America, U.S.A, Spain and Portugal to name just a few of the sources.

However, in recent years, waste paper has become one of the major sources of raw-material. Because of this, all the problems associated with virgin pulp are further aggravated, and extra chemical treatments are often necessary.

The preparation of stock from virgin pulp is unlikely to give problems to the paper-maker, other than the resinous material, which can be carried forward, depending on the type of pulping process that it has received. These resinous substances can agglomerate and can give deposition and processing problems later on in the system. This agglomeration can be prevented by the use of talc or dispersants, such as Naphthalene Sulphonates.

Waste paper, on the otherhand presents many more problems which need to be addressed at the stock preparation stage. In some cases the waste paper is used as is (eg packaging materials) but in other cases, it is necessary to include a de-inking process, in order to upgrade the quality of the end-product (eg newsprint and tissue manufacture).

For untreated wastes, the only additives required are wetting agents, which facilitate the breakdown of fibres in the hydrapulper; this is particularly applicable to waste paper with high levels of size, and will result in increased pulping speeds and associated savings in energy costs.

For better quality papers, it is often necessary to incorporate a de-inking process. As the name implies the purpose of this process is to remove printing ink.

Two methods are available for this purpose. In the early days this was done using a wash de-inking process, where the ink was dispersed into very small particles, which were then washed out, as the pulp went through the thickening processes. In order that the dispersed particles were not entrapped in the fibrous web, nonyl phenol ethoxylates were used to ensure that particle size was kept to a minimum.

Most new de-inking plants now use froth flotation. In this system, fatty acid soaps react with available Calcium ions. These calcium soaps attract the ink particles, which are collected and transported to the surface of the vessel by air and subsequently skimmed off. Whichever system of de-inking is employed, the net effect is a far cleaner pulp.

After a variety of mechanical cleaning and screening processes, the pulp is now ready to be chemically or mechanically treated in such a manner as befits its intended end-use.

Typical chemical additions at this stage would be;-

Sizing Aids
Dry strength additives
Wet strength additives
Dyes
Fillers

Each of these additives, although producing the desired end-results, can lead to problems in the system. eg:

1. Sticky sizes can cause agglomeration of the filler leading to unwanted deposits, which can build-up and later break-away. This will lead to problems with both paper production and quality.

2. Starch is an excellent nutrient for biological growth. This growth can also lead to deposits in the system, which build-up and again break away.

In order to overcome problems such as this, dispersants may be used to prevent agglomeration and hence maintain a clean system. Biocides are used to prevent biological growth.

A novel new method for the removal and prevention of slimy deposits is the use of specially cultivated enzymes. The enzyme breaks down the slime and so allows more intimate contact between the bacteria and the biocide.

The presence of many of the above-mentioned chemicals in the stock are conducive to the formation of foam. To the paper-maker, foam represents a problem, not only with the smooth operation of the paper-machine, but can also lead to surface defects in the end-product.

Operational problems could include air locking in the circulation pumps and retardation of the drainage rate.

Foam can also contribute to the build up of deposits.

In order to overcome the foaming problem, the use of a defoamer is essential. Initially oil-based defoamers were used and indeed they still enjoy wide application. Nonionic surfactants and silicones would normally be added to the formulation. However, environmental

pressures accelerated the change to water-based emulsions, and the added economic advantage of dosing 40% active material in place of 100% oils have probably been an even greater incentive; accidental overdosing does not lead to the same degree of waste. These water-based defoamers are usually based on long-chain alcohols and esters;

more recent developments have included EO/PO copolymers and blocked alkoxylates.

Inorganic fillers are added to improve the surface characteristics of the paper and to increase the opacity. However, to retain these fillers within the matrix, it is necessary to use flocculating agents such as polyacrylamides.

PAPER MACHINE

A paper machine is a dynamic system. The addition of all these chemicals can give both added benefits and also some disadvantages. The selection of the right chemical for any given application is of prime importance, eg:

1. Defoamers not only control the amount of foam, but also reduce the amount of entrained air in the stock.

   This results in an increased rate of drainage of water through the sheet; a drier web will then proceed to the dryers and hence reduce energy costs.

2. An adverse effect is that the emulsifiers, present in the defoamer can act as wetting agents, which have a detrimental effect on the sizing process.

3. It is also obvious that flocculants and dispersants are, in many respects, working directly against each other. It is important, therefore, that the correct balance is maintained.

No matter how effective the dispersants are in keeping the system clean, the paper-machine will eventually need to be shut down for a thorough cleaning. This can be done with high pressure hoses, but the process is made easier if the 'soils' are first softened and wetted. Strong acid or alkali cleaners are used in conjunction with conventional wetting agents and detergents. Hydrotropes are often necessary to hold the formulations together. These chemicals are used to clean both the wire and the felts in the press section, which may have been blinded by a gradual build-up of the materials used in the system. Without this regular housekeeping, operational problems will occur and loss of production will result.

PAPER COATING

Not all papers are coated but many are, and coating is a significant part of the industry.

There are two systems for doing this;-

1. On-machine coating, where the coating unit is an integral part of the paper machine, located part way down the drying section.

2. Off-machine coating, where the plant is independent of the paper machine and may obtain its base paper from any supplier.

Whichever method is used, the objective is the same - to put a smooth even finish coat onto the surface of the paper. The nature of the coating depends on the end-use of the paper and each will have its own problems. The most common coating is a pigmented coating, which is used primarily to improve the printability of the paper. Typically, china clay or calcium carbonate are adhered to the surface with a binder such as latex or starch.

To ensure a smooth coating, the pigments must be very finely divided and dispersants are once again necessary. In order to obtain the required solids levels, whilst still maintaining fluidity, it is essential that these dispersants should also have plasticising properties.

The chemicals used in the coating process include metal soaps and polyacrylates.

Again, as at earlier stages in the paper-making process, the addition of these chemicals can lead to foaming problems. To ensure that the surface quality is maintained defoamers are used extensively in the coating process; compatability of the various chemicals is once again of primary importance.

It can be seen that surfactants or surface - active agents play a very important role in the manufacture of paper.

1. They enable waste paper to be used in applications hitherto not thought possible.

2. They enable the paper-maker to produce the quality of paper he desires.

3. They keep his system clean and improve the efficiency of his plant.

APPLICATION	TYPE OF SURFACTANT
DEFOAMERS	Long-chain fatty alcohols. EO-PO co-polymers Silicone based surfactants. PEG esters.
DRAINAGE AIDS	Sulphosuccinates Nonyl Phenol Ethoxylates EO/PO copolymers
PITCH CONTROL AGENTS	Naphthalene Sulphonates
COATING LUBRICANTS	Calcium Stearate
WETTING AGENTS	Nonyl Phenol Ethoxylates Alcohol Ethoxylates
CLEANERS	Typical anionic and nonionic cleaning formulations, but since the system is generally alkali, hydrotropes may also be necessary. Alkyl Benzene Sulphonates Nonyl Phenol Ethoxylates Alcohol Ethoxylates Sodium Xylene Sulphonate Phosphate Esters.

FLOCCULANTS	Polyacrylamides
SIZING AIDS	Emulsified waxes
BIOCIDES	Cationics, Amphoterics
DISPERSANTS	Naphthalene Sulphonates Polyacrylates Alcohol, Ethoxylates
DEINKING	Nonyl Phenol ethoxylates Fatty acid soaps

FUTURE?

As more and more waste paper is used in the industry, the papermaker will be looking for more effective ways of solving old problems. He will also be looking to create new qualities of paper, and surfactants will necessarily be at the forefront of any developments.

For these reasons a close working relationship between the papermaker and the suppliers of chemical processing aids is essential.

Use of Surfactants in Mineral Flotation

F. J. Kenny
ANAMET SERVICES, ST. ANDREWS ROAD, AVONMOUTH, BRISTOL

Introduction

Flotation was first used commercially in the mining industry at the turn of this century with the first significant application involving flotation of a sulphide ore at Broken Hill Block 14 mine in Australia. Since these modest beginnings flotation has become the predominant process in mineral separation. The main reasons for this have been the treatment of lower grade and more finely disseminated ores, the broad applicability of the process to a whole range of different ore types, and the comparative lack of limitations.

The application of flotation for mineral separation and concentration fits into the overall mining and metallurgical process between the comminution of an ore and the production of metal or metals from the valuable constituents of the orebody. It has an advantage over operations such as magnetic, electro-static and gravity separation because the small differences in surface chemical properties can, with an ever increasing range of reagents, be enhanced to selectively separate a wide range of minerals.

The flotation mechanism is easily understood by consideration of the simplest system. This system consists of a solid phase, a gas phase, and a liquid phase. The solid phase consists of two different and totally liberated minerals, one of which is hydrophilic (water wetted) and the other hydrophobic (water repellent). These are mixed into a slurry with the liquid phase.

The slurry is agitated and the gas phase is introduced by air injection which causes fine bubbles to form. The agitation maintains the solid minerals in suspension and ensures good mixing of the bubbles and also helps good bubble to mineral contact.

The hydrophobic minerals adhere readily to the gas bubbles on contact which then float to form a removable mineralised froth on the surface of the slurry. The hydrophilic minerals remain behind in the slurry thereby effecting a separation of the two minerals. However, flotation is always more involved than this in the practical application. Combinations of reagent additions are therefore designed to arrive at the required selectivity and plant circuits are designed accordingly.

Preparation of the minerals' surfaces is of utmost importance. This is accomplished by crushing and grinding the ore to an optimum feed size distribution followed by chemical conditioning.

Particle size is a major factor in mineral separation and concentration and may vary widely between different orebodies. Flotation is readily applied to this wide range of size distributions, with the maximum particle size being between 1.5 to 2.5 millimetres and a lower limit of about 10 microns. This minimum particle size varies with the nature of the material being treated and can be a limitation to the process. One of the more recent advances in flotation has been selective flocculation which, in combination with flotation, improves the recovery of the fine minerals.

Chemical conditioning is usually required to impart hydrophobicity to the required mineral surfaces and is strongly dependent on parameters such as pH. A few minerals, e.g. molybdenite, native sulphur, graphite, stibnite, and coal float naturally because their surfaces are naturally hydrophobic and do not necessarily require an artificially prepared hydrophobic surface.

In the majority of cases, however, hydrophobicity is imparted by treatment with specific reagents, usually in advance of flotation. This conditioning may start in the grinding section, but is usually performed continuously in mixers prior to flotation, with further reagent additions to the flotation circuit itself.

Numerous chemicals are used in froth flotation and can be classified as:

1) Collectors
2) Frothers
3) Modifiers
   i) Activators
   ii) Depressants/ Dispersants
   iii) pH modifiers

1 COLLECTORS

The purpose of the collector (or promoter) is to render the required mineral particle with a hydrophobic surface so that the particle attaches itself to an air bubble. The collector molecule, without exception, is made up of a polar group and one or more non-polar hydrocarbon groups.

The polar group is hydrophilic in character and is the group by which the collectors attach to the mineral. It is usually ionic and determines properties such as collector hydrolysis, chemisorption to, and electrostatic interaction with the mineral surface.

The non-polar group is the hydrophobic portion by which bubble attachment occurs. The hydrophobic tail is usually a hydrocarbon, but recent research has shown that fluorocarbon and silane tails have a potential application. The most important feature of the hydrocarbon chains is their ability to form micelles, due to their non-ionic nature and incompatibility with polar water molecules which render the collected mineral particle floatable.

Collectors can be divided into anionic and cationic groups, with further sub-division of the anionic group into sulphydryl and oxyhydryl groups.

Anionic Collectors

Sulphydryl. Sulphydryl or thiol collectors are used for the flotation of sulphide minerals. The most common of these are the xanthates and dithiophosphates which are all characterised by a functional group that contains a sulphur atom bonded to a carbon or phosphorus atom. Figure 1 shows various sulphydryl collectors. Less widely used groups such as dithiocarbamate, dithionocarbamate, thiocarbanilide, mercaptan, and xanthate derivatives are also available.

Much evidence from work with xanthates performed on sulphide minerals such as galena, sphalerite, pyrite, chalcocite,and chalcopyrite has shown the presence of either dixanthogens formed by the electrochemical oxidation of xanthate (see equation 1 and figure 2), or chemisorbed xanthate ions , or in the case of galena, bulk precipitated lead xanthate at the mineral surface. The presence of whichever form or forms of xanthate at the surface gives the surface its hydrophobic property. The more commonly used xanthates include ethyl-, isopropyl-, isobutyl-, and amyl- derivatives.

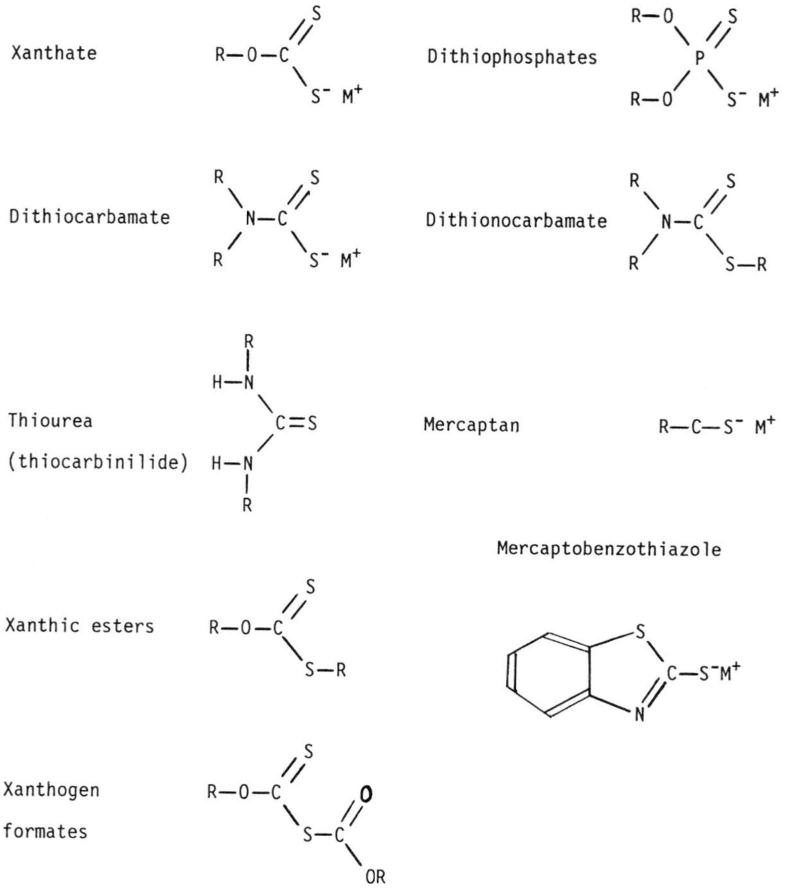

Figure 1 Structures of Various Sulphydryl (Thiol) Collectors

$$2X^- + {}^1/_2O_2 + H_2O \rightleftharpoons X_2 + 2OH^- \qquad \text{Equation 1}$$

where X represents xanthate and the structure is illustrated in figure 2.

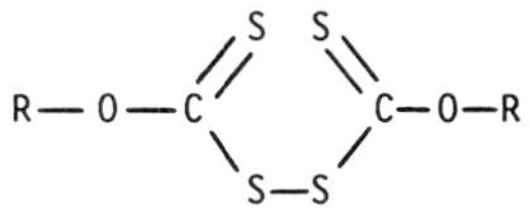

Figure 2 Structure of Dixanthogen

In the past xanthates have been chosen for their selectivity. As a rule selectivity decreases as the non-polar group homologue lengthens. For example, use of ethyl xanthate produces greater selectivity in the flotation of one sulphide mineral such as galena, (PbS). On the other hand amyl xanthate is widely used as a general sulphide mineral collector when selectivity against other sulphide minerals is not required.

This apparent increase of selectivity from amyl to ethyl xanthate correlates well with the corresponding solubility products. It is further correlated with the solubility products of zinc, lead, and cuprous ethyl xanthate which is important in the selective flotation of sphalerite (ZnS), galena (PbS), and chalcopyrite ($CuFeS_2$). This trend in solubility products is illustrated in table 1.

Sodium diethyl dithiophosphate is a commonly used dithiophosphate which has been found to be very selective for copper sulphides, and particularly pyrite (FeS_2). The solubility products of the dithiophosphates follow the same trend as those of the xanthates but the metal dithiophosphate solubility products are greater than those of the corresponding xanthates, implying a greater degree of selectivity than the xanthates.

The nonpolar groups associated with sulphydryl collectors include ethyl (C_2H_5-) through to hexyl (C_6H_{13}-) alkyl groups, phenyl (C_6H_5-), and cyclohexyl (C_6H_{11}-) groups.

	Zn		Pb		Cu	
	X	DTP	X	DTP	X	DTP
ethyl	$4.9x10^{-7}$	$1.5x10^{-2}$	$2.1x10^{-17}$	$7.5x10^{-12}$	$5.2x10^{-20}$	$1.4x10^{-16}$
propyl	$3.4x10^{-10}$	-	-	$6.0x10^{-14}$	-	-
butyl	$3.7x10^{-11}$	-	-	$6.1x10^{-16}$	$4.7x10^{-21}$	$2.1x10^{-18}$
amyl	$1.6x10^{-12}$	$1.0x10^{-8}$	$1.0x10^{-24}$	$4.2x10^{-18}$	-	-

X = xanthate, DTP = dithiophosphate

Table 1 Solubility Products of various metal xanthates and dithiophosphates (after Kakovsky[2], 1957)

Oxyhydryl. For nonsulphide minerals, variety is greater and oxyhydryl collectors are generally used. The polar groups include; carboxylic (fatty) acid group, sulphonates, and less commonly, the sulphates, phosphonates, arsonates, sulphosuccinates and sulphosuccinamates. These collectors dissociate to yield negatively charged surfactant ions which adsorb onto the surface of a mineral. They adsorb either physically by electrostatic interaction or chemically. They are used to collect a wide range of oxide, silicate, and salt-type minerals. The hydrocarbon radicals of the oxyhydryl collectors have to be more hydrophobic than the sulphydryl collectors because of the hydrophilic nature of these minerals, usually resulting in carbon chains of between 12 and 22 carbon atoms. Figure 3 shows various examples oxyhydryl collectors.

Collector chemisorption to oxide and silicate mineral surfaces is most probably achieved by a slight dissolution of mineral at the surface and adsorption of hydroxy complexes to the surface, resulting in a collector-hydroxy-complex reaction. This has been illustrated by observing the effects of $MgOH^+$ and $FeOH^+$ in solution on the pH of point of zero charge, PZC, of quartz shown in figure 4[3,4]. These pH values are 8 and 11 and correspond with the positive flotation responses of chromite at the same pH values shown in figure 5[5]. This is attributed to the formation of the hydroxy complexes of Mg^{2+} and Fe^{3+}.

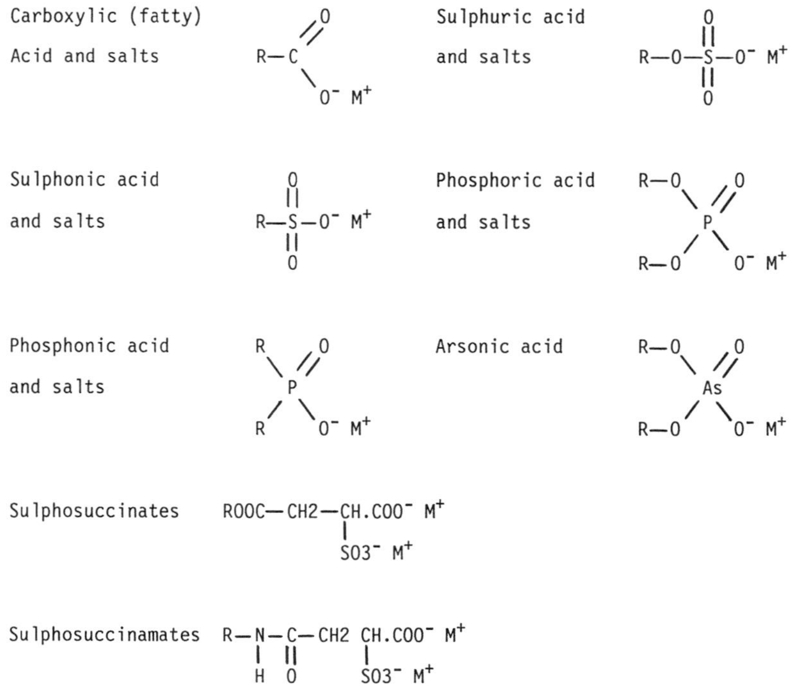

Figure 3 Structures of Various Oxyhydryl Collectors

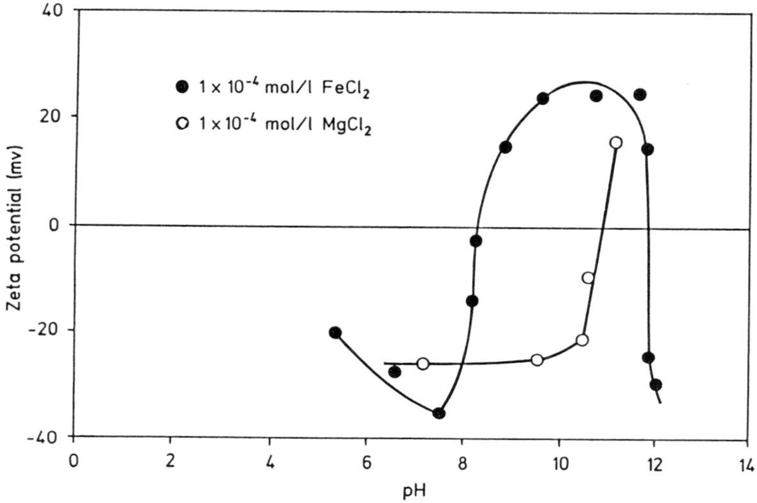

Figure 4 Zeta potential of quartz as a function of pH with additions of $MgCl_2$ (after Fuerstenau, Elgillani and Miller, 1970) and $FeCl_2$ (after Fuerstenau, 1975)

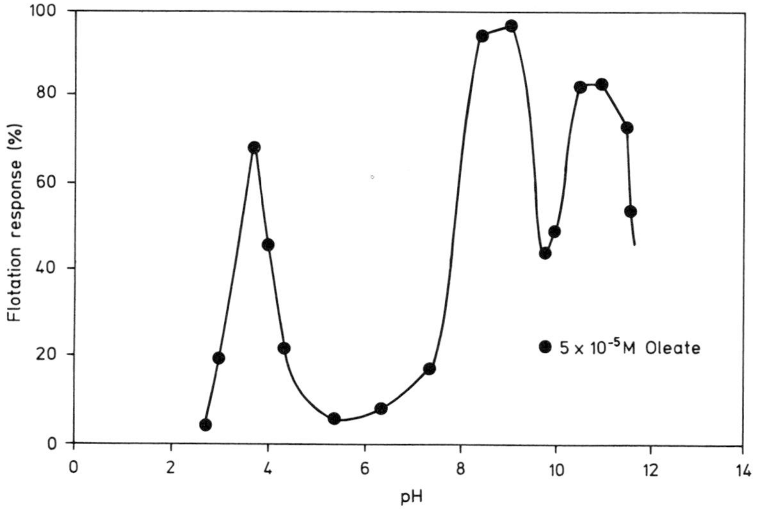

Figure 5 Flotation recovery of chromite as a function of pH (after Palmer, Fuerstenau and Aplan, 1975)

Physical adsorption is a balancing of electrostatic forces between the mineral surface, polar, and non-polar groups and is dependant on collector concentration at the surface. This collector concentration at the mineral surface determines the formation of hemi-micelles. Hemi-micelle is a term used by Gaudin and Fuerstenau (1955)[6] to describe the Van der Waal's type bonding of the non-polar groups (as with micellation) with the polar group adsorbing to the mineral surface. Wakamatsu and Fuerstenau (1973)[7] have shown that the hydrophobicity of the mineral surface increases rapidly once the collector concentration at the mineral surface has reached the critical concentration. This, in turn, increases the flotation rate. The formation of hemi-micelles is illustrated in figure 6.

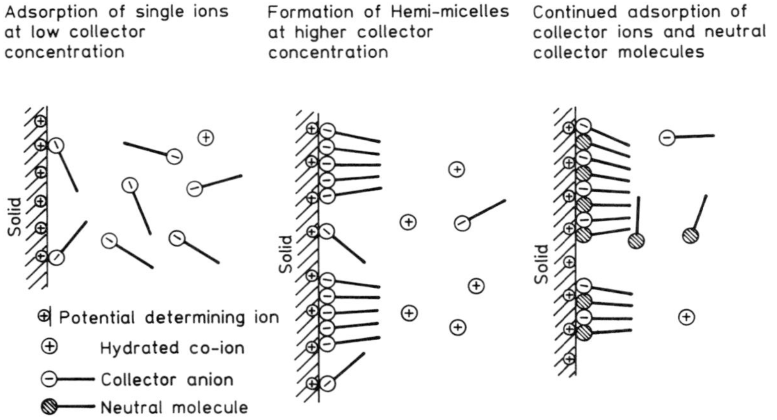

Figure 6 Formation of Hemi-micelles

The carbon chain length of the non-polar group determines the collector solubility and the concentration at which hemi-micellation occurs. The solubility and the concentration at which hemi-micellation occurs decreases as the carbon chain length of the non-polar group increases.

Cationic Collectors

The amine-type collectors are the only cationic collectors used. They are classified into primary, secondary, tertiary, and quarternary amines. The primary, secondary, and tertiary amines are weak bases and the quarternary amines are strong bases. The quaternary amines ionize completely at any pH, but the ionization of the primary, secondary, and tertiary forms is pH dependant and is controlled by addition of usually either acetic or hydrochloric acid. Figure 7 shows the primary, secondary, tertiary, and quaternary amine (cationic) collectors and their ionization products.

Primary amines	$R.NH_2 + H_2O \rightleftharpoons [R.NH_3]^+ + OH^-$
Secondary amines	$(R)_2.NH + H_2O \rightleftharpoons [(R)_2.NH_2]^+ + OH^-$
Tertiary amines	$(R)_3.N + H_2O \rightleftharpoons [(R)_3.NH]^+ + OH^-$
Quaternary ammonium salt	$[(R)_4.N]^+ \; Cl^-$

Figure 7 Cationic (amine) Collectors and Their Ionization Products

The collecting action of these collectors is not specific since they attach to mineral surfaces of opposite charge. They are generally used in conjunction with modifiers to ensure that the selected mineral has a negatively charged surface. The surface charge can be determined by electrokinetic experiment and the pH at the point of zero charge be determined. Iwasaki et al. (1960)[8] showed that the surface potential of geothite (FeOOH) decreases and reverses as pH increases, and the pH at the point of zero charge, PZC, is pH 6.7.

The flotation response using cationic collector occurs in the pH region corresponding to negatively charged surfaces. This is because pH is the most important variable effecting the potential-determining ions, such as H^+ and OH^- for oxide mineral surfaces and vice-versa with an anionic collector. In the case of geothite for example, a pH below 6.7 renders the mineral surface positive and the anionic collectors, sodium dodecyl sulphate and sodium dodecyl sulphonate render the mineral floatable, whereas at a higher pH the cationic collector dodecylammonium chloride renders the mineral floatable. Figure 8 illustrates the flotation responses of cationic and anionic collectors compared to the surface charge of the mineral geothite.

Examples of their use are in the flotation of sylvite (KCl) from halite (NaCl), where primary and secondary amine salts are applied. The quaternary amines are increasingly being applied to the flotation of silica, (e.g. flotation of silica from phosphate rock concentrates). They are also used for flotation of sulphidized zinc carbonates and silicates. In general, good separations can usually be achieved between acidic (silicate) and basic (oxide or carbonate) minerals.

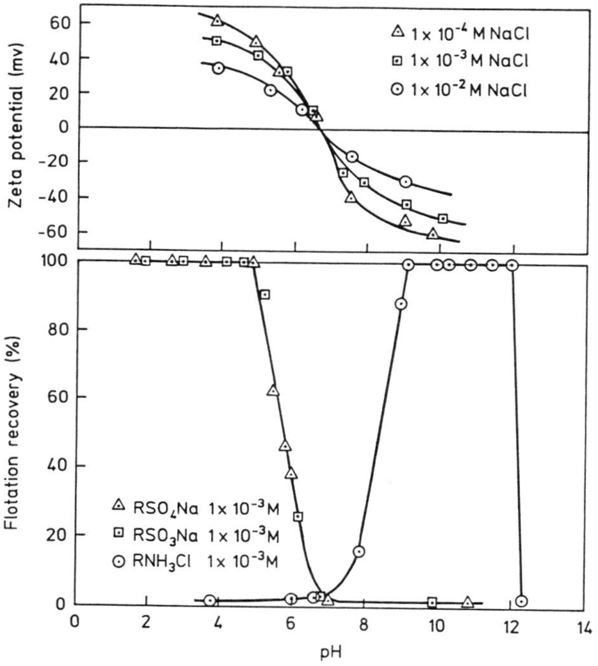

Figure 8 Flotation recovery of geothite as a function of pH with collector additions compared with mineral surface charge at different concentrations of NaCl (after Iwasaki <u>et al.</u>, 1960)

<u>Other collectors</u>

<u>Hydrocarbon Oils.</u> Other reagents that do not readily fit into the above classification are used as collectors and are used to promote flotation of readily or naturally floating minerals. They include fuel oil, kerosine, and others and are effective in the flotation of coal, graphite, molybdenite,

phosphate, and sulphur. Since these reagents have collector properties some may be used as collectors or in conjunction with another collector to enhance the hydrophobic effect of the first collector.

Chelating Agents. The chelating agents represent another group of reagents that have some application in flotation, but are not used commercially. Certain of these have been shown to be very selective with respect to certain metal ions in the flotation pulp and to individual minerals. Examples of these are the hydroxamates, oximes, 8-hydroxyquinoline, diphenylquainidine, and benzotriazole amongst others. Little is known about their flotation mechanism, but the selective nature of the chelating agents has been found applicable to the flotation of certain minerals. It is thought that mineral hydrophobicity is only obtained when the metal chelation is insoluble and hydrophobic. The extent of the surface reaction and surface compound stability is unknown. Similar reagents have been used in Russia in the commercial flotation of tin and other complex ores.

2 FROTHERS

The purpose of a frother is to provide a stable and controllable froth. The bubble diameter is controlled to an average of approximately 0.5 millimetre in the pulp. The bubble size in the froth is highly variable and is dependent on many variables such as type and concentration of frother, type of collector, mineral loading, and froth height.

Frothers are surface active and usually non-ionic, but also influence the kinetics of the attachment of particles to bubbles. In non-sulphide flotation the collectors used are usually sufficiently surface active to provide their own frothing action. In sulphide flotation, the sulphydryl collectors are generally not, and therefore require a reagent specifically designed to produce a froth.

It would be simplistic to view frother, collector, and modifiers all separately in their individual roles, and the type of froth produced is undoubtedly affected by the other chemical species present. Frothers also have an effect on selectivity, due to certain minor collecting properties.

The range of frothers has increased over the years, from the use of pine oil and cresylic acid which were also used as collectors, to a wide selection of frothers, with emphasis usually on their non-selective properties. There are three main groups; the alcohol, the alkoxy, and the polyglycol types.

Alcohol-type

This group is subdivided into;

Aliphatic alcohols. The most commonly used of all frothers is methyl isobutyl carbinol, MIBC, which produces fine textured , fairly selective, and brittle froth. It is low cost and effective and so has found a wide commercial application in sulphide, fatty acid, soap, sulphonate, amine, and coal flotation.

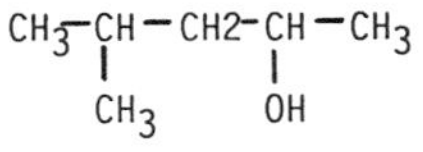

Figure 9 Structure of Methyl Isobutyl Carbinol

Cyclic alcohols. These are the pine and eucalyptus oils of which terpene alcohols, often alpha- terpineol, are the active constituents. They are frequently used in combination with another frother and also show some collecting capabilities.

Aromatic alcohols. The most important aromatic alcohol frother is cresylic acid which is also used for its collecting properties. It is made from distillation of coal tar using the middle and heavy oil fractions and is mainly a mixture of cresols and xylenols. Different grades are available which are boiling point range dependant. The lower ranges produce lighter, less persistent froths than the higher range grades.

Alkoxy-type

These are a further development of the alcohol type and tend to be more powerful and more selective. An example of this type is 1,1,3-triethoxybutane.

Polyglycol-type

The polypropylene glycol and methyl esters of polyethylene are widely used in flotation and have a similar application as MIBC. They produce fine, fragile froths that are very selective and show no collecting properties. The frothing characteristics produced are individual to this type of frother and are not classified with the alcohol-type frothers although they contain the OH group. The general formulae of these is shown as follows:

CH_3--(O-C_2H_4)n-OH

and CH_3--(O-C_3H_6)n-OH

where n can vary between from 4 and 10.

3 MODIFIERS

It is often necessary to develop a latent difference between the various mineral species before the collector is adsorbed preferentially by a selected mineral since it is rare for this to be acheived by the collector alone. This is usually carried out in the conditioning stage prior to flotation. At this stage, some minerals lose their attraction to collector and/or inherent hydrophobicity, while in others their attraction to collector or inherent hydrophobicity is increased.

Modifiers can be divided into three types; the activators, depressants, and pH regulators.

Activators

This is a group of chemicals which give a mineral surface the required surface charge and environment for collector adsorption. There are several important instances of this:

a) Copper Sulphate

The Cu^{2+} ions have been shown by Gaudin et al. (1959)[9] to displace Zn^{2+} from the lattice of zinc sulphide mineral surface, resulting in a layer of copper sulphide. This is represented in equation 2.

$$Zn^{2+}S^{2-}(s) + Cu^{2+}(aq) \rightarrow Cu^{2+}S^{2-}(s) + Zn^{2+}(aq) \qquad \text{Equation 2}$$

The copper coated zinc sulphide is then open to conventional copper sulphide flotation conditions and reagents. Copper sulphate is used extensively to activate sphalerite/marmatite for flotation with a xanthate, or dithiophosphate collector.

b) Sodium Sulphide

The activation of oxidised copper and lead minerals by Na_2S or NaHS for flotation by sulphydryl (thiol) collectors has been proposed to occur via the chemisorption of a metallic sulphide layer from the aqueous sulphide ions.

$$H_2S \rightarrow HS^- + H^+ \qquad \text{Equation (3)}$$

then $$HS^- \rightarrow H^+ + S^{2-} \qquad \text{Equation (4)}$$

Abromov[10] suggested that the sulphidised layer produces an increase in hydrophobicity, but flotation only occurs when a collector is added. This sulphidised layer also reduces the high solubilities of these minerals and diminishes the dissolution of cations from the mineral surface. The result is a stronger bonding of collector ions to the mineral surface and improved mineral flotation.

This has been used commmercially on oxidised copper and lead ores, particularly the carbonates (malachite, azurite, and cerussite).

c) Hydrofluoric Acid

Fluoride has been used to activate feldspar, an alumino-silicate for selective flotation in the presence of quartz. It has been suggested by Smith (1965)[11] that activation is caused by the formation of SiF_6^{2-} ion which adsorbs on aluminium sites leaving a negative surface charge open for chemisorption of a cationic collector, dodecylamine.

$$Al..SiF_6^- + RNH_3^+ \longrightarrow Al..SiF_6.NH_3R \qquad \text{Equation (5)}$$

d) Inorganic ion activation of oxide and silicate minerals

The formation and adsorption of hydroxy complexes on mineral surfaces has been used to explain the mechanisms of collector adsorption to oxide and silicate minerals. Cations such as Cu^{2+} have been shown to activate chrysocolla by Palmer et al. (1975)[12] and Scott and Poling (1973)[13]. Quartz has been shown by Cooke and Digre (1949)[14] to be activated by Ca^{2+} ions and made floatable by fatty acid collectors. Kraeber and Boppel (1934)[15] showed quartz activation by certain metal cations at pH ranges similar to those listed by Fuerstenau and Palmer (1976)[16]. This type of activation has also been shown by Modi and Fuerstenau (1960)[17] to take place by adsorption of anionic ions such as $SO4^{2-}$ on positively charged alumina using alkylammonium chlorides for flotation.

Depressants

These are reagents which help ensure that minerals other than the 'selected' valuable mineral do not float. This is done by enhancing the hydrophilic character of the gangue and the 'non-selected' valuable minerals if present. Flocculation and dispersion are both properties that many of these depressant reagents exhibit and since both depression and flocculation are dependant on an increase of surface charge or hydrophilicity, the role of the reagent may be difficult to ascertain.

The depressants can be sub-divided into two further groups: inorganic and organic

Inorganic. In the flotation of sulphide minerals , inorganic salts are widely used as depressants. The active ions of several of the important ones include sulphite (and HSO_3^-), sulphide, cyanide, and chromate amongst others. The depressing action comes from the formation of a more stable metal depressant surface layer on the mineral surface which either prevents or replaces the formation of collector layers on the mineral surface. Another of these is lime, which is widely used as a pH modifier and also exhibits a flocculation characteristic. When lime is used to control pH the formation of surface hydroxides on sulphide minerals at various values of pH result in a hydrophilic layer which depresses pyrite and some other sulphides in the same way as the other ions. According to Fuerstenau et al. (1968)[18] depression of pyrite is also caused by the adsorption of Ca^{2+} ions which change the surface charge, and at above pH 11 the oxidation of xanthate ion to dixanthogen is inhibited. The metal hydroxides formed by oxidation and/or the addition of metal salts show a significant effect on flotation responses.

The sulphide oxidation rate has been shown to be greatly increased in the presence of pyrite and is due to a galvanic interaction between pyrite and other sulphide particles (Kocabag and Smith, 1985[19], Rao and Finch 1988[20]). The pyrite particle acts as a cathode and electron transfer, due to sulphide conductivity, is such that in the presence of oxygen, hydroxide is formed. This results in a complex scenario when selective flotations of a polysulphide orebody is considered. Conflicting responses may occur unless the system is accurately analysed and carefully controlled. A common feature of these orebodies is that they remain un-mined, in preference to the simpler orebodies which are amenable to flotation.

In some cases, it has been shown that inorganic ions such as K^+ have a depressing effect on non-sulphide minerals such as quartz by effectively competing for adsorption onto the mineral surface with a cationic collector such as dodecylamine (Onoda and Fuerstenau 1964)[21]. The activation of inorganic ions for anionic collector flotation could be considered as depressants for cationic flotation and vice-versa.

Sodium carbonate and sodium silicate are used to depress semi-soluble minerals such as calcite and fluorite. According to Miller and Hiskey[22] carbonate from sodium carbonate or the air reacts with the surface of fluorite to form a hydrophilic surface of calcium carbonate. Various forms of sodium silicate at particular pH ranges and dosages show depression on calcite and no depression effect on fluorite (Fuerstenau *et al.* 1968)[23].

Organic. Comparatively little work has been done to investigate the mechanisms of mineral depression with these compounds, and they are used almost exclusively on the silicate, oxide, and semi-soluble minerals. They are large complex compounds with high molecular weights of 10,000 and more. The organic depressants can be split into three types: polyglycol ethers, polysaccharides, and polyphenols.

a) Polyglycol Ethers are from the same family of reagents used as frothers and by changing the radical group and the number of ethylene oxide groups degrees of hydrophilicity and hydrophobicity can be generated.

$$R\text{-}(OC_2H_4)n\ OH$$

The use of these reagents therefore is bi-functional, acting as both depressant and frother simultaneously. A typical example of this type is where R is a nonylphenyl group and n is 4, resulting in nonylphenyltetraglycol ether. It is used in the depression of calcite and dolomite.

b) Polysaccharide depressants are generally derived from the natural products of starch, cellulose, and natural gums.

Starches are used most commonly unmodified, but are also used in the form of dextrins. Their action has been to depress talcecous gangue and calcite. The chemical structure of starches and dextrins is based on the dextrose molecule and the 1 to 6, and 1 to 4 glycosidic linkages of the dextrose molecules form branched chains made up of dextrose units (Caesar, 1968)[24].

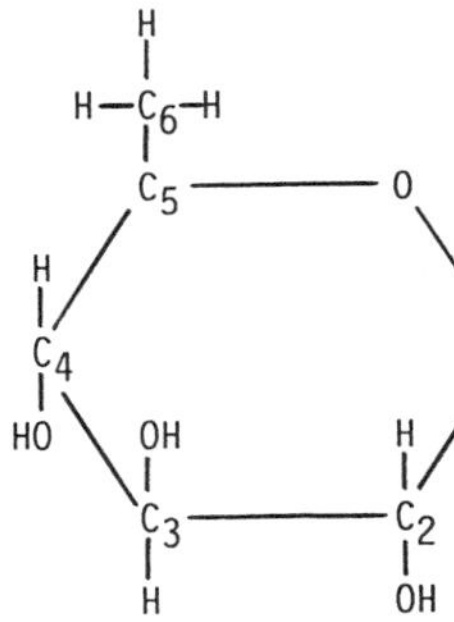

Figure 10 Structure of Dextrose

The action of starches is due to their anionic nature and adsorption to positively charged surfaces. Specific starches contain specific anionic types and chemical interaction of these anionic groups with particular metal sites in the mineral surface result in a hydrophilic starch covered mineral surface. This anionic property is also utilised for flocculation and the possibility of selective flocculation is apparent, as in the case of hematite flocculation.

The use of cellulose is only in the modified form of which carboxymethylcellulose is predominant and is used in talc depression.

The natural gums are large molecules such as guar gum used to depress talc and siliceous minerals. Due to the size and anionic nature of these compounds they also act as flocculants.

c) Polyphenols are large and complex compounds. The tannins are the most commonly used and of these the chemistry of quebracho has been extensively studied, but not directly to the application of flotation. It is used in the depression of calcite and several plausible mechanisms are used to explain its adsorption to the surface of calcite. Figure 11 shows the structure of Tannin[25].

pH modifiers

Control of pH is the most crucial operating parameter for most flotation operations.

The most commonly used pH modifiers that are used specifically for pH control are lime, slaked lime, sodium carbonate, sodium hydroxide, and sulphuric acid. They also have other effects, as already mentioned, such as activation and depression. Many other flotation reagents have an effect on pH. This effect on pH may be compatible with the operation but if not further pH control is required.

The above classification should be considered as being flexible since the action of the modifier is specific to the minerals present in the pulp and can result in the same reagent acting as a depressant for one mineral, an activator for another, and a pH modifier. For example, sodium sulphite is a depressant for sulphides, an activator for oxidized minerals, and affects the pulp pH.

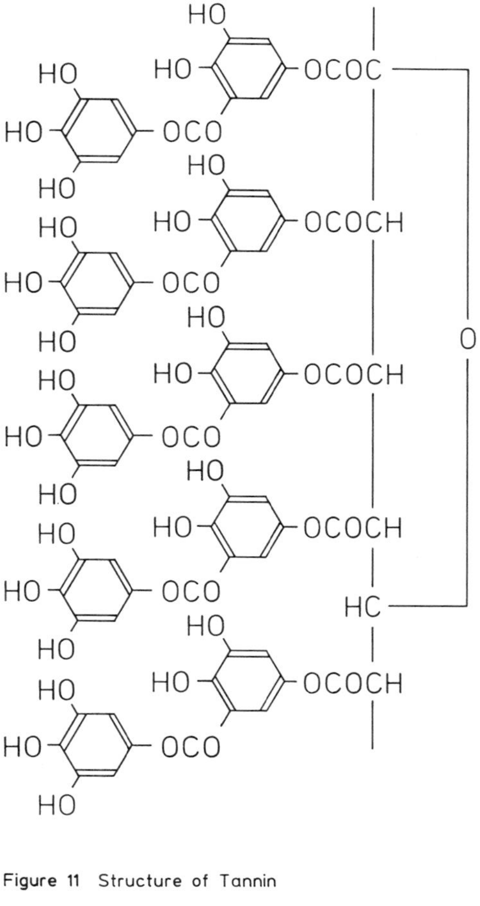

Figure 11 Structure of Tannin

This cross over of roles is apparent in the flocculation and depression by starches, flocculation, dispersion, depression, activation, and frothing of the polyglycol ethers, dispersion and depression by sodium silicate, amongst others. The level of understanding about the interaction of these reagents with the mineral systems that they are applied to is limited.

CONCLUSIONS

The onset of flotation technology heralded a mineral separation technique which has proved applicable to almost all mineral separations. Nevertheless it is still considered by many operators to be a skill or an art opposed to a technological process. The resultant performance of many plants is heavily dependant on the aquired knowledge of the plant floor operators and their art of applying that knowledge. It is a knowledge that is gained specifically to a particular orebody and is usually acquired over many years.

Flotation responses of all minerals are usually very specific to the characteristics of the individual orebody, which are in the form of:

a) the types of minerals present,
b) the reactivity of the minerals present,
c) the dissolvable ions present in the flotation pulp,
d) the selectivity required, and
e) the economic performance.

Design of plant flowsheets has therefore to be based on extensive testing of samples from throughout the orebody if economically viable. The prediction of potential plant performance with reagents and conditions tested can only be as representative as the samples tested.

The results in the more academic field of research investigating the chemical and physical mechanisms of flotation is still a long way from direct application to the practical operations and should be carefully considered when direct comparision with the plant operation is made. Often this type of research is carried out on artificial, usually simple, and known constituent ores which enable otherwise complex interpretation of results. This kind of research is essential though to further the understanding of plant operations. Recent work on sulphide flotation by Shannon and Trahar (1986)[26], particularily the complex (polymetallic) sulphide ores, has suggested that the conventional role of a xanthate collector is not one solely of coating a mineral surface in collector ions, but more as a reagent to react with the mineral oxidation products and ions in solution which prevent self-induced potential for flotation. The counteraction effect of these mineral oxidation products can be displaced using other chemicals such as ethylene diamine tetra-acetic acid (EDTA), and soluble sulphides. The mechanisms to produce this self-induced flotation may not be identical as for xanthate but are very similar. This work also revealed that the order of flotation for various sulphide minerals was the same as that by sulphydryl collector induced flotation, which suggests that the selective sulphide flotation reponses are predominantly a function of the minerals' properties and not those of the collector.

The specific nature of orebodies has made their classification with respect to flotation very complex and due to this the development of flotation chemicals over the past 80 years has been largely an empirical process.

As mineral separations become more complex due to depletion of the more simply treated ores our understanding of the pulp and surface chemistry in the flotation environment will have to respond to enable continued treatment by flotation. This is likely since the development of new technological processes to date still require the involvement of a pre-concentration stage and flotation remains in the forefront of the processes used.

References

1. V.M.Lovell, "Principles of Flotation", edit. R.P. King, SAIMM, J'burg, 1982, Monograph Series 3, Chapter 5, 73.

2. I.A.Kakovsky, Vol 4, Proc. 2nd Int. Cong. Surf. Act., 1957, London, 223.

3. M.C.Fuerstenau, D.A.Elgillani and J.D.Miller, Trans AIME, 1970, 247, 11.

4. M.C.Fuertenau, AICHE Symp. Series no. 150, edit. P.Somasundaran and R.B.Grieves, 1975, 71, 16.

5. B.R.Palmer, M.C.Fuerstenau and F.F.Aplan, Trans. AIME, 1975, 258, 261.

6. A.M.Gaudin, and D.W.Fuerstenau, Trans. AIME, 1955, 202, 958.

7. T.Wakamatsu and D.W.Fuerstenau, Trans. AIME, 1973 254, 123.

8. I.Iwasaki, S.R.B.Cooke and A.F.Colombo, U.S Bureau of Mines, 1960, R.I.5593.

9. A.M.Gaudin, D.W.Fuerstenau and G.W.Mao, Trans. AIME, 1959, 214, 430.

10. A.A.Abramov, Jev. Vyssh. Ucheb. Zaved. Tsvetn. Metall., 1969, 12(5), 7.

11. R.W.Smith, Trans. AIME, 1965 232, 160.

12. B.R.Palmer, G.B.Gutierrez and M.C.Fuerstenau, Trans. AIME, 1975, 258, 257.

13. J.W.Scott and G.W.Poling, Canadian Met. Quarterly, 1973 12, 1.

14. S.R.B.Cooke and M.Digre, Trans. AIME, 1949, 184, 299.

15. L.Kraeber and A.Boppel, Metall. und Erz. 1934 31, 412.

16. M.C.Fuerstenau and B.R.Palmer, Flotation - A.M.Gaudin Memorial Volume, edit. M.C.Fuerstenau, AIME, New York, 1976, vol 1, Chapter 7, 148.

17. H.J.Modi and D.W.Fuerstenau, Trans. AIME, 1960, 217, 381.

18. M.C.Fuerstenau, M.C.Kuhn and D.A.Elgillani, Trans. AIME, 1968, 241, 148.

19. D.Kocabag and M.R.Smith, Complex Sulphides, edit. A.D.Zunkel, R.S.Boorman, A.E.Morris and R.J.Wesley, The Met. Soc., New York, 1985, 55.

20. S.R.Rao and J.A.Finch, Canadian Met. Quarterly, 1988, 27, 253.

21. G.Y.Onoda and D.W.Fuerstenau, Proc. 7th Int. Min. Processing Cong., New York, 1964, 301.

22. J.D.Miller and J.B.Hiskey, J. Colloid Interface Sc., 1972, 41, 567.

23. M.C.Fuerstenau, G.B.Gutierrez and D.A.Elgillani, Trans. AIME, 1968, 241,

319.

24. G.V.Caesar, Starch and its derivatives, edit. J.A.Radley, Chapman and Hall, London, 1968.

25. H.S.Hanna, 3rd Arab. Chem. Conf., Cairo, 1972.

26. L.K.Shannon and W.J.Trahar, Advances in mineral processing; a half century of progress in application of theory to practice, SME-AIME Arbiter Symp., SME, New Orleans, 1986, 408.

Subject Index